A BIM Professional's Guide to Learning Archicad

Boost your design workflow by efficiently visualizing, documenting, and delivering BIM projects

Stefan Boeykens

Ruben Van de Walle

BIRMINGHAM—MUMBAI

A BIM Professional's Guide to Learning Archicad

Group Product Manager: Rohit Rajkumar

Publishing Product Manager: Kaustubh Manglurkar

Senior Editor: Keagan Carneiro

Senior Content Development Editor: Debolina Acharyya

Technical Editor: Joseph Aloocaran

Copy Editor: Safis Editing

Project Coordinator: Sonam Pandey

Proofreader: Safis Editing

Indexer: Rekha Nair

Production Designer: Jyoti Chauhan

Marketing Coordinators: Nivedita Pandey, Namita Velgekar, and Anamika Singh

First published: May 2023

Production reference: 1210423

Published by Packt Publishing Ltd.
Livery Place
35 Livery Street
Birmingham
B3 2PB, UK.

ISBN 978-1-80324-657-4

www.packtpub.com

After my first book with Packt Publishing (Unity for Architectural Visualization), I'm so glad this book became a collaborative work with Ruben. He brought in many different approaches to productivity and finishing the model and drawings. Although we couldn't cover everything there is to say about Archicad, which would have taken at least two more books, I hope you will experience the same joy of learning about a software that makes so much sense, as it did to me in 1997, even before we started calling it BIM.

It captures knowledge from manuals, from years of using the software, initially in architectural practice but later more in research, teaching, and currently, consulting with D-studio. It was put into perspective by a wide variety of internet posts, videos, and blogs and by meeting so many enthusiastic users of Archicad (and other BIM software): if you are passionate about your work, wonderful things can happen.

Finally, thanks (again) to my wife, Kathleen, and my three boys, Bram, Wannes, and Jonas, for allowing me some time to dive into these "side projects," which often take over my thoughts and focus.

– Stefan Boeykens

It's hard to believe I can officially call myself an author now. If it weren't for my wife, Machteld, and my kids, Noor, Niene, and Wannes, this book would have never been possible. They were kind enough to give me the time and patience to come up with another one of my crazy ideas, so a massive thank you to them!

I also have to give a shoutout to my co-writer and friend, Stefan. Thanks to Pieter Pauwels' request for us to teach "something about Archicad" at the University of Ghent in a post-grad program on Building Information Management (BIM), we ended up bonding over our shared love for Graphisoft Archicad. And that, my friends, is the main reason we collaborated on this book. So, thanks for asking, Stefan, it has been a wonderful experience in which I have learned a lot from you (yet again)!

Let's not forget my partner in crime at studiov2, Pieter Vandewalle (no, not a relative!), for believing in me and allowing me to embark on this wonderful journey on behalf of our company.

And last but not least, a massive thanks to all the Howest students in applied architecture and every studiov2 or KUBUS client who has challenged me over the last decade or so. Your questions, critiques, and (seemingly) impossible ideas related to Archicad have all contributed enormously to what I know about the software today and how I can share this knowledge with you in the most understandable way. Without your struggles, this book would for sure have been a lot thinner…

– Ruben Van de Walle

Foreword

I have known *Ruben van de Walle* and *Stefan Boeykens* for many years now, as active, enthusiastic, and knowledgeable BIM practitioners, BIM educators, and overall digital content creators. Both Ruben and Stefan work primarily in Belgium and Europe and represent the Belgian community in 3D modeling, BIM processes and software, and digital architectural design practice at large.

With this book, Ruben and Stefan present a practical and hands-on introduction and get-started book on the topic of Archicad, which is one of the major software tools for **Building Information Modeling (BIM)**. BIM software such as Archicad allows modeling building designs into comprehensive 3D models that can be used to guide the design and engineering process, and to organize any construction site. As such, BIM in itself is an indispensable tool in the toolchain of the design and engineering specialist. Archicad in particular is known as the first commercially available BIM software in the world, and it is a tool of preference in architectural design offices primarily.

This book is a very complete and all-inclusive book, which starts by explaining what Archicad is and how you can get started with this software. The book continues with the creation of a basic residential model, which is used as an example model throughout the book, to show example functionality. Full geometric modeling techniques are explained for all types of objects (stairs, beams, spaces, etc.). Furthermore, different types of drafting, linework, and hatching techniques are visually explained.

The more advanced sections of this book deal with composite elements (floors, walls, layers, etc.), the use of the **Renovation** tool, the use of Complex Profiles, Detailing, and Viewpoints, as well as data extraction and visualization procedures (schedules). Full detail is given on the available publication workflows from the 3D BIM model to various outputs, such as PDF drawings, plots, hyper-models, and similar. Finally, the book explains in detail how to make compelling and functional visualizations in Archicad.

As such, this book is your go-to resource for practical insights and hands-on tips and tricks that will boost your design workflow using Archicad.

Dr. Ir.-Arch. Pieter Pauwels

Associate Professor, Eindhoven University of Technology, Netherlands.

Guest Professor, Ghent University, Belgium.

Contributors

About the authors

Stefan Boeykens is a Belgian architect-engineer.

Starting out as a professional architect in several offices, he returned to KU Leuven in 2000, at the Department of Architecture, for teaching and research, completing his PhD. in BIM in 2007.

Stefan is an experienced researcher with a variety of IT skills. Aside from teaching BIM at KU Leuven and a few post-graduate courses, he mainly works as senior innovation and BIM manager for D-studio, focusing on BIM middleware and consultancy. He is a frequent speaker at BIM-related events and is actively involved in BIM standardization groups, including CEN/TC 442 (Europe), buildingSMART Benelux, and Belgian technical committees on digital construction.

He is the father of three boys and enjoys musical composition, guitar playing, reading, cycling, and life-long learning.

Ruben Van de Walle (Belgium) also has a master's degree in architecture and engineering.

Having worked for both small and large architectural firms, he has dedicated himself to teaching and researching BIM at the Howest University in Bruges, briefly combining this with activities at KUBUS, the local Archicad reseller, for whom he helped develop the Belgian Template.

Ruben is a BIM and Archicad expert at Howest's bachelor program in applied architecture and post-graduate programs on BIM. He is also a co-founder and BIM expert at studiov2, a Belgian BIM consultant with a focus on architects and SMEs in general. He is a frequent speaker at BIM-related events for the construction industry.

He is married and is the father of three children (two girls and one boy) and enjoys playing bass guitar, reading comics, and continuing to learn.

About the reviewer

Nathan Hildebrandt is a globally recognized Archicad expert. His templates are world-renowned. He presented at Graphisoft's KKC in Budapest in 2014 on the development of FTA's Archicad Template. In 2017, his Skewed template was downloaded over 1,000 times.

He is the founder of several Archicad user conferences, including the world's first 24-hour Archicad user conference, ARCHINTENSIVE, which was also the world's largest Archicad user-run event.

His expertise in Archicad is driven by the successful delivery of over 100 projects across his 20-year career. Nathan currently leads his own architectural practice, *Skewed*, where he provides architectural services alongside Archicad and BIM implementation services and digital advisory services for the government.

Table of Contents

3

Building a Basic Residential Model: Modeling the Construction Elements   35

4

Building a Basic Residential Model: SPACEAdding Roofs, Zones, Beams, and Columns   83

5

Building a Basic Residential Model: Modeling Openings, Stairs, and Objects   117

6

Basic Drafting and 2D Views   175

7

Adding Annotations and Creating 2D Output   219

Part 2: Becoming an Archicad Professional – Learn About Archicad Tools and Settings to Create and Publish Any Type of Project in Full Detail

8

Using Advanced Modeling Tools for Developed Design   261

9

Using Advanced System Tools for Designing Stairs and Curtain Walls
295

10

Using the Mesh tool and Wizards to Finalize a Design   331

11

Using Advanced Attributes and the Renovation Tool for a Wider Design Range   357

12

2D Construction Drawings and 3D Views with Linked Annotations   379

13

Data Extraction and Visualization   429

14

Automating the Publication of BIM Extracts   461

15

The Various Visualization Techniques in Archicad   511

Appendix: Some Final Tips and Tricks   563

Index   571

Other Books You May Enjoy   586

Preface

A BIM Professional's Guide to Learning Archicad is a comprehensive introduction to all that Archicad has to offer for the creation of 3D models, 2D document extracts, and related output. This book is not a click-by-click series of recipes but rather focuses on understanding why and how Archicad works the way it does by providing realistic examples for you to apply, and how you can apply this knowledge to improve your own BIM design workflow.

The book introduces you gradually to Archicad tools, using ample examples, and mastering its complexity through several clear parts. This allows you to start a first project quickly, obtain useful skills in succeeding projects, and keep using the book at a later stage as a source for real insights into the software and useful expert tips and tricks. We'll start with the basic modeling of construction elements and then gradually add roofs, stairs, and objects to the project. We'll then dive into basic drafting and 2D views for creating 2D output, and explain how to use attributes and more advanced modeling tools for designing curtain walls and sites. You'll also learn how to extract and visualize your data and how to automate the process of publishing your extracts and 2D documents into a variety of output formats.

By the end of the book, you will have a profound understanding of Archicad and how to implement it efficiently in your own architectural projects, and how BIM can improve your overall design workflow.

Who this book is for

The book is for architectural designers, engineers, residential designers, BIM professionals, and anyone working in construction, manufacturing, or similar fields. Whether you're an absolute beginner or a professional looking to upgrade your architectural or engineering design or urban planning skills, you'll find this book useful.

What this book covers

Chapter 1, What Is Archicad and How Can You Learn It?, introduces Archicad and its developer, Graphisoft, with a short overview of its history, from the early 1980s. We also explain license types and what to expect from our approach to learning about the software.

Chapter 2, Getting Started with Archicad, covers the first steps with the software, introducing the interface, the major concepts and interaction methods, and the onscreen feedback and data entry you will encounter. We end with starting and saving our first project file, ready to start modeling in the next chapter.

Chapter 3, Building a Basic Residential Model – Modeling the Construction Elements, finally introduces modeling inside Archicad, by creating a small house step by step. This involves learning how to navigate in 2D and 3D and shows you how to model accurately. We will start with walls and slabs, but the main interactions apply to many other types of elements too.

Chapter 4, Building a Basic Residential Model – Adding Roofs, Zones, Beams, and Columns, continues the basic model by introducing more construction elements that make up the core shell and structure of our project, but also its spatial content, using zones.

Chapter 5, Building a Basic Residential Model – Modeling Openings, Stairs, and Objects, closes our first three chapters on modeling, with parametric objects such as doors and windows, and versatile system tools, such as stairs and curtain walls. We also tell you how to create a freeform shape, to create basically any object you can imagine.

Chapter 6, Basic Drafting and 2D Views, uses the 3D model to derive 2D views, which can be enhanced with 2D drafting methods. We introduce different attributes that control how lines and fills look.

Chapter 7, Adding Annotations and Creating 2D Output, closes the first part of the book, by adding 2D annotations onto the different views and preparing the final output documents, ready to be printed or shared as PDF.

With this chapter, we end the first part by creating a basic residential model and using it to extract drawings, ready to be printed or digitally shared.

Chapter 8, Using Advanced Modeling Tools for Developed Design, expands the model with more advanced modeling tools and techniques and further refinement of the elements, including their composite structure and how Archicad creates connections between elements.

Chapter 9, Using Advanced System Tools for Designing Stairs and Curtain Walls, returns to the stairs and curtain walls for more refinement and complete editing in detail.

Chapter 10, Using the Mesh Tool and Wizards to Finalize a Design, explains how to model our terrain and introduces two wizards to generate a complete truss and roof structure, starting from basic geometry.

Chapter 11, Using Advanced Attributes and the Renovation Tool for a Wider Design Range, explains how to control element sections with parametric, complex profiles and also introduces a few attributes to indicate the renovation status of objects, their structural function, and whether they are positioned as part of the building envelope or shell.

Chapter 12, 2D Construction Drawings and 3D Views with Linked Annotations, dives deep into the Archicad attributes that help control the look and feel of lines, fills, and other attributes in more detail. We will also return to our sections to control how they can dramatically change the display of the model. Finally, we will learn a few methods to organize construction detailing and how to display building fragments, to help you communicate the intent of the design.

Chapter 13, Data Extraction and Visualization, leverages the model to extract information into tabular schedules and to override the graphics of elements, to create alternative, thematic views, all model-based.

Chapter 14, Automating the Publication of BIM Extracts, elaborates on the publication workflow, from the initial viewpoint to defining views, collecting them in layouts, and finally, generating all the output as a single publisher set. This unlocks the power of being able to update the model and have all related documents and output updated automatically. Let the system work for you!

Chapter 15, The Various Visualization Techniques in Archicad, completes the book by generating different compelling and interesting visual output, including photorealistic rendering and how to bring the model into an interactive application.

The *Appendix* section gives a summary of what the book covered, but also a few pointers to additional features and options that could not be covered in the book, but which can help you to grow to the next level of Archicad proficiency.

To get the most out of this book

No prior knowledge of Archicad is necessary. You are expected to have construction knowledge (architectural design), and a basic understanding of 2D CAD drafting is recommended. The book is based on the international version of the software and uses metric units. The hardware requirements should be fine for a basic project (single-family or residential), but larger projects will need more RAM and higher-end CPUs and GPUs.

Software/hardware covered in the book	Operating system requirements
Archicad 25 or Archicad 26	Windows 10 64-bit or macOS 10.15
Intel Core i5 or AMD Ryzen 5 or Apple M1	
8+ GB RAM, SSD Hard Disk, 2+ GB VRAM, Full HD Display Resolution	

While model files are provided from the download link in the next section, we advise you to go through all steps one by one to familiarize yourself with input and the software interaction. This experience is essential to become productive in your own projects. When you get stuck, you can take a look at the provided models to get a clean start or compare them with your own model.

You can follow most of the book using versions that are up to 3 or 4 years old. Earlier versions may lack some of the current features, and the interface has also been updated considerably since older versions. If you don't have access to the full version of the software, a trial or educational version works just fine. Contact a local reseller for further support.

Download the example model files

You can download the example model files for this book from GitHub at `https://github.com/PacktPublishing/A-BIM-Professionals-Guide-to-Learning-Archicad`. If there's an update to the models, it will be updated in the GitHub repository.

We also have other code bundles from our rich catalog of books and videos available at `https://github.com/PacktPublishing/`. Check them out!

Download the color images

We also provide a PDF file that has color images of the screenshots and diagrams used in this book. You can download it here: `https://packt.link/JIoo1`

Conventions used

There are a number of text conventions used throughout this book.

`Code in text`: Used where a user needs to enter a value or text string in the user interface. You can use any other value if you want, but if you want to follow the examples, please use the provided text entry. Here is an example: "We'll keep it simple and set our layout number in **Custom ID** to `01` and **Layout Name** to `Quick Layout`."

Keywords: Used to introduce or emphasize new concepts, typically the first time they are encountered. These are mostly terms for concepts and features that the software introduces. Here is an example: "As an alternative, we sometimes attach a drawing to an **Independent Worksheet** and use **Trace and Reference** to display it in a section or on a ground floor plan, which gives more flexibility."

Bold text: Indicates important words or any term you see onscreen, in a dialog box, or anywhere else in the user interface. Here is an example: "So, in this case, the **A3 Landscape** master layout."

Italics: Indicates emphasis on certain words, product names, or brands and also menu entries and keyboard shortcuts. It is also used to indicate references to figures or other chapters. Here are two examples: "This may seem quite a lot, but we will return to the published workflow in *Chapter 14*."

> **Tips or important notes**
> Appear like this.

Get in touch

Feedback from our readers is always welcome.

General feedback: If you have questions about any aspect of this book, email us at customercare@packtpub.com and mention the book title in the subject of your message.

Errata: Although we have taken every care to ensure the accuracy of our content, mistakes do happen. If you have found a mistake in this book, we would be grateful if you would report this to us. Please visit www.packtpub.com/support/errata and fill in the form.

Piracy: If you come across any illegal copies of our works in any form on the internet, we would be grateful if you would provide us with the location address or website name. Please contact us at copyright@packtpub.com with a link to the material.

If you are interested in becoming an author: If there is a topic that you have expertise in and you are interested in either writing or contributing to a book, please visit authors.packtpub.com.

Share Your Thoughts

Once you've read, we'd love to hear your thoughts! Scan the QR code below to go straight to the Amazon review page for this book and share your feedback.

https://packt.link/r/180324657X

Your review is important to us and the tech community and will help us make sure we're delivering excellent quality content.

Download a free PDF copy of this book

Thanks for purchasing this book!

Do you like to read on the go but are unable to carry your print books everywhere?

Is your eBook purchase not compatible with the device of your choice?

Don't worry, now with every Packt book you get a DRM-free PDF version of that book at no cost.

Read anywhere, any place, on any device. Search, copy, and paste code from your favorite technical books directly into your application.

The perks don't stop there, you can get exclusive access to discounts, newsletters, and great free content in your inbox daily

Follow these simple steps to get the benefits:

1. Scan the QR code or visit the link below

https://packt.link/free-ebook/9781803246574

2. Submit your proof of purchase
3. That's it! We'll send your free PDF and other benefits to your email directly

Part 1:
Getting Started with Archicad – Project Setup and Essential Modeling Tools for Your First Residential Project

The objective of this part is to familiarize you with Archicad, the BIM concepts used in the software, the user interface, and the essential modeling tools that are commonly used when designing a small residential project. Throughout this part, you will learn how to use a variety of tools by creating a first project step by step. At the end of this first part, you will be able to model a basic residential project, at a preliminary stage, including using simple drafting tools and preparing printed output.

This section comprises the following chapters:

- *Chapter 1, What Is Archicad and How Can You Learn About It?*
- *Chapter 2, Getting Started with Archicad*
- *Chapter 3, Building a Basic Residential Model – Modeling the Construction Elements*
- *Chapter 4, Building a Basic Residential Model – Adding Roofs, Zones, Beams, and Columns*
- *Chapter 5, Building a Basic Residential Model – Modeling Openings, Stairs, and Objects*
- *Chapter 6, Basic Drafting and 2D Views*
- *Chapter 7, Adding Annotations and Creating 2D Output*

1
What Is Archicad and How Can You Learn It?

In this first chapter, we will introduce you to the book and its main content. We will get some background information about the Archicad software and its development and how it is installed and licensed. We will also learn about the main objectives and approach of this book, setting expectations right from the start. This is not meant to be a cookbook with step-by-step recipes showing buttons to click but, instead, aims to give you a proper understanding of how the software fits into a professional workflow.

The main topics we'll cover in this chapter are as follows:

- An introduction to Archicad

- A brief history of Archicad

- Setting up Archicad

- Licensing

- The approach of the book

By the end of this chapter, you will know what Archicad is, how you can install it, and what license types exist. You will also have a clear understanding of the approach toward learning Archicad in this book.

Introducing BIM and Archicad

This book introduces **Graphisoft® Archicad®**, advanced software for architects and other designers, which follows a methodology that is known as **Building Information Modeling** (**BIM**). It is an approach to developing digital models for buildings or other constructions that forms the basis for all kinds of documents and reports that are typically created by architects, such as drawings, renderings, and schedules, among many others.

BIM is gaining worldwide popularity and is quickly becoming the best practice in construction projects of all sizes. It helps you to better manage information and all related graphical documents, while at the same time avoiding mistakes, inconsistencies, or inefficient workflows.

You will use the Archicad software as one of the main tools in your BIM toolkit, not only to model and draft but also to develop a design, help you understand the spatial layout and qualities of your concepts, and help you in collaboration with other project partners, such as your client, other designers or engineers, the contractor, and manufacturers. The models you develop will become one of the main sources of information for a project, and ensuring they can be created efficiently is essential to become productive in a project where BIM is applied.

While we don't expect you to already have experience with BIM, it would be helpful if you have an overall understanding of the construction process and, possibly, some experience with making drawings using general CAD software, such as **AutoCAD**®.

The objective of using software such as Archicad is not simply to create 2D drawings but to really develop a model, with 3D geometry and embedded information, that allows drawings to still be extracted from it. This is also what we will cover in this book.

About Graphisoft and Archicad – a short history

Graphisoft was founded in 1982 by *Gábor Bojár* and his business partner *Ulrich Zimmer*, with *Archicad* being the first program they developed, specifically for solving coordination issues in a project by the Hungarian government to construct a new nuclear power plant. More specifically, the coordination of the complex 3D geometry of the building's structure and its piping didn't go as well as it should have done.

Graphisoft was able to develop its software in time to win the bid for this project and used the earnings to further develop the company. It's worth mentioning that Gábor Bojár and *Steve Jobs* met in 1984 at a trade show in Germany. Jobs saw opportunities in the software, to prove what his machines were capable of, and Bojár was at the time looking for good hardware to continue development. We should not forget that in the 1980s, the Iron Curtain was still in place, and allegedly, four *Apple*® *Macintosh*® machines were smuggled into Hungary, and Graphisoft was able to further develop Archicad. Although Apple's *Lisa* was a commercial failure, reviews of the Archicad software at the Hannover Fair that year were excellent, and Bojár and his team managed to survive the next few years. Macintosh machines improved and became a success in their own right, along with Graphisoft steadily growing as a company and Archicad being developed into a more mature product.

The good relationship between the founders of Graphisoft and Apple is worth mentioning because it probably explains to a large extent their somewhat shared vision of how software should work for a user. Like Steve Jobs, Gábor Bojár believes in the benefits of "intuitive" software, a compact and "logical" user interface, and a focus on the user experience, rather than on the software itself and all of its marvelous capabilities.

It also explains why this Hungarian company was the first to unveil a statue of the late Steve Jobs – it was a commission by Bojár for his old friend, about whom he expressed that *Graphisoft might not be around today if it weren't for him.*

Since 1993, Archicad has also been available for *Microsoft® Windows®.* The software has gradually expanded, including more advanced modeling and collaboration tools, high performance using multi-threading and hardware acceleration, and going beyond architecture with its structural and mechanical toolsets. One of its strong points, since the beginning, has been its **Geometric Description Language** (**GDL**), which is at the core of the parametric objects in the Archicad library, presenting highly efficient and flexible objects. Another point of note is the **BIMcloud** module, which allows multiple users to connect to the same Archicad model (stored in the cloud) and make (coordinated) edits simultaneously.

The *Graphisoft Help Center* (`https://community.graphisoft.com`) contains an overview of all versions of the software, giving you an insight into the evolution of it and the inclusion of major features and changes (`https://community.graphisoft.com/t5/Let-s-get-started/Archicad-versions/ta-p/304207`).

In 2007, Graphisoft was acquired by the German *Nemetschek™* group, where it now is part of a large group, including the likes of *Vectorworks®, Allplan®, Scia® Engineer®, Solibri®,* and *Cinema 4D®,* all mainly targeted at 3D modeling or construction.

Archicad is currently one of the major BIM-authoring software tools, alongside competitors such as *Autodesk® Revit®,* Vectorworks, and Allplan. It has a good reputation for being user-friendly and supporting a wide variety of 2D and 3D file formats for collaboration with others. It is worth mentioning that *Graphisoft* is a strong supporter of the *openBIM* approach by supporting the exchange of data across BIM software, using open standards and related file formats.

Installing Archicad

While we assume that installing software should not be a problem in general, there are a few aspects of the Archicad installation that should be considered.

Platform requirements

Archicad is **multi-platform** software, which can be used on Microsoft Windows and Apple macOS®. The exact requirements of the software are indicated on the Graphisoft website (`https://graphisoft.com/resources-and-support/system-requirements`), so for fully up-to-date specifications, you should check there first.

In general, Archicad supports the current iterations of both operating systems. However, when a major update of the platform is released, it is recommended to consult the Graphisoft site prior to upgrading. macOS users especially have encountered issues by jumping too soon when Apple releases an upgrade to its operating system, which is about every year around autumn. There are often large changes that impact graphically demanding software such as Archicad. Conversely, Graphisoft always tries to benefit as much as possible from software and hardware improvements, and not only fully takes advantage of multiple cores but also, more recently, embraced Apple's *M1*® chipset.

In general, there are a few requirements that are of primary importance, which you can see in the following list. Archicad is demanding software, both in terms of storage and in terms of system requirements:

- Having a fast *CPU* is but one of the requirements. Archicad makes heavy use of **multi-threading**, so having a multi-core processor (four cores or more) is recommended for smooth operation. If your system is still 32-bit, it won't work, but in that case, your computer is too old to run such intensive applications anyway.

- You need plenty of *RAM* available, both in the main memory and the video memory. As the software is graphics-intensive, if at all possible, use a system with a **dedicated GPU** or graphics adapter. This is often difficult with office laptops, which tend to have a limited GPU. This may limit navigation and 3D display. CAD and 3D software ideally would run on a workstation-class desktop or laptop. Some users also get good performance with gaming-oriented systems, since they tend to have powerful graphic cards.

- It is highly recommended to use **solid-state drives** (**SSDs**) for the fast loading of projects and library files with sufficient storage space. The installer is almost 2 GB, and when installed, the software uses several GB of storage. This is largely due to the object libraries, including the default plugins and textures.

- Depending on the size of the projects you develop, you also have to understand that Archicad projects tend to generate large files. Smaller projects may need less than 100 MB, but larger and more complex projects can go above 1 GB for a single project file.

- As with most professional applications, having a **large monitor** (or two or three) also helps with the setup of your workspace. Using a full HD or 4K screen resolution is recommended. Professional users typically select a dual-screen setup or a very large widescreen to get the best working environment, but if needed, you can use the software on a laptop and bring the whole authoring environment with you, on-site or to a client.

- Archicad is mainly controlled with a mouse and keyboard, so touchscreens do not provide any advantage, but an advanced 3D mouse can be beneficial.

And finally, as discussed in the following section, Archicad is commercial software that uses a protection system. This can be in the form of a hardware key, requiring a USB port, or a software key, in combination with your personal **Graphisoft ID**.

Older versions of the software

It is still possible to use older versions of the software, which remain available for download from the Graphisoft website. While your license entitles you to use any older version of the software, the actual version that will work depends on the operating system you use. Older versions are not further developed nor supported and may not work in current versions of your operating system.

For people using an older computer with an older operating system, this is sometimes a solution, as they can run a previous Archicad version, with lower hardware requirements.

Another reason to use older versions may be related to the project – when construction projects last several years, it is not always acceptable or advisable to upgrade software tools throughout the project, unless this is agreed upon with other people you collaborate with. In that case, offices sometimes tend to continue a project using the software version that was used when setting up the first models. In other cases, upgrading to a newer release has to be planned carefully to not disturb agreed workflows. It is not uncommon to wait a few months after a new release before upgrading.

A final reason for keeping the libraries of older installations available is related to the project and library management – although there is the option to save a project in an archive format (which contains any Library Part used in the project), occasionally some Library Parts are not correctly migrated or are even completely missing. In such a scenario, still having copies of older versions of the Library Container Files can be a lifesaver. You do not even need to have the software fully installed for this, and in any version of the software, Migration Libraries are included, but this does not always solve every migration issue that could arise...

By now, you have learned how Archicad came into existence, what it is meant for, and which hardware and software is required to use Archicad. Next, we will learn what license types are available and what type and version of Archicad is required for this book.

Licensing

Archicad is developed and marketed as commercial software for professional users. As with most advanced and specialized software, this requires a user to select one of the **different license models**, depending on their activity.

License types

The primary license model is a **full commercial license**. The user gets access to the full software using either a hardware lock (*dongle*) or a software key, to prevent abuse of the license or sharing a license. Graphisoft still offers this as a perpetual license, which never expires, but it also advises users to get a maintenance contract, which adds support and all future Archicad updates at a yearly cost.

For people not requiring the full feature set of the commercial license, a cutdown **STAR(T) edition** is also available. This comes at a lower license cost and is primarily targeted at smaller offices, where collaboration and integration features are not required.

Alternatively, a **pay-per-use license** is also available when you need one or more additional licenses for a limited time. This can also be of interest to users who use the software just occasionally, although you may need a full commercial license to qualify for a pay-per-use license. It should also be noted that this license type may not be available in your market.

If you intend to develop add-ins for Archicad, you can opt for a **developer license**, which is valid for 1 year and gives you full access to Archicad and support from the development team. Note that such a license cannot be used for commercial purposes, such as architectural or engineering services.

More recently, Graphisoft and its resellers have started to provide a **subscription license**, which, as the name implies, gives you access to the software and all updates for a fixed, monthly, or yearly fee.

For actual license pricing, you are advised to contact a local reseller, as there may be differences between not only markets but also the offered services and additional benefits, such as localized templates or libraries.

While you are still discovering the software, you can request a **temporary trial license**, which gives you full access to the software, including saving and printing, but with a fixed expiration date and locked to the computer where the trial was activated. If you decide after the trial to continue with a commercial license, Graphisoft can assist you with upgrading the files saved during the trial.

Finally, if you are a student or teacher, you have free access to the **educational license**. This has no functional limitations, but the output of documents and exports will be watermarked to avoid commercial use. You can access educational licenses at `https://myarchicad.graphisoft.com`. You can start immediately with a temporary 30-day license, which is extended for 1 year after approval by the regional reseller, upon proof of educational enrolment. This offer can be extended as long as you still comply with the educational requirements.

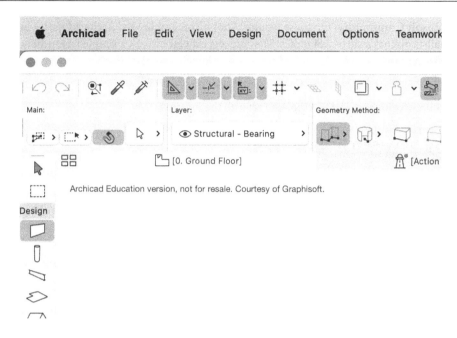

Figure 1.1: An example of the educational watermark in the Viewport

In all cases, you must activate the software using a (free) Graphisoft ID to ensure you are entitled to the license you use. This Graphisoft ID also gives you access to multiple Graphisoft products and services, such as the following:

- **Graphisoft BIMcloud**: This server technology lets you collaborate with several team members on one central model, using Teamwork technology. It is available in three license types, of which *BIMcloud Basic* is included with any commercial license.

- **Graphisoft Learn**: This online training platform can be a great accompaniment to this book. Tutorial videos are offered on a variety of topics, some of which are discounted or even free for certain license types.

- **BIMx and BIMx Model Transfer**: BIMx is available as a separate mobile app and as a desktop application (the free version is included with every Archicad installation – you can easily upgrade to the paid **BIMx Pro** for some extra features). The app combines fluent 3D navigation and visualization with interactive hyperlinks to 2D-derived documentation. This transfers a model and a full set of (chosen) data from your modeling tool to your mobile presentation device, making it well suited for client meetings and visits to a construction site.

- `https://bimcomponents.com`: Here, you can find additional Library Parts for Archicad, checked and approved by Graphisoft.

- **Graphisoft Community**: On these forums, Graphisoft employees, Archicad gurus, and BIM specialists discuss a variety of topics. They are a great place to ask a simple technical question, propose an improvement to the software, lose yourself in a discussion on a GDL parameter, or just make new friends.

For more on these resources and links, please refer to the *appendix*.

Migrating between license types

Exchanging projects between Archicad versions with different license types is somewhat limited. You can perfectly exchange Archicad projects created with a full commercial license, a subscription license, or a pay-per-use license. A full version can open projects created with any STAR(T) edition, but that edition does not allow you to open files other than the ones created with this license type.

Should you decide to acquire a fully licensed version of Archicad, after your trial period has ended, you will be able to continue to work on the projects created within this trial version of the software, as long as you **upgrade** the trial and did not uninstall it.

Projects created with an educational license can be opened in a fully licensed version of Archicad, but Archicad will (temporarily) switch to Educational Mode and show the **educational watermark**. Opening a file made with a commercial version in an educationally licensed version of Archicad will transform the file irreversibly to an educational version. Copying from educational projects to commercial ones is also prevented.

Localized versions

Archicad is developed for a worldwide audience. However, to cater to the large difference in languages, terminology, and construction methods, it is distributed in **different localized versions**. This not only defines the language of the user interface but also gives Graphisoft, or the local reseller, the opportunity to adapt the content that is installed with the software.

There is one main version, which is called **International** (**INT**). This uses the English language and contains the default Archicad library and template.

Various countries or regions have a localized version available, and they can be discovered on the Archicad website. Be sure to select the region that is associated with your Graphisoft ID. Check with your reseller to see which of the more than 20 localized versions is available to you and whether your license gives you access to different localized versions. If your region is not available, you can use the INT version instead.

Requirements for this book

For this book, it doesn't really matter which license you have access to. Considering the depth of the subject, a time-limited trial version may not last long enough to work through the whole book, but you can use any commercial or educational version without issue.

The book is written primarily using **version 26** of the software, but as the software is quite mature, most of the content still applies to previous versions. However, we don't recommend trying to follow along with versions older than 3 or 4 years, since the interface and the features have evolved considerably.

The approach in this book

This book is written with the assumption that you are starting from scratch, with only minor experience in architectural drafting and general IT skills. We don't want to bother you with countless trivial details about software in general but, rather, focus on Archicad and where it can differ from other systems.

There is a logical order in the chapters, starting from the humble beginnings in *Part 1*, where you are introduced to the software and learn just enough so you can create a basic house, sufficiently developed for the preliminary design. This will help you at the start of a project.

In *Part 2*, we will increase the level to an intermediate stage, where we will learn about many of the tools we haven't covered yet and dive deeper into the ones we already learned about. You may start to experience the sheer depth of Archicad and how flexible it is. It must be, because it is used worldwide, in a variety of countries with vastly different construction methods and materials.

After these two parts, you should be able to set up and complete your first few projects, at least if you are willing to practice and dive deeper into the documentation or the online communities when some details need more explanation.

While there is much more to Archicad, we had to leave out some of the more advanced and complex methods in it. The *Appendix* gives a brief overview of many of the additional features Archicad has to offer, such as collaborative BIM workflows, both with other Archicad users and with people running other software, by using *openBIM* methods and formats. There are also engineering-focused structural and MEP modules (used for modeling Mechanical, Electrical and Plumbing elements). To cover all of this, we would probably need one or two additional books.

What to expect as a reader

This is not a cookbook of point-and-click recipes. We want to explain the different tools, but with a strong focus on methodology and the reasoning behind the different choices that are made. The aim is for you to *understand what you are doing*, rather than memorizing tricks that only work well in a fixed scenario.

Every project is different, and the many tools in Archicad can all be used in a variety of ways. It is a flexible system, but with that comes certain complexity. So, we are not able to cover every little detail, setting, or configuration. After all, the whole manual, which hasn't been a printed book for a long time, is over 4,000 pages! It is digitally available from within the software, and properly searchable and indexed.

> **Note**
> The Archicad reference guide is the full catalog of every tool and setting. This learning book is a guideline for using the software efficiently. They are complementary.

If you need more resources, the *Appendix* will give you more references to the various communities and other channels of information and knowledge related to Archicad.

Our experience, as users and teachers of Archicad, is that starting with core training (such as working through this book) before you properly get used to finding your way around the software and have a good grasp on the terminology and tools at your disposal works best. From that point on, it becomes much easier to look for answers, to search through the documentation, and, in general, to solve your problems yourself.

This workflow approach could have gone in different directions; one approach is going through all the tools, one by one, to ensure that you don't miss a spot. That is not the approach we like to bring you, as a reader, since it takes a long time and is, frankly, very boring.

The opposite approach would be to take a single project from start to finish. There are many arguments for this, so we will follow this approach to a certain extent. We will move mostly in chronological order of what you would encounter in a project, but there are a couple of catches:

- Developing a full project works differently the first time around! Once you have completed a few projects, you know where you will end and, thus, can better prepare for it. You cannot do this when you first start. So, we must jump into a project with no assumptions about what we already know or understand. We will introduce tools that fit at that point in the process, but we may not cover them completely to not overwhelm you with details that don't matter initially.

- We haven't come across a single project, yet, that requires every tool in Archicad. Or, to put it another way, the architects in us are not too keen on arbitrary designs that don't have any architectural value and are just a mixed bag of all the tools you may encounter. They tend to be more like Frankenstein designs.

So, we opted instead to work with a few basic reference projects to guide you through the process and introduce smaller, singular examples if needed, or if the tool that is discussed requires it.

Version and units in this book

There is another choice we had to make in this book. To cater to the widest possible audience, we don't use one of the many splendid, localized versions of the software or one of the extensive specialized libraries available. We stick with the basic **INT version** and its default library and template. It was created by Graphisoft to be globally applicable for many project types. Since every localized version is derived from it, you could state that this version is at the core of every other version.

We understand your personal situation may differ from that, and you are strongly advised to adopt first and foremost the localized version and related templates and libraries for your language or region. This is especially advisable when your local reseller has prepared this generic template to better cater to local building practices, regulations, documentation habits, or anything else that Graphisoft could not prepare for you.

Another choice here is to stick with the **metric system** throughout the book, which is valid for all but a few countries in the world. We assume that those who prefer to work in imperial units can adapt easily, since the display of units is but a setting in the software. Since you will work at full scale, Archicad can display units as you see fit, using feet and inches or, rather, meters and their derived units.

A final note should be made about the screenshots included in this book. These are taken from projects opened with Archicad 25 or 26 INT, running on **macOS**. There are only a few differences with Archicad running on Windows, which will be mentioned throughout the book where applicable. Also, both the Mac and Windows variants of shortcut keys will be mentioned (e.g., *Cmd + D*/*Ctrl + D*).

So, with these explanations and disclaimers out of the way, we are ready to start our journey into Archicad.

Summary

In this chapter, we introduced you to Graphisoft Archicad, its variety of license types, and how you can select the applicable one for you. In the next chapter, we will learn a little bit more about the installation and explore the user interface of the software before we set up our first Archicad project!

Further reading

If you are interested in the history of Graphisoft and Archicad, here are some interesting links we based our summary on:

- `https://hbr.org/1991/01/micro-capitalism-eastern-europes-computer-future`
- `https://www.cultofmac.com/428277/software-firm-commemorates-golden-steve-jobs-moment-with-bronze-statue/`
- `https://graphisoft.com/it/press-releases/jobsstatueunveiled`

2

Getting Started with Archicad

When getting started with complex software such as Archicad, we must learn about its interface and how the software can be configured. First, we will be introduced to the way Archicad provides onscreen feedback. Then, we will get an overview of what gets installed with the software and why it is taking up so much room on your computer. And finally, we will start our first project and set it up properly.

Here are the topics covered in this chapter:

- Understanding the Archicad installation
- Understanding the Archicad **user interface** (**UI**)
- Setting up your first Archicad project

Technical requirements

For this chapter, you will need a working copy of Archicad—ideally, version 26 or 25. Any of the license options mentioned in *Chapter 1,* are fine. Choose the version that suits you best through the Graphisoft website (`https://graphisoft.com/try-archicad`).

Understanding the Archicad installation

Let's look at what gets installed onto your computer, what this all contains, and what the impact of a localized version on all of this could be.

What is installed?

The Archicad installation takes up quite some space... not only for the software and its components but also because of the **libraries** with content that get installed. The installation process is straightforward but may take a while. Meanwhile, you are greeted with a few nice pictures of projects designed by international Archicad users.

After the installation wizard finishes, you are asked to reboot your computer. This is caused by the drivers for the software protection system and the filesystem integration. Some security messages may also pop up to give Archicad the required permissions for the filesystem.

All in all, the installations on macOS and Windows are *practically identical*. They have the same folder structure and contain the same library files. Only the executables (the main programs) and the add-ons are platform specific.

Apart from minor differences between platforms, such as the availability of fonts or some platform-specific features, it should not pose any issue to switch between platforms or to collaborate with others.

Other languages or versions

If you use a **localized version** of Archicad in a *different language*, some of the folders may have a translated name, but their function remains the same.

For example, when we install the local *Key Member edition* for Belgium or the Netherlands, it gets installed over the current INT version and adds translated default libraries and some additional libraries, plugins, and resources.

By way of example, if you have a French, German, or other language edition of Archicad, the folder names and libraries are fully translated, and you end up with a separate installation folder for Archicad instead of the INT version. You can have multiple versions of Archicad installed at the same time: either different releases (24, 25, and 26) or languages (INT, USA, GER, NED...).

Now that we have a clear understanding of what is installed, it is time to fire up the Archicad application for the first time and explore its UI!

Understanding the Archicad UI

When you open Archicad for the first time, you are presented with the default interface layout or **work environment**:

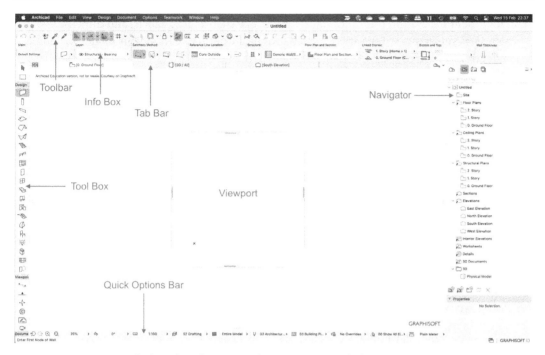

Figure 2.1: Default Archicad 26 INT work environment with the main UI components

Depending on your installed version of Archicad, things may look a little (or a lot) different, but it always works the same way.

Getting to know the main UI components

As with most software, Archicad has a **main menu** (or **menu bar**) that sits at the top of the screen on macOS and above the toolbars on Windows. The menu is organized logically from left to right:

- The **File** menu gives access to starting new projects, opening, closing, and saving files, and the menus for managing libraries, reference files, and interoperability tools.

- The **Edit** menu focuses on elements in the model, including commands for moving and transformations, alongside the typical **Undo/Redo** and **Copy/Cut/Paste** commands.

- The **View** menu groups many commands related to how Archicad displays the model, with a variety of options that can be toggled, such as the display of the grid, line thickness, and onscreen options to help when modeling.

- The **Design** menu helps you manage the Virtual Building, including accessing all modeling commands, extensions for design analysis, and the management of model information.

- **Document** focuses on the annotation commands, such as text, labels, and dimensions. It also gives you access to the variety of views Archicad can create from your model, including renderings and schedules.

- **Options** is where you manage attributes, properties, and classifications. We'll return to attributes in detail later in the book.

- **Teamwork** groups the **BIMcloud** collaboration features, to share and synchronize models with other co-workers.

- **Window** allows you to show or hide all UI toolbars, panels, palettes, and other windows that define the Archicad UI.

- Finally, the **Help** menu, the last one, gives access to the Archicad documentation, including online and offline resources and license information, and checks for updates to the software.

Depending on your Archicad installation, additional menus may also show up (between **Window** and **Help**), but we won't discuss them here.

Toolbars and palettes

Around the main graphical window, we get a series of docked toolbars and palettes.

Toolbars contain commands you can click, whereas **palettes** contain multiple widgets, such as buttons, lists, checkboxes, or other UI elements:

- **Viewport**—The largest area on your screen, meant for modeling, drafting, drawing… in other words: creating your design.

- **Toolbar**—By default at the top, this contains various non-modeling tools that are frequently needed but can also be accessed through the menus at the top. Some of the buttons in the toolbar can have underlying drop-down menus, indicated by a downward-pointing small arrow.

- **Tool box**—By default on the left, contains all the modeling tools, grouped in a logical way. From top to bottom, we have the following:

  - **Selection** tools (**Arrow** tool and **Marquee**)

  - **Design** tools (**Wall**, **Door**, **Window**, **Column**, **Zone,** and **Mesh** tools)

  - **Document** tools (**Dimension**, **Spot Dimension**, **Figure,** and **Drawing** tools)

  - **Viewpoint** tools (**Section**, **Elevation**, **Interior Elevation**, **Detail**, **Worksheet**, and **Camera** tool)

- **Info Box**—By default at the top, beneath the toolbar, this displays (condensed) settings for the currently selected command or tool. It contains a selection of the complete collection of settings that can be found by accessing **Element** (**Default** or **Selection**) **Setting**s, which is explained in *Chapter 3,* and further.

- **Navigator**—By default at the right, this contains the structure of your model and its different views, sheets, layouts, and publisher sets. This "table of contents" structures everything you create in a project and keeps track of how you want to view the model, create layouts, and publish all the content.

- **Tab Bar**—By default at the top of the **Viewport**, this shows all open views or viewpoints and lets you easily switch between active views or viewpoints without scrolling through the Navigator. Open tabs are given priority when the model is being updated in the background and when a change in the model is being applied.

- **Quick Options Bar**—At the bottom of the screen, below the **Viewport**, this shows the current display settings of the active tab. These settings can be used to quickly change the displayed content and the appearance of the currently shown **Viewport**, using the pop-up lists showing the available options. How a certain combination of settings is saved into a view is discussed later.

Toolbars and palettes can be docked so that they stick to the sides of the Archicad window, or they can float where you want. You can also close them if you don't need them. To reopen toolbars and palettes, go to the **Window** menu, and look for the desired toolbar or palette.

As with most software, Archicad also has countless *dialogs*, which can be opened from menus, from commands on a toolbar, or from palettes. In contrast with a palette, a dialog is *modal*, which means that it demands your full attention: the rest of the interface is blocked until you close the dialog, either by accepting its input or by canceling it.

Understanding work environments

The whole UI layout can be configured in detail and stored in a so-called **Work Environment**. You can store and recall work environments at any time. With your Archicad installation, a few preset environments are included, allowing you to switch between them.

A **work environment profile** is used to store countless settings and configurations, so you can apply this when you need it. We apply the default **Architectural Profile** from the menu (**Options** > **Work Environment** > **Apply Profile** > **Architectural Profile**).

When you apply the **Structural Engineering Profile** (**Options** > **Work Environment** > **Apply Profile** > **Structural Engineering Profile**), you get a different interface layout, with a few toolbars getting hidden and the set of available tools in the Toolbox limited to structural elements. To see what gets included in these profiles, we open the **Work Environment** dialog (**Options** > **Work Environment** > **Work Environment…**), which can be a bit overwhelming at first! Here in *Figure 2.2*, you get a full overview of the different profiles and what is controlled by them.

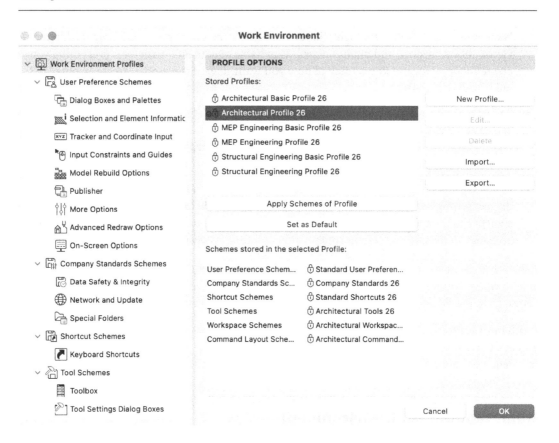

Figure 2.2: Work Environment dialog

This dialog is very, very extensive, as it allows you to completely reconfigure the Archicad interface, including showing or hiding tools and commands, changing shortcuts, and changing how Archicad works. It even includes options to let Archicad behave like an older version, which may be required when opening some older projects.

Advanced users often create *custom profiles*, to include a series of related toolbars and palettes that are focused on the task at hand. You already have profiles for architectural, structural, and **mechanical, electrical, and plumbing** (**MEP**) design, but you could also configure one profile that is more oriented toward drafting, one for visualization, and maybe another when training new colleagues, such as a "getting started" profile. When you install other versions of Archicad or extensions, new profiles may be installed as well.

As for this book, we will stick to the default **Architectural Profile** and, if needed, we tell you when to switch to another one.

Understanding Archicad onscreen feedback

When we model in Archicad (starting in the next chapter!), the software provides a large variety of onscreen feedback. We will dig deeper into how to use this feedback and associated interfaces to our benefit in later chapters, but it is important to define the types of interfaces we will encounter early on.

Status bar

At the bottom left of the screen, just below the **Quick Options** bar, Archicad gives feedback text and instructions on what we should do next when we activate a particular tool—for example, **Enter First Node of Line** when starting the **Line** tool, or **Complete Line** when the first node of a line is already defined. The **status bar** therefore can be seen as a dynamic *quick tooltip*, always telling you what Archicad "needs" to be able to perform an action, such as creating a wall or adding a dimension line:

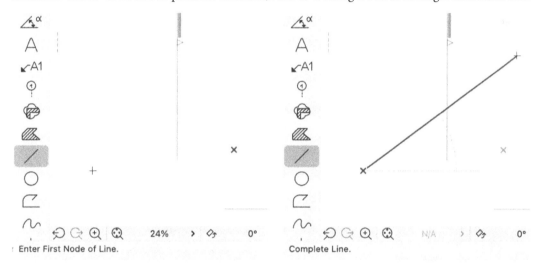

Figure 2.3: The status bar in action

Pet Palette

The **Pet Palette** pops up next to your cursor when selecting an element in the **Viewport**, displaying several relevant options on how you can edit the selection. The selection of icons shown is based on where you click on the element (for example, clicking the endpoints of a line gives different options than clicking the line itself). This is called a **contextual menu**. Learning how to use this Pet Palette efficiently is key to becoming an Archicad modeling expert, and a lot of its options will be covered in later chapters:

	Endpoint (2D)	Edge/Reference Line (2D)	Edge/Reference Line (3D)
Line			
Wall			
Shell			

Figure 2.4: Some versions of the Pet Palette

By default, the Pet Palette follows your cursor around the **Viewport** like a pet, hence its name. If this behavior bothers you, it can also be set at a fixed position through **Options** > **Work Environment** > **Dialog Boxes and Palettes** and setting **Pet Palette movement** to **Jump to preferred position**:

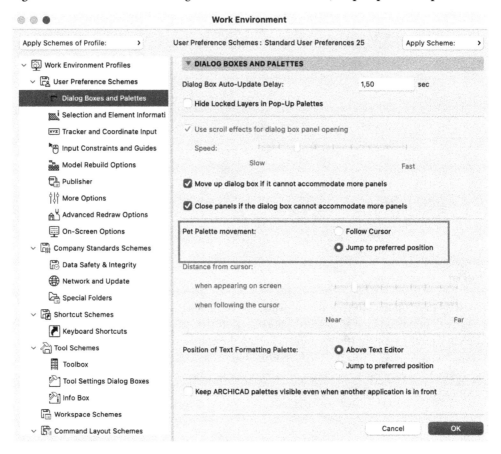

Figure 2.5: Changing the Pet Palette behavior

> **Getting to know the Pet Palette options**
>
> Any Pet Palette can contain a multitude of options/icons, but it is not necessary to know all of them immediately. Most will be introduced throughout this book when we start modeling. If you want to know what a certain option in the palette does, however, you can always hover over the icon and wait for a moment until a tooltip appears, describing the function of any icon.

Cursor states

The mouse is the most important instrument for input in the Archicad viewport. To help the user to *model accurately and precisely*, without having to zoom in and out all the time, the *shape of the pointer* changes to provide the user with useful and meaningful feedback. This way, we can see immediately if we are clicking a point, edge, or face of an element. This is called the **Intelligent Cursor** by Graphisoft. It can tell us that we are going to cut/trim an element (by pressing *Cmd/Ctrl* while clicking), that we are picking up parameters of an element, or that we are injecting those picked-up parameters into something else (by pressing *Opt/Alt* or *Cmd + Opt/Alt* while clicking).

An overview of all cursor states can be found in the Graphisoft Help Center:

Figure 2.6: Different cursor shapes (Pencil Shape) during modeling input

As an example, *Figure 2.6* shows the cursor shape during modeling input, when pointing at empty space, a reference line, other edges, an intersection of edges, a node of a reference line, or other nodes.

Tracker and other onscreen feedback

Besides all this *visual onscreen feedback*, there are even more tools and feedback close to the cursor when we are modeling in Archicad:

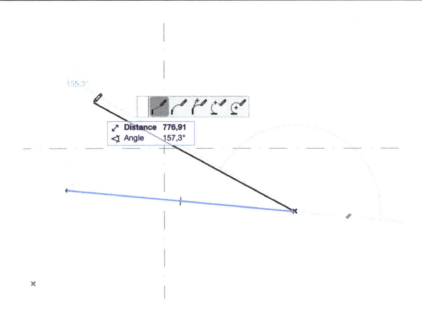

Figure 2.7: Tracker showing distance and angle, Pet Palette, pencil icon, and Snap Guides

The **Tracker** gives numeric feedback—for example, the "length" or distance of a line and its (relative) angle to other elements. It can be used for exact numeric input, which we will explain in later chapters. On top of this, there are also all sorts of blue lines appearing in the **Viewport**, called **Snap Guides**, further helping the user in modeling accurately and efficiently—for example, parallel or perpendicular to other elements in the current view.

All this feedback is a lot to take in as a new user, but this book will get you familiar with each of these modeling aids and how to use them to your benefit step by step. They will be explained in combination with the modeling tools, starting from *the next chapter* onward. Before we start modeling, though, we have to set up our project correctly, as we will see in the next section.

Setting up your first Archicad project

When you launch Archicad, you are presented with a welcome screen, which gives you a few options to open a project. You can open one of your recent previous projects, using the tiles, browse for a local file, connect to a BIMcloud project, or you can start a new one from scratch based on one of the available templates:

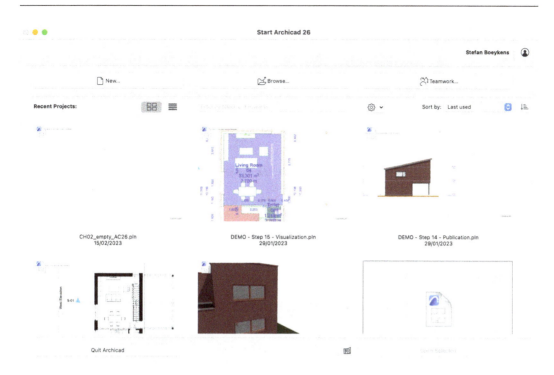

Figure 2.8: Start Archicad dialog, showing a preview of recent projects

When you do select to create a new project, you can choose the template and work environment profile you want to use for this project. Since at this point in this book, you lack the experience to develop a template for yourself, you are advised to stick to the template(s) installed by default. This, again, depends on the language version of Archicad you have access to:

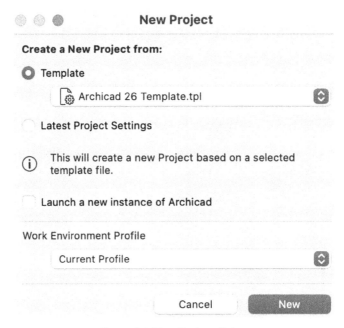

Figure 2.9: New Project dialog

We start here with the **Archicad 25 Template.tpl** template, which gives a generic but usable setup that can be applied in many situations. It prepares an empty project, with a few views and a full configuration set up for you to work in. Since there are so many aspects of the software we can control and configure, it would be impossible to really start from an empty file, with nothing configured, as it will hinder you in every single step. It is more efficient to work in a template and tweak the configuration if needed—for example, by adding new types or adjusting some of the display settings and view settings. We'll show you how.

Project Info

While not immediately required, it is a good habit to fill in the project information early on. These are properties at the project level, and they can be used later, such as when automatically filling in information for a title block.

You can access the project information from the menu (**File** > **Info** > **Project Info...**) or by *right-clicking* **untitled** at the top of the Navigator and getting a dialog where you can set all the relevant information about the project:

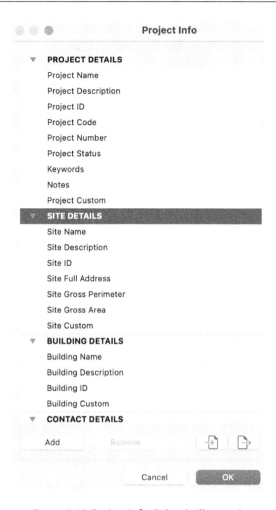

Figure 2.10: Project Info dialog (still empty)

There are sections with details on the project, site, building, contact, and client.

It may not be immediately obvious, but this information is also included when you exchange files with other project stakeholders: one of the many examples where **Building Information Modeling (BIM)** is more than geometry—where information is at the core of everything.

Building stories

Since Archicad is clearly targeted at the modeling of buildings, it uses terminology and a setup that assumes you are creating a **Virtual Building**, which is the term Graphisoft launched as a concept in 1987, describing what Archicad stands for, before the BIM acronym became the accepted term in the industry around 2003, more than a decade later!

When you design a building, *multiple stories* could be required. Stories in Archicad define the different *floor plans* you can enter, and they are vertically stacked: they are positioned at a specific elevation, which defines the height to the next story above. Typically, the elevation value of an Archicad building story corresponds with the real-life elevation value of a finished floor in a building.

Access the **Story Settings** dialog in the menu (**Design** > **Story Settings...**), through the contextual menu in the Navigator (*right-click* on a story), or press *Cmd + 7/Ctrl + 7*. You get a tabular overview of all stories in your project:

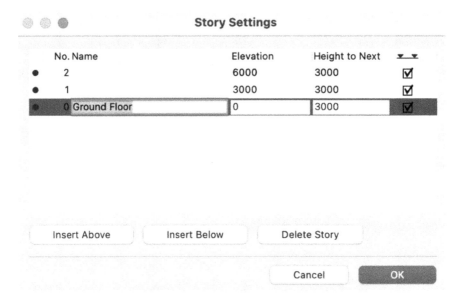

Figure 2.11: Story Settings dialog with default values

The default template has already created a **Ground Floor** story and two unnamed stories above. You can rename them as you wish and insert new stories above or below the selected one. Note that **Elevation** and **Height to Next** are *linked*: changing the **Height to Next** setting also modifies the **Elevation** setting from the story above and vice versa.

> **Pro tip**
>
> Here, you may encounter the *auto-apply* behavior of Archicad: whenever you fill in a value in one of the text or number boxes, Archicad automatically applies this after a small delay. You don't have to press *Enter* or close the dialog to apply the value. This gives you the option to continue making changes and it also works in palettes, where there is no **OK**, **Apply**, or **Close** button available. Just enter a value and wait half a second.

Origin and display grid

Since an Archicad project is meant to host a construction, you must position it somewhere. As with any **computer-aided design** (**CAD**) or BIM software, you must place elements by defining the position of their main reference points against a *reference* **coordinate system**.

Origin

Archicad uses a *Cartesian* coordinate system, with a small cross indicating the **origin** or center of the Cartesian world. While you can start modeling anywhere you want, it is highly recommended to use the Archicad origin as a central point of reference for anything you do—consider it as the origin of the project, with *all elements* of the projects modeled *in the close vicinity of this origin point*:

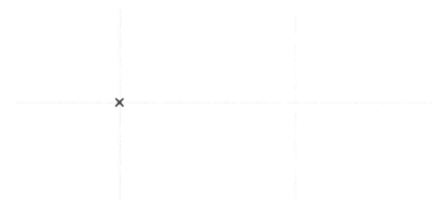

Figure 2.12: Archicad origin cross – the "origin" of the project

In many cases, this point will be located at or close to a corner of the building or at some other reference point, such as the crossing of two grid lines.

Even though Archicad is very flexible in navigating around projects, a good project origin is important for file exchange and collaboration. In fact, it is so important that it is one of those conventions you need to agree upon in collaborative BIM projects.

Grid

Another aid when modeling is the **grid** (**View** > **Grid & Editing Plane Options** > **Grids & Background**): a visual series of lines at a fixed distance, to give a frame of reference toward *sizes and scale*. Without the grid, you may start to lose grip on the size of the project, especially since you can zoom in or out almost infinitely:

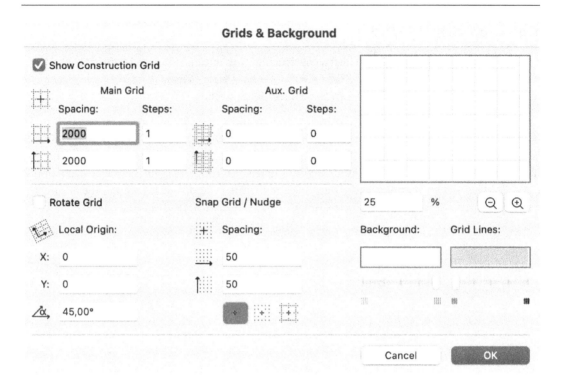

Figure 2.13: Grids & Background dialog, to configure Snap Grid, Construction Grid, and grid line colors

Archicad has two main options for the grid:

- The **Construction Grid** is intended as a larger-scale grid, aligned to typical construction distances, such as 6 m or 7 . 2 m (typically a multiple of 60 cm, based on human proportion). You can set both a **Main Grid** and an **Auxiliary Grid**, with two different repeated grid sizes. This is sometimes used in architectural design, such as Japanese *ken/tatami* grids or Dutch *structuralism* when, as an example, you want to have a repetition of two main grid steps and one auxiliary smaller step.

- The **Snap Grid** is intended as a modeling aid, with rather small snap sizes—for example, set at every 10 cm. It is also applied when "*nudging*" selected elements, by shifting them by one grid step (*Shift* + arrow key).

The grid in Archicad can be used for two different purposes:

- You can use the grid to snap the position of the cursor, enforcing you to always *work at a fixed grid size*. This can be helpful when modeling, as the position of walls, slabs, and other elements will be guaranteed upon clicking on the grid.

- You can also use the grid as a *visual aid*, by displaying lines on the screen. They don't impact snapping but give a visual clue. This is helpful as the size and proportion of a project are not always easy to interpret on a computer screen, especially when compared to working on paper, where you traditionally had a better feel for the scale of a project. You can set the line color and weight as you want. The default uses thin gray lines on a white background color.

Both options can be combined as you see fit. You can display the grid or not, and you can choose whether it influences snapping or not. For the initial sketch design, snapping to a Construction Grid is convenient, but once you start developing the design in more detail, the Construction Grid is too obtrusive. In that case, you'd probably use the Snap Grid or rely on automatic snapping to the geometry and not the grid.

We learn in the following chapters how other snapping and guiding methods are available for precise modeling.

Saving a project file

Once you have properly configured the basics of your project, do not forget to save the project file in a suitable location, ideally dedicated to this project. This will help to keep things together and ensure that all output can be kept in an easy-to-find location.

When you save the project, Archicad stores it as a **PLN file**. This is a single, compressed file, containing everything you have created and configured in the project. It is specific to the version of Archicad you are using, but if needed, you can export the project to a previous version (limited to *one* earlier version), albeit with a warning that new concepts may not transfer properly.

You are advised to *save frequently*, to ensure you don't lose too much work should anything unexpected happen, such as a freeze of the computer or a crash of the software. Archicad is a complex system and has many features, and while the software is mature and goes through extensive testing, it isn't entirely possible to rule out bugs that may lead to the software crashing. If that occurs, two things will happen:

- Archicad attempts to create a **recovery file** (created by an auto-save function), allowing you to continue your work.

- Archicad also launches a **bug reporter**, which gives you the opportunity to send the crash information to Graphisoft, helping their engineers to investigate and solve the issue for future updates or hotfixes.

> **Pro tip**
>
> In our experience, the recovery file has proven to be quite reliable, and we seldom lose more than a few minutes of work, mainly thanks to the **Data Safety** settings of the software. They can be accessed through **Options** > **Work Environment** > **Data Safety and Integrity…**. By default, this is set to **Ultra Safe**, saving every step. This may cause performance issues in more complex projects (for example, where editing a large, complex **Curtain Wall** takes quite some calculation time for every step), so you may consider setting it to **Safe**.

As an additional precaution, Archicad also saves your previous PLN file as a backup **BPN file**, containing the last saved version of a project, should you want to return to the previous state. However, this is more dependent on how often you save your project manually.

One final point of attention, should you want to open your project on another computer: the PLN file contains the whole project, but it only keeps *references to any library object or texture* that was used in the project. These library objects and textures are stored in **Library Container Files** (.lcf as the extension). This helps to keep file sizes down, but it also implies that the library content is not transferred with your project file. When you just use the default library from your Archicad installation, another user can open the PLN file, and everything gets restored with no issue from that library.

If, however, they use a different setup that lacks the library you used, this does not get transferred. In that case, you should make an *archive* copy of your project, by saving it as a **PLA file**. Here, you have the option to embed the library into the project, for archival usage or for transferring the whole project as a self-contained file. You may opt to only embed the referenced objects, to keep the PLA file size as low as possible. If you embed the full library, your project will be really huge, and this is not needed.

We can summarize the different Archicad file types as follows:

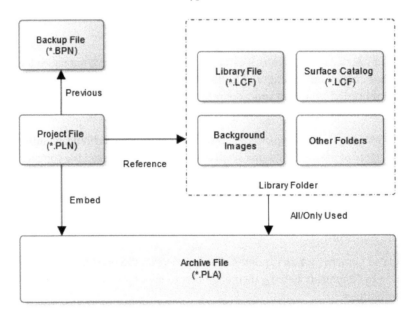

Figure 2.14: Diagram of project, backup, library, and archive files

Summary

So, there you have it. We have looked at the Archicad interface and selected a template and work environment. Then, we started a new project from a template and set up a basic configuration for our project.

We are now finally ready to get started creating a model!

3

Building a Basic Residential Model: Modeling the Construction Elements

In this chapter, we will cover the basic modeling concepts of Archicad, needed for a first small-scale residential model. Many of the tools and techniques introduced in this chapter will be repeated and expanded later in a second, more complicated model. This is an extensive but necessary chapter. Here are the topics covered in this chapter:

- Navigating in 2D/3D and learning the basics of modeling
- Modeling tools in general
- Core construction—walls and slabs
- Using drawing aids in Archicad

Technical requirements

For this chapter, you will need our demonstration model, which can be downloaded from GitHub (`https://github.com/PacktPublishing/A-BIM-Professionals-Guide-to-Learning-Archicad/blob/main/DEMO.pln`). Save it to a local folder on your system.

Navigating in 2D/3D and learning the basics of modeling

In this section, we will learn how to fluently navigate through and interact with a model in both 2D and 3D.

Opening your first model

Before we can learn how to navigate in 2D and 3D, we need a model. For this, we will use the demo model that you just downloaded. We will use this example throughout this chapter to learn how to navigate in Archicad.

You can open the model by either double-clicking it in your Finder/Explorer (hence starting the Archicad application) or by first starting up Archicad and browsing for the file through **File** > **Open** > **Open...** (*Cmd + O/Ctrl + O*), clicking **Open** once you have selected the correct project within the right folder in the following dialog screen:

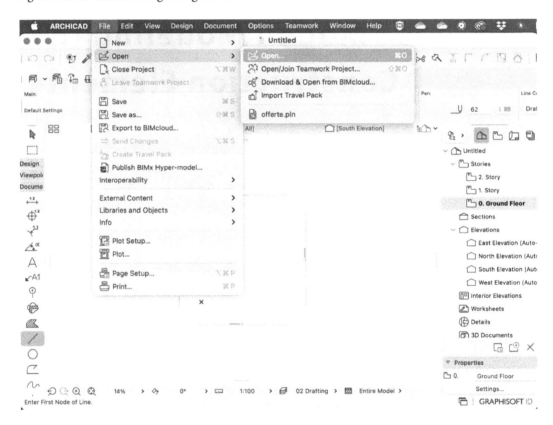

Figure 3.1: Opening a project from within a running Archicad session

Archicad models in the cloud

Storing models in the cloud is not advised in most cases. The backup process of Archicad (creating the `.BPN` file) may conflict with the synchronization processes of your cloud service of choice. Some cloud storage solutions offer an option to keep a file or folder on your hard disk permanently. In such a case, using this file/folder for Archicad might work better. More information is available at *Graphisoft Community* (`https://community.graphisoft.com/t5/Project-Management-articles/Sync-utilities-and-issues-with-Save-operation-in-ARCHICAD/ta-p/303937`). Note this is not the same as using **BIMcloud** from Graphisoft, which is designed specifically for storing and sharing collaborative Archicad projects.

The main part of the Archicad interface (the Viewport) is where you will interact with the building. Since buildings are much larger than your screen, you will need to learn how to navigate around the model through the various views.

Navigating in 2D

In this section, we go through the basic operations and explain the related mouse movements and shortcuts.

Panning

Panning is the equivalent of moving your display, drawing, or another view sideways, without changing its size. You can pan up, down, left, and right. This is one of the core navigation methods of Archicad, and there are a variety of methods to pan.

Technical prerequisite

Open the demonstration model—**1. Story** should be the default view when opened. This is also shown in the **Navigator** on the right-hand side, marking this story in bold text.

When using the mouse, you press and hold the **middle mouse button** (**MMB**—*the scroll wheel*) to drag the view left, right, up, and down. The cursor changes to a hand to indicate panning:

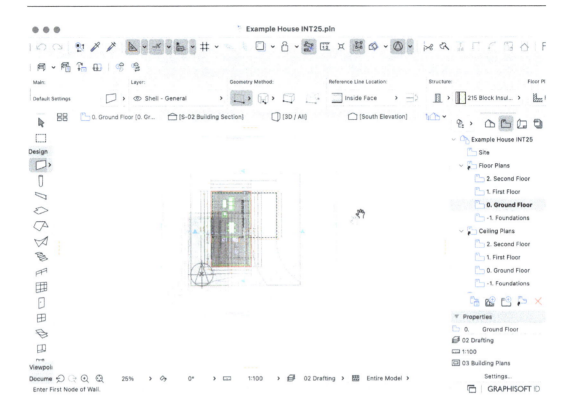

Figure 3.2: Panning a floor plan diagonally (right + upward) using the MMB

With the *Apple Magic Mouse*, there is no *MMB*, so you slide your finger left, right, up, and down instead.

Using the keyboard, you can use the arrow keys to navigate the view in small steps. The arrow indicates the direction in which you want to move, which is the opposite of the direction in which the view itself will move. It sounds more complicated than it is, really.

You can also pan using the trackpad, by swiping with two fingers at the same time.

Zooming

You can display the model at any scale, from an overview of the whole project up to the finest detail. You use the **Zoom** navigation command to enlarge or shrink the display of the model. However, the model is always at the same absolute size! Zoom in to look in more detail; zoom out to get a large overview.

When using the mouse, you scroll the mouse wheel to zoom in and out. Archicad will smoothly enlarge or shrink the display. Beware that the position of your mouse will remain the focal point for the zoom navigation. When you hold down *Shift*, this will increase the speed of zooming.

With the Apple Magic Mouse, there is no scroll wheel, so you use a finger swipe combined with the *Option* key. It works well but requires two hands.

Using the keyboard, you can press - (minus) to zoom out and + (plus) to zoom in. This works best when you have a separate numeric keypad, which is not always available if you work on a laptop.

Finally, you can also zoom using the trackpad, by pinching two fingers. This works quite intuitively, but only if you have a large trackpad, such as on a MacBook Pro.

Navigating around like a pro

Take some time to learn these two navigation methods. With the mouse, you can immediately switch between zooming and panning, which you'll have to do continuously when working in Archicad. Try to move around, pick a small region of interest, and see if you can bring it up close, in the middle of the display. Switch between the different shortcuts (mouse, keyboard, and trackpad) to get a good grip on navigation.

When you want to bring the whole project into the center of the view, you can use the **Fit in Window** command. You find this command in the menu (**View** > **Zoom** > **Fit in Window**), using the shortcut displayed in the menu or just double-clicking your *MMB*. The same command is also available in the context menu (*Ctrl + left-click* or *right-click*):

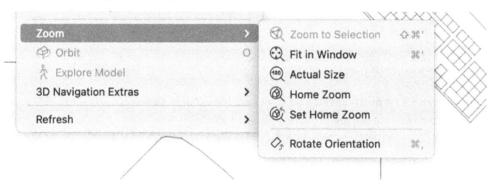

Figure 3.3: Zoom commands in the View > Zoom submenu

In the same menu, you also get the option of zooming the view to its actual size, at 100%. What does this mean? This is an approximation of the view at print size, for the current *display scale* (for example, `1:100; 1:50`).

Both **Fit in Window** and **Actual Size** (through **Current Zoom Factor** > 100%) can also be activated from the buttons in the **Quick Options** bar below the Viewport:

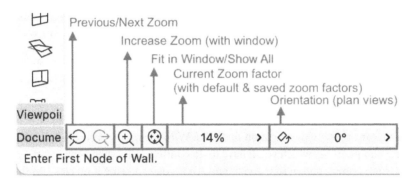

Figure 3.4: Zoom options in the Quick Options bar

> **Contextual menu**
>
> The contextual menu will be used throughout the book. You can always access it, in any view or layout, by using the **right mouse button** (**RMB**) or the *Ctrl* + **left mouse button** (**LMB**) combination if you are a Mac user. We already encountered it briefly in *Chapter 2*, in the *Building stories* section.

And even further navigation

After you select an object by left clicking on it, an additional **Zoom to Selection** command becomes available. It can be found in the menu (**View** > **Zoom** > **Zoom to Selection**) or in the context menu. This adjusts the zoom so that the object gets framed as large as possible in the center of the view.

> **Note**
>
> You may have noticed by now that Archicad is a professional application, with loads of commands and variants to cater to a wide variety of situations. You are advised to start with the basics but also be attentive to the shortcuts displayed in the menus so that you can pick them up gradually. We are not going to overwhelm you with every possible command, variant, and shortcut. You'll remember them best when using them repeatedly.

Configuring mouse behavior

If you don't like the default behavior of the mouse wheel or Magic Mouse, you can configure them differently in the work environment (**Options** > **Work Environment** > **Input Constraints and Guides**). Here, you can alter the way the mouse wheel, Magic Mouse, and trackpad react. The options are slightly different between Windows and macOS, obviously:

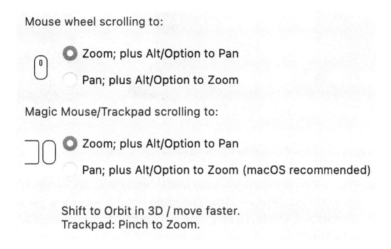

Mouse wheel scrolling to:

◉ Zoom; plus Alt/Option to Pan

○ Pan; plus Alt/Option to Zoom

Magic Mouse/Trackpad scrolling to:

◉ Zoom; plus Alt/Option to Pan

○ Pan; plus Alt/Option to Zoom (macOS recommended)

Shift to Orbit in 3D / move faster.
Trackpad: Pinch to Zoom.

Figure 3.5: Configuring mouse and trackpad behavior via Input Constraints and Guides

Navigating in 3D

So far, we have stayed in the 2D floor plan window. At any given moment, you can switch to a 3D window. This can be done from the menu (**Window** > **Floor Plan** / **3D Window**) or using the *F2*/*Fn + F2* and *F3*/*Fn + F3* shortcuts respectively. If you have a *Touch Bar* on your Mac, these two options are shown using an icon and text.

From within the 3D window, the main methods of navigation (pan and zoom) still apply, and they work mostly the same: scroll to zoom, drag with the *MMB* to pan, and the related swipes from the Magic Mouse or trackpad. You also have the **Fit to Window** and **Zoom to Selection** methods, as in a 2D window.

However, two additional navigation methods help you to easily navigate around the model in a 3D window.

Orbit

When you want to look at the model from a different angle, you can rotate around the model with the **Orbit** command. This is performed by holding down *Shift* and dragging with the *MMB*. On the Magic Mouse, you also press *Shift* but will have to swipe instead. *Shift* works even when you swipe the Trackpad with two fingers. You can see a visual representation of **Orbit** being used here:

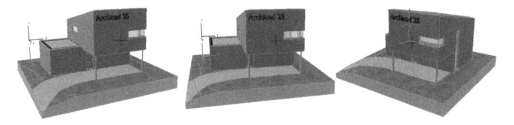

Figure 3.6: Different viewpoints during Orbit

All these navigation gestures are temporary, while you hold down *Shift*. If you want to remain in **Orbit** mode, you can press *o* as a shortcut. Now, the **Orbit** command remains active, and you can orbit using the regular *LMB* or the trackpad. **Orbit** is a very important method, so practice it a bit before you continue. You will use it all the time.

Explore (or the "game" mode)

When you are in a 3D window, the **Quick Options** provide a navigation method that is inspired by gaming. **Explore** uses a mouse/keyboard combination, just like typical 3D shooting or exploration games:

Figure 3.7: Accessing the Explore navigation mode

It introduces a whole new set of shortcuts and navigation gestures that will feel very familiar to gamers. To help you, Archicad displays them in a dialog, so take a few seconds to read through the options.

Remember to press *esc*, or click with the mouse, to exit the **Explore** mode. Archicad reminds you about this at the bottom of the main window, where a few additional options are presented, such as switching between walking and flying.

Take some time to explore the model and imagine how it would feel when roaming your own project in real-time 3D. This is very helpful, not only for presentations but also as a designer.

Modeling tools in general

We are almost ready to start modeling on our own. While Archicad has a huge set of modeling commands, they all behave quite similarly and are structured the same way. Once you can model a linear element, such as a wall, it is not difficult to apply the same methods to create a beam, and you will see similarities between both tools with the basic 2D Line Tool. And after you have learned how to draw a slab, roofs and fills use the same approach, as their geometry is all contour-based.

Here are a few common behaviors or methods, which are available for many tools:

- Archicad distinguishes between **Default settings** and **Selection settings** in the Info Box. The properties displayed in the Info Box are either those from the currently selected element (**Selection settings**) or, if no element is selected, the setting that will be used for the next element you create (**Default settings**). Each tool remembers the last settings that were used, and these become the default settings for that tool in the project.

- Elements are typically created in a floor plan window, closely resembling 2D drafting on a drawing board or **computer-aided design** (**CAD**) software. This defines the position and horizontal extent of elements. Each floor plan is related to a building story, so whenever you place an element, it will be *linked* to that story, which becomes its **home story**. Some objects can also link their height to a story, such as walls and columns. This allows you to adjust the story settings while retaining the vertical linkage of the elements in the stories.

> **Home story**
>
> The home story of any element in Archicad is the story where this element "belongs"—just like in the real world (when the project is constructed). Setting the home story correctly is key to getting good 2D visualizations (for example, showing all elements on a floor plan of a certain story) but also helps with tracking quantities in schedules (for example, grouping elements and their quantities per story), so keep an eye on this setting!

- Some elements, such as walls or slabs, have an internal **structure**. This defines their composition, by assigning *building materials* (discussed later). There are three main types of structure: **Basic**, **Composite**, and **Complex** profiles. We will encounter all of them throughout the book.

- When creating a linear element, it is typically defined by its reference line: this is the imaginary line between its start and end points or *nodes*. The actual geometry is then set relative to the reference line. This reference line also defines the way elements can automatically connect—at corners, T-joints, or other combinations.

- Many tools provide different **geometry methods** during modeling. This defines how the user will enter the main reference points, lines, or contours of the element. The available geometry methods differ between tools but are quite similar: linear, rectangular, and polyline methods are used for many tools. The geometry method is typically selected before entering points, but it can be toggled during input.

- Last, but certainly not least, each element is placed on a **layer**, which is very similar to other CAD software. Layers are used to control the visibility of elements. When you hide a layer, all elements on that layer get hidden too. When you start a project, you are presented with a default set of layers. The visibility of the different layers is managed as a **layer combination** so that they can be retrieved easily.

> **Note**
>
> We return to all these concepts several times, so don't worry if you didn't fully understand their purpose right now. If you pay close attention, you'll quickly figure out that the different tools in Archicad are very similar: once you have learned a few techniques, they can be applied to many other commands as well, so you don't have to relearn them for other elements.

Finally, we are now ready to start modeling our very first project. Through the upcoming parts of the book, we will learn the skills to model a complete small residential building—the example house you used to learn how to navigate in Archicad. It will take a lot of steps and a variety of techniques, but we will take it slow and explain every step along the way.

Core construction – walls and slabs

In a **Building Information Modeling** (BIM) context, we should mention that we are always drawing or modeling *on a 1:1 scale*. In CAD systems, this is not always the case, as they apply drawing units linked to a certain real-life measurement (for example, 1 drawing unit = 1 mm). Since Archicad lets you create a virtual building, it makes sense that we use the real-life size to model this. We will learn later how to create different views, representing our design at different scales and with dimensions displayed in the units we want for a specific view.

> **Starting a new project**
>
> Start a new project (**File** > **New** > **New...** or *Cmd + N/Ctrl + N*), using the **Archicad 26 Template. tpl** and **Architectural Profile 26** templates as the work environment. Default story settings will suffice for now.

Modeling the basic shape of the ground floor

We will start creating our project by modeling a rectangular shape on the *ground floor* using the **Wall Tool**:

1. Activate the **Wall Tool** in the **Tool Box**.

2. Choose **Rectangle** for **Geometry Method**, by clicking the first option in the **Info Box** and holding the *LMB* for a moment (four options appear: **Single**, **Chained**, **Rectangle**, **Rotated Rectangle**) and then choose the third option:

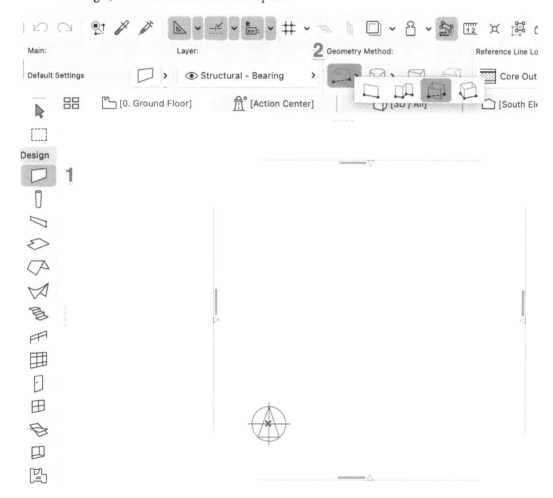

Figure 3.8: Starting the Rectangle tool and choosing the appropriate geometry method

> **Note**
>
> When clicking buttons in the Info Box marked with an arrow for at least 0.5 seconds, that button will expand and offer more applicable options for that setting. This way, the work environment is kept as efficient as possible.

3. Set **Reference Line Location** to **Outside Face** as we will edit the wall structure later, and the reference line is the fixed part of our wall contour—we come back to this concept in *Chapter 4*:

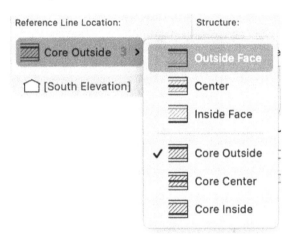

Figure 3.9: Choosing the best location for the reference line

4. Enter your first node at the bottom left, close to (but not on) the *Archicad* origin. The exact location is not important but try to mimic the image.

5. At this point, the **Tracker** appears, showing us the current *dimensions* (not coordinates!) of the rectangle that we are modeling (**Dimension 1** along the *x* axis; **Dimension 2** along the *y* axis). These numbers will change while moving around the cursor until we enter the second node (at the opposite corner of the rectangle)—but don't do this yet!

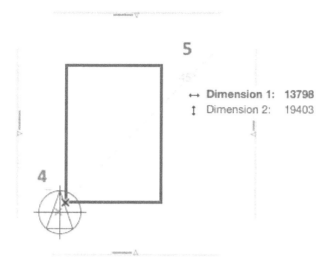

Figure 3.10: The Tracker displaying our current dimensions

6. The outer perimeter of our building at the ground floor should measure 6 by 10 meters. Although we can model this approximately by using the Tracker, in combination with a grid, we can also directly enter the dimensions into the Tracker:

    · Press *Tab* to turn the first dimension into an input field. Give a value of 6000 (working units are mm!), followed by another *Tab* press (not *Enter*!), to jump to the next field:

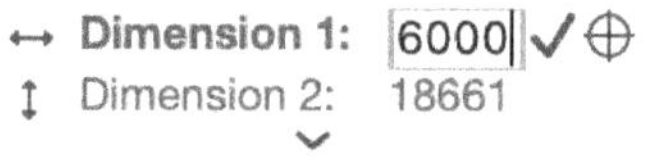

    → **Dimension 1:** 6000 ✓ ⊕
    ↕ Dimension 2:   18661

Figure 3.11: Input values in the Tracker using Tab

    · You can now input a value of 10000 for **Dimension 2**.

    · Watch how the shape adapts to your input.

    · Finish and confirm your input by pressing *Enter*.

Tip

Using *numeric input*—mostly through the Tracker—hugely increases your precision and speed. This topic is specifically covered in more detail further on in this section.

7. We have created four walls in a rectangular layout, as shown here. Archicad keeps the **Wall** command active, so if you click again, you'll start creating new walls. To prevent this, you should return to the main selection command by pressing *esc*:

Figure 3.12: Walls in the 2D window

8. Now is a good time to switch to the 3D window (*Fn + F3/F3*, as explained earlier), to confirm that the rectangle we just created effectively contains walls in 3D, with a 2D representation on the ground floor:

    ▪ You can see the four walls, but also a blue grid, which is our current working plane. The origin is indicated by three small axis lines, labeled **X**, **Y**, and **Z**.

    ▪ Remember to pan, zoom, and orbit around the model to confirm the layout:

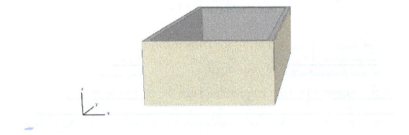

Figure 3.13: Walls in the 3D window

9. Return to the 2D window (*Fn + F2/F2*).

Grouping

When you model walls as a chain of wall segments using either **Chained Rectangle** or **Rotated Rectangle** as the geometry method, Archicad applies **grouping**: the four walls we just created using the **Rectangular** geometry method are still individual walls, each with their own starting and ending point, but have been automatically grouped for convenience because they were created at the same time.

When you toggle **Suspend Groups** (*Option + G/Alt + G*), elements remain grouped, but you can temporarily ignore them as a group, allowing you to edit individual elements. This is needed when you want to move a single wall in the group or if you want to edit its shape or properties independently from the others:

Figure 3.14: Suspend Groups toggle button

The status of Suspend Groups influences the options available in your Pet Palette, and this works as follows:

- **Groups Active**: Elements can only be moved or rotated (only **Move** options available in the Pet Palette)

- **Groups Suspended** + one element selected: You can fully edit the selected element (**Move** and **Adjust** options are available through the Pet Palette—options depend on where you click on the element: node, edge, reference line)

- **Groups Suspended** + all elements of the group selected: You can edit the shape of the group as a whole (**Move** and **Adjust** options are available through the Pet Palette— specific options such as **Offset one/all edge(s)** are also available)

Try this out for yourself: select the four walls and switch on/off **Suspend Groups**, each time clicking nodes/edges of the current selection. Hover over the options of the Pet Palette that pops up. See if you can find **Offset all edges**. Try some of the other options in the Pet Palette to get familiar with it. Restore the original shape afterward by undoing your modifications (**Edit** > **Undo** or *Cmd + Z/Ctrl + Z*).

As an example of the Pet Palette appearance, *Figure 3.15* shows various versions of how it looks for a wall selection, depending on whether or not groups are suspended and on how many grouped elements are selected:

Group "status" selection	Endpoint (2D)	Edge/Reference Line (2D)
Active whole group	⊞ ⟳ ⅍ ⚒ ⟲	⊞ ⟳ ⅍ ⚒ ⟲
Suspended 1 element	╱⊹ ⊞ ⟳ ⅍ ⚒ ⟲	◁ ◇ ⌁ ⊟ ⬭ ⊞ ⟳ ⅍ ⚒ ⟲
Suspended whole group	╱⊹ ⌁ ▢ ⊞ ⟳ ⅍ ⚒ ⟲	◁ ◇ ⌁ ⊟ ▢ ⊟ ⬭ ⊞ ⟳ ⅍ ⚒ ⟲

Figure 3.15: Overview of how the Pet Palette looks in various situations

Let's edit the size of our rectangle—it now measures 6 by 10 meters, but the size in the *y* direction should be 11 meters. Do this as follows:

1. Make sure groups are suspended.

2. Select the short wall at the top.

3. Toggle **Suspend Groups** twice (hit *Alt* + *G* twice)—this first selects the whole group, then makes the group editable as a whole.

4. Click the reference line of the short wall at the top and choose **Offset edge** in the Pet Palette.

5. Move upward and enter a value of 1000 in the **Tracker** for **Distance**:

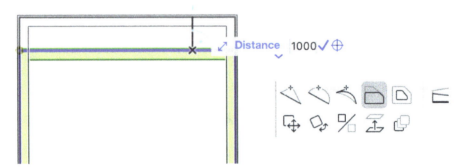

Figure 3.16: Adding 1000 mm

6. Press *esc* if the group of walls is still selected, to deselect them. When you hover the mouse over one of the walls, an info popup is displayed to only show basic information about a single wall:

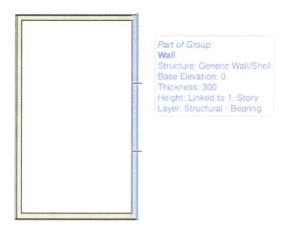

Figure 3.17: One wall as part of the group

Looking at the many settings of a wall

When we started modeling our first four walls, we (deliberately) did not discuss the material or the structure for the walls, nor did we consciously set the height or offsets for these walls—our first steps were already complicated enough! Of course, the material, height, and other properties of any element in our virtual building are of great importance, and we should take the time to set these up appropriately.

At any given moment, you can change the settings of a selected element in the **Info Box**, which displays a series of widgets to configure it.

You can also open the dedicated **Selection Settings** dialog (**Edit** > **Element Settings** > **Wall Selection Settings**, or click *Cmd + T/Ctrl + T*). This opens an extensive dialog, with a series of foldable panels to access groups of settings:

- **GEOMETRY AND POSITIONING** is used to set the size, elevation/offset(s), current home story, and structure of the element.

- **FLOOR PLAN AND SECTION** is used to configure how the element is displayed in 2D.

- **MODEL** is used to configure how the element is displayed in 3D.

- **STRUCTURAL ANALYTICAL PARAMETERS** is used for exchanging architectural models with structural engineers. This is outside of the scope of this book.

- **CLASSIFICATION AND PROPERTIES** is used to manage element "information" and other data or attributes.

Depending on the element, other foldable panels are provided. Luckily, the Archicad interface is filled with icons, and if you hover the mouse over a setting, you get a tooltip with additional information to help you understand each setting (as explained in *Chapter 2* for the Pet Palette):

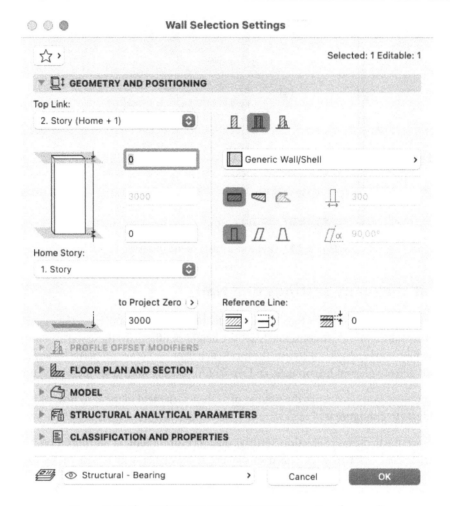

Figure 3.18: The GEOMETRY AND POSITIONING section for a wall

Adding a slab

After the main walls, we'll add a few slabs. Contrary to a typical drawing, in BIM modeling software, you do need slabs in the 2D floor plan views—for example, when you want to display information about the floor or ceiling in this 2D view. They can be modeled in either a 2D floor plan view or in 3D. You may not see them easily on the floor plan, but they are part of the building and thus required for them to show up in a section or 3D view. Here are the steps:

1. Activate the **Slab** tool. We'll stick to the default settings for now.

2. Just as with our walls, we select the **Rectangular** geometry method to create them in just two clicks. You can press *G* to toggle between the available methods.

3. Since we already have our walls in place, we'll use the snapping tools and guides, to work accurately without having to enter exact positions or dimensions. Click once on the lower-left interior corner of the rectangle to place the starting node of our slab. Archicad shows a preview during your cursor movement.

4. Click on the top-right interior corner of the rectangular walls to place the opposite node. Archicad creates a rectangular slab. The slab is positioned inside the walls. Depending on local design conventions and construction practice, you could also choose to model the slab edges at the outside of the walls. As we use this project throughout the book, we think it is best to follow our example, though. Press *esc* to return to the **Arrow** tool:

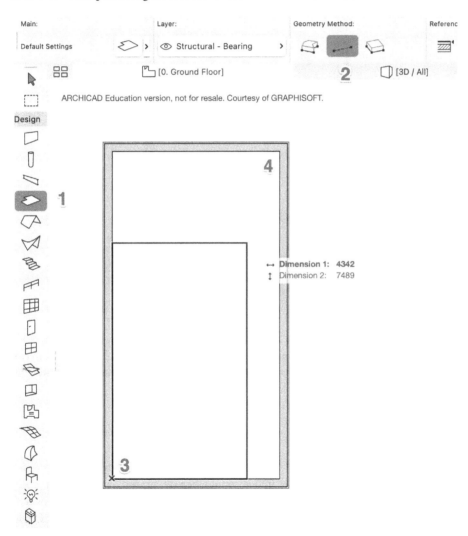

Figure 3.19: Creating a slab between the walls

5. Hover with the mouse over the slab to confirm that a slab is effectively placed. You cannot easily see a slab in a floor plan if its contour is covered by walls:

Figure 3.20: Showing an info popup for a slab

> **Note**
>
> Be careful not to model the slab a second time. Having two slabs in the same location may be hard to see but will lead to mistakes later, for example when making a quantity estimate. Press *esc* when you are done creating elements.

Trace & Reference

When you open a story for the first time, it doesn't show any elements. However, walls need support, so you should take lower (and upper) stories into account when designing. Archicad took inspiration from a manual drafting board, where different semitransparent sheets were overlayed to ensure walls, columns, and staircases were aligned. This is reimagined as **Trace & Reference**: in any 2D window, you can request any other 2D window to be referenced. It is shown, by default, in a slightly transparent single color, to be used as a modeling reference.

We will use it here to display the ground floor when modeling the preceding story. The same snapping tools apply, so you can pick corners, midpoints, or perpendicular references easily:

1. From the **Navigator** palette on the right, double-click on **1. Story** to make that story active in our 2D window. The **View** tab will now display the name of the story. Notice that you are looking at an empty window. The default elevation markers are visible, but your walls seem to have disappeared. They are still there on the ground floor, but you can't see them here.

2. Activate **Trace & Reference** from the menu (**View** > **Trace**) or press *Opt + Fn + F2/Alt + F2*. You should see the walls appear again, but they are tinted in light beige now.

3. To open the **Trace & Reference** palette, pick the pop-up menu on the right of the trace icon (two overlapping rectangles) on the **Toolbar**. Pick the last option in this menu (**Trace & Reference**). This palette brings together a set of convenient controls to adjust how trace works:

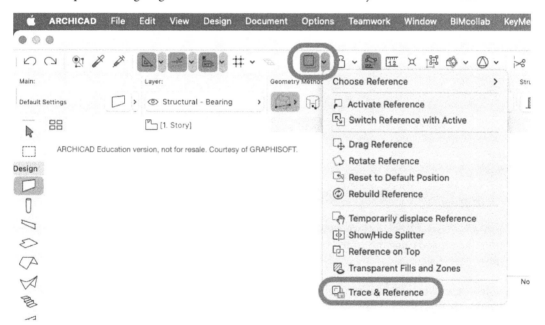

Figure 3.21: Trace & Reference pop-up menu to open its palette

4. By default, Archicad will display the ground floor below the current active story, which is exactly what we need. This is controlled from the top-right button, which displays **Below Current Story**. Keep it like that for now:

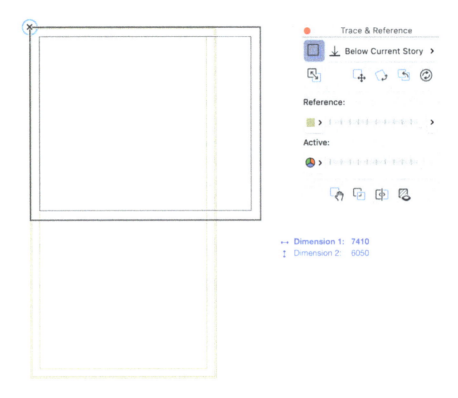

Figure 3.22: Trace & Reference showing Below Current Story

5. Model the upper level of the design, by picking the upper-left corner of the walls from the ground floor and using the same Tracker method as before to set the dimensions: *Tab*, 10000, *Tab*, 7000, *Enter*.

Now, you have created a second rectangular set of walls on story 1.

> **Note**
>
> The **Trace & Reference** system brings you many options, including controlling transparency of both current and traced views, overriding colors, and even ways to interact with the trace—for example, when you need to check the coordination between stories, sections, elevations, and other views.

Using the Pet Palette during modeling

We will now show you a few ways that the Pet Palette can be used during modeling.

Pet Palette in 2D views

When you are modeling walls or slabs, keep a close eye on the Pet Palette, as this is where the modeling magic happens. Follow these steps:

1. Select one of the walls you have created.

2. Now, click on its *end node*. The Pet Palette appears with options that make sense for adjusting the nodes of the wall.

3. Pick the **Drag** icon, which is the first tool on the bottom row of the Pet Palette. Now, you can move the wall in its entirety. Archicad displays a preview of the element while moving, to make it clear where it will be moved to:

Figure 3.23: Dragging one of the walls using the Pet Palette

4. Click again to confirm the new position of the wall.

5. When you click on the *reference line*, between the start and end nodes, you get a few new tools. Pick the first tool on the top row: **Insert new node**.

6. When you move the cursor, you'll notice that the wall is now split into two walls, connected at the cursor position:

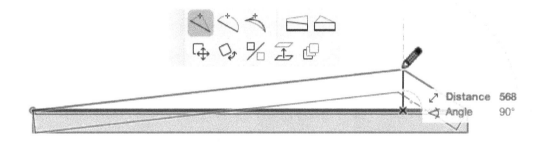

Figure 3.24: Adding a node to the wall using the Pet Palette

7. Click anywhere to finish this adjustment.

8. Click on one of the wall's reference lines again but select the second tool: **Curve Edge**. This turns the straight wall into a curved wall.

9. Move the cursor to interactively set the curvature radius of the wall:

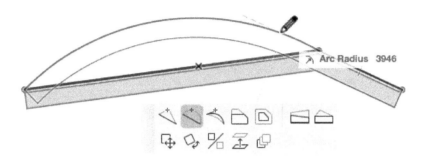

Figure 3.25: Curving the edge of a wall, tangential to another wall, using the Pet Palette and a snap guide

10. Click one final time to stop the adjustment.

As this is not the desired result for the wall, undo all previous steps until you reach the initial state before editing.

The same approach can be followed with a slab:

1. Select a slab.

2. Click on one of its contour nodes. The Pet Palette for slabs appears. It looks like the Pet Palette for walls, but a few new options are available. By default, the **Move node** tool is activated (the first tool on the top row). This is used to refine the contour. Don't do this now:

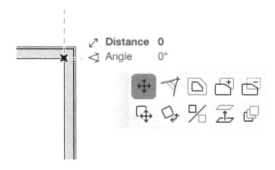

Figure 3.26: Pet Palette for nodes of a slab

3. Pick the **Fillet/Chamfer** tool (the second tool on the top row). A small dialog opens, which requests a fillet or rounding radius to be applied to the selected node. Set the radius to 500 (mm) and click **OK** to confirm the dialog:

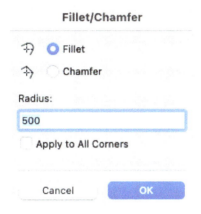

Figure 3.27: Setting the radius

4. The contour of the slab is adjusted by inserting a circular segment in between the two straight segments connected to the selected node. Beware that there is no separate node anymore at the original position:

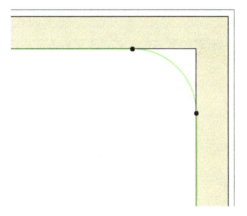

Figure 3.28: Fillet applied to a slab corner

5. Now, click on one of the contour edges. The Pet Palette presents a few new options:

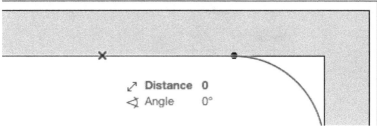

Figure 3.29: Pet Palette for slab contour

6. Pick the **Offset all edges** tool (the fifth tool on the top row). You get a preview of where the contour will arrive. You can either pick a point, using snapping, or press *Tab* to enter an exact **Distance** value. Or you can press *esc* to exit the command:

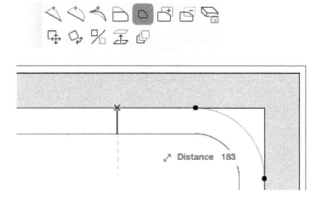

Figure 3.30: Pet Palette during Offset all edges

7. Undo these last few operations, since we only needed the rectangular slab.

Pet Palette in a 3D window

When you perform the same operation in a 3D window, additional transform commands become available. Feel free to experiment a bit with what's available. You can always undo when the results are not what you need:

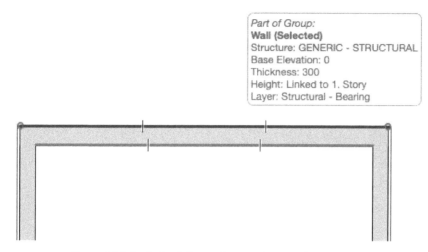

Figure 3.31: Basic Pet Palette when moving an element in 3D

Materials and composite structures (basics)

Every element in a building is made from certain materials. Some elements have a single material, while others may be comprised of many materials. For walls and slabs and a few other elements in Archicad, you can choose between different compositions:

- **Basic structure** is for elements that have a single material.
- **Composite structure** is for elements made from layers or skins of materials.
- **Complex profile** is for elements that are defined from their cross-section, possibly comprising multiple materials. This type of structure is only available for walls, beams, and columns.

In the examples of walls and slabs we created so far; we didn't bother with their composition. However, at some point, we must set and configure their composition, as it defines not only the materials that are used but also their thickness. You can access the type of structure for a selected element from the Info Box, as shown here:

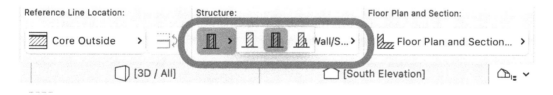

Figure 3.32: Selecting the type of structure of a wall

Follow these steps:

1. Select a wall. Notice that our default wall has a composite structure called **Generic Wall/Shell**. Don't worry about it for now but notice that our wall has two skins of material.

2. When you press and hold the **Structure** button, you can switch between the structure types (Figure 3.32).

3. Switch the structure of this wall to **Basic** (the first option on the left).

4. The Info Box now shows **GENERIC - STRUCTURAL**. You can click on this name to gain access to other available materials. Take a look around, but don't worry about configuring this at the moment:

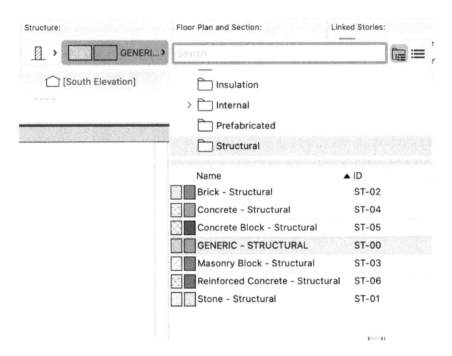

Figure 3.33: Basic structures selection list

5. You'll notice in the floor plan that the wall now has a single layer, and if you hover over the wall, this is also shown in the pop-up info panel:

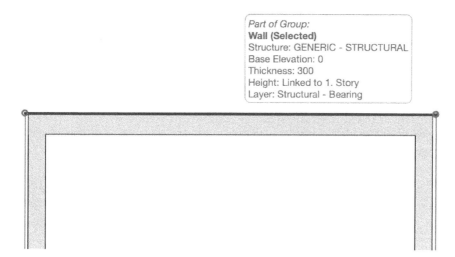

Figure 3.34: Wall information popup

You can do the same thing for the slab but beware that only basic and composite structures are available, as slabs cannot have a complex profile. Before continuing, make sure that you set the wall composition back to its original definition!

Using drawing aids in Archicad

We have noted before that Archicad provides plenty of onscreen feedback. A large part of this feedback consists of several drawing aids, aimed at modeling more precisely and efficiently. All of these drawing aids can be turned on and off through the menu or by using the buttons in the **Toolbar**. In this section, we will get to know the following aids and how to use them: **Snap Guides**, **Guide Lines**, and the **Tracker** we already mentioned. They are available both in 2D and 3D (although the screenshots in what follows have been taken from 2D views for clarity). The **Grid** is also considered a drawing aid but was already covered in *Chapter 2*:

Figure 3.35: Drawing aids in the Toolbar (left to right)

The drawing aids can be toggled, and their options can be set through the buttons shown in *Figure 3.35*. From left to right, this screenshot shows **Guide Lines**, **Snap Guides**, **Tracker**, and **Grid**, each with a drop-down menu for more options.

Using Snap Guides and Guide References

Snap Guides are a temporary drawing aid that can be referenced while modeling an element. Snap Guides are shown as a *blue dotted line* while **Snap References** use a *blue solid line*. They are turned on by default but can be disabled through **View > Snap Guides and Points**.

Activating Snap Guides

Snap Guides pop up automatically during modeling, but they also disappear just as quickly. Seeing numerous guides at the same time can be a bit overwhelming at first. However, once you know how to use them, they are a very powerful and versatile drawing aid. They are temporary and disappear after completing a command, but you can **pin** one (or more) of many of the appearing options to make it easier to focus on the appropriate one. You do this via the *context menu* > **Pin Snap Guide**:

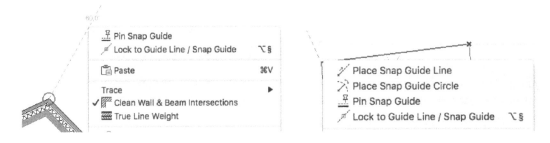

Figure 3.36: Using the Pin Snap Guide feature

When modeling a new element without referring to an already modeled one, the context menu is a bit more extensive. Then, you also have the possibility to place a specific snap line in the direction in which you draw the object (**Place Snap Guide Line**) or a snap circle at the distance of the cursor from the starting point of the element (**Place Snap Guide Circle**). Finally, when you choose the **Lock to Guide Line / Snap Guide** option, the cursor will only move along the line that you have clicked with the *RMB* to access the *context menu*.

> **Important note**
>
> The angle at which Snap Guides appear is set to **every 45°** by default. This can be changed under **Options > Work Environment > Input Constraints and Guides... > Incremental angle**. At **Input Constraints and Guides...** you can adjust many settings for **Snap Guides**, **Snap References**, and **Guide Lines**, by the way.

Activating Snap References

You will, however, often start from or refer to elements that were modeled earlier. Then, you will notice that **Snap References** (dots or lines) are generated with which you can start creating/building Snap Guides. To activate these Snap References, you have the following options:

- Hover over the end point/node of an element or a corner of two elements: a blue circle appears
- Hover over an edge (straight or bent): a (full) blue line appears

You can see an illustration of this here:

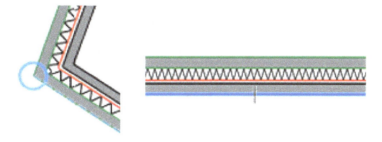

Figure 3.37: Node- and edge-type activation

Pro tip

You can activate **Snap References** very quickly by pressing the *Q* key, skipping the default 1-second delay. Once the desired Snap References are visualized, different Snap Guides appear by moving the cursor. New Snap References can be activated at intersections between Snap Guides that have been created.

Figure 3.38 shows how a Snap Reference can be created at the intersection of two Guide Lines created with a node reference and an edge reference.

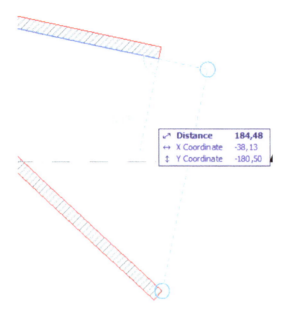

Figure 3.38: While modeling, complex references can be used

This shows it can get quite complex quite fast, but after this section, you should feel a lot more comfortable with these drawing aids.

Using Guide Lines

Sometimes, it can be useful to keep guiding lines permanently visible/active in a project. This can be done in Archicad in both 2D and 3D, using permanent **Guide Lines**. Unlike the Snap Guides, they always remain in a project (unless deleted), even though they are also not printed. They can be made invisible through **View** > **Guide Lines** or by pressing *L*. They can be distinguished from Snap Guides by their (default) orange color. Settings can be altered through **Options** > **Work Environment** > **Input Constraints and Guides…**.

Adding and removing Guide Lines

Guide Lines are always generated in the same way:

1. From one of the "buttons" on the four sides of your drawing board, drag a Guide Line into the window using the *LMB* (note that these buttons are only visible when **Guide Lines** is turned on...):

Figure 3.39: Guide Line "buttons" visible at the top and to the left of the Viewport

2. Place the line at the desired location:

   - **Parallel to the direction from which you are dragging**: Release the *LMB* where you want to place the Guide Line.

   - **Referring to an existing element in any direction and with any shape (even circular)**: Release the *LMB* at the reference element.

Some of the most common options for Guide Lines are shown in *Figure 3.40*.

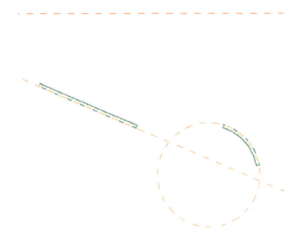

Figure 3.40: Three types of Snap Guides (no reference, linear and circular reference)

Through the drop-down menu in the Toolbar (or *Shift + L*), you can create a Guide Line segment by clicking two points.

You can delete Guide Lines using the context menu or drop-down menu in the Toolbar:

- **Remove All Guide Lines**: Delete all permanent guide lines in the current view
- **Remove Guide Line**: This requires you to right-click the guide line in question (only available in the context menu)
- **Erase Guide Lines**: Click on superfluous lines (with an eraser)

The options are illustrated here:

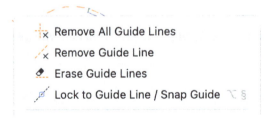

Figure 3.41: Removing Guide Lines context menu

Learning to use numeric input

We have used the **Tracker** for accurate modeling a few times already, and it is now time to get to know all the options. Mastering this part of drafting in Archicad is key to obtaining an efficient and precise modeling technique!

Among the many drawing aids Archicad provides, the Tracker gives us the clearest control over the dimensions of the elements we are modeling. It allows us to enter exact values for length, angle, and coordinates as variables. The "active" variable is always shown in bold. The value can easily be changed by either directly typing in the value (using a numeric keypad) or by "entering" the Tracker by tapping the *first letter* of a variable (*D* for **Distance**, *A* for **Angle**). Using *Tab* (or *Shift + Tab*) cycles forward (or backward) through the available parameters.

In the following examples/use cases, we will discover the possibilities of the Tracker using the **Wall** tool (in 2D), but of course, this works for all tools (and in 3D)—although the available parameters are not always the same!

Entering length and angle using Tracker and Guide Lines

We already know how to use **Snap Guides** to make guiding lines appear at certain angles and/or in certain directions. In this way, you can use the cursor to refer to existing objects or logical "connections" between elements. By combining this function with the Tracker, we can draw objects at a certain angle, with a fixed length:

1. Activate the **Wall** tool and click to define the starting node of the element.
2. Move the cursor in the desired direction until the Tracker appears.
3. Refer to a Snap Guide to determine the desired angle (note that the cursor changes into a "pencil with a line")—only the defined angles (see previously) are available.
4. Enter the value numerically (use a numeric keypad or use the *Shift* key in combination with the numbers on the top row of keys)—you may release the mouse at this moment.
5. Confirm the entered values with *Enter*.

You should see an output similar to this:

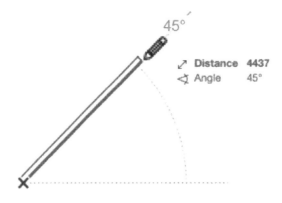

Figure 3.42: Guide Line for an angle

> **Note**
>
> The input is always relative, meaning that it is positive for the direction you are moving the cursor toward. Entering a negative value (for example, -1000 instead of 1000) results in an element being drawn in the opposite direction of where you were heading.

Entering coordinates using the Tracker

Although it is not explicitly made visible, we already explained that there is a coordinate system active in Archicad (when we talked about the origin in *Chapter 2*). Consequently, we can also apply (relative) coordinates when modeling:

1. Activate the **Wall** tool and click for the first node of the element.

2. Move the cursor in the desired direction. When the Tracker appears, press the letter of the desired coordinate (*x*, *y*, or (in 3D) *z*) or use the *Tab* key to cycle—you may release the mouse at this point.

3. Enter the desired numerical value—if you move left or down, this will be a negative *x* or *y* value respectively, preceded by a - sign. You can also derive this from the feedback provided by the tracker. The value entered is relative to the point, not linked to the Archicad origin...

4. Notice how the other variables (**Distance**, **Angle**) in the Tracker change accordingly.

5. Confirm with *Enter*.

You should see an output similar to this:

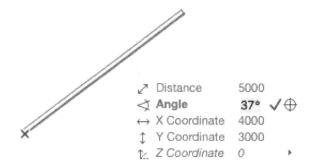

Figure 3.43: Values in the Tracker demonstrating the Pythagorean theorem

Inputting a value in the Tracker graphically

The values in the Tracker can also be entered "graphically", by referring to already modeled elements, or even by measuring certain reference elements in a project (without interrupting the running command for this). Referring to an existing element (when modeling, but also when moving or copying elements) is done as follows:

1. Activate the **Wall** tool and define the first node (using Snap Guides if necessary).
2. Move the cursor in the desired direction (orthogonally, or along a Guide Line).
3. Hold down the *Shift* key to "lock" the selected direction
4. Move the cursor to the desired reference point—the pointer will show a "black pencil" or a "check mark".
5. Click on the desired reference point to confirm the input.

You should see an output similar to this:

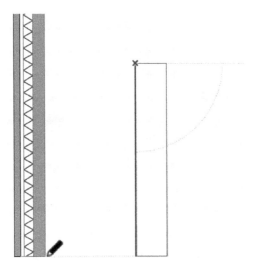

Figure 3.44: Tracking the endpoint of another wall

Using the *Shift* key, the cursor (black pencil) can be moved away from the element that is being modeled, to refer to an existing element in the project.

If you wish to track a value from another element that is already modeled, but to which you cannot refer graphically, proceed as follows:

1. Activate the **Wall** tool and define the first node (using Snap Guides if necessary).
2. Move the cursor in the desired direction (orthogonally, or along a Guide Line).
3. Type the letter of the Tracker value that you want to measure (or cycle through the Tracker with *Tab*)—you can let go of the mouse at this point.
4. Press the shortcut key for the **Measure** tool (*M*)—a **Transfer Value to Distance** message appears in the Info Box near the cursor.
5. Measure the desired value/length with two clicks—the measured value appears in the Tracker.
6. Confirm by pressing *Enter*.

Measure tool

The **Measure Tool** is a handy aid on its own. You can use it to measure distances, cumulative distances (perimeters), angles, (relative) coordinates, and areas in a temporary way. The measured values are displayed in the Tracker. No dimension lines or other annotations are placed by this tool. It can be activated via the Toolbar or using the *M* shortcut key.

How the **Measure Tool** is used is shown in *Figure 3.45*. The measured distance(s) appear in the Tracker (in blue).

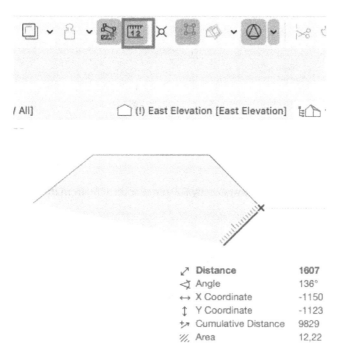

Figure 3.45: The Measure tool in action

Defining the last node of an element by referring

Using the tracker, in combination with graphical input and feedback, you can model an element up to a certain distance from another element. It is also possible to make an element an exact number of mm longer or shorter with respect to another element. This technique can even be combined with entering *x* and *y* coordinates. Proceed as follows:

1. Draw a certain element, such as a wall—the Tracker always and continuously displays the length (distance) of the object being drawn.

2. Hover over the edge of the object from which you want to keep a certain distance—pay attention to the cursor feedback ("pencil with a line", "black pencil", "perpendicular angle").

3. Press *D* or navigate to **Distance** in the Tracker with *Tab*.

4. Enter numerically the desired spacing between the two objects, followed by a - sign—a Snap Guide will be created at the entered distance.

5. Confirm with a click on the Snap Guide or press *Enter*.

You should see an output similar to this:

Figure 3.46: The wall stops at exactly 500 mm (input: 500-) from the other one

Calculations with the Tracker

Doing "calculations" in the Tracker can be used for any variable. The "operator" (+ or -) is always at the end of the input. You can continue calculating as long as you have not pressed *Enter*. Remember that a – sign in front of the **Distance** variable or any of the coordinates gives a different result!

To enter the length of an element relative to an existing element, proceed as follows:

1. Start modeling the desired element.

2. Select the desired direction and hold down the *Shift* key to lock the direction.

3. Refer to the point relative to which you want to define the relative length.

4. Press *D* or cycle with the *Tab* key to **Distance** in the Tracker—*Shift* can be released at this point.

5. Enter the desired relative distance, followed by a + sign if the element is to be longer than the reference, and a - sign if the element should be shorter than the reference—appropriate Guide Lines will appear.

6. Confirm by pressing *Enter*.

You should see an output similar to this:

Figure 3.47: The new wall is 500 mm (input: 500+) longer than the parallel wall

Defining the first node relative to a reference point

Using the (relative) coordinate system, we can also define a starting point relative to an existing corner or end point/node of an element. To do this, proceed as follows:

1. Activate the desired modeling tool.

2. Do *NOT* click for the first node, but hover over the desired reference point (or use *Q*—a circular Guide Reference should appear.

3. Enter the desired relative distance to this point in the horizontal and vertical direction using the *x* and *y* keys. Add a + sign after the value for positive coordinates: (to the right (*x*) or up (*y*)) and a - sign for negative coordinates (to the left (*x*) or down (*y*)).

4. After entering all desired values, confirm with *Enter* and start modeling from the point you just defined.

Locking coordinates or directions

While modeling, we can lock certain directions/coordinates by using the *Alt* key. In combination with the first letter of the variable from the Tracker, this variable is considered "fixed" during further modeling—a ("pinned") Snap Guide is created on the "fixed" value each time:

- *Opt + x*/*Alt + x*: Fix *x* value—vertical Snap Guide

- *Opt + y*/*Alt + y*: Snap y value—horizontal Snap Guide

- *Opt + a*/*Alt + a*: Snap angle—angled Snap Guide

- *Opt + d*/*Alt + d*: Snap to length—circular Snap Guide

After locking, you can refer to a second point to model the object. To disconnect or detach, the same key combination is pressed again. Notice that the created Snap Guide is not removed as long as it's locked.

Adding interior walls

We will now add a few interior walls, which will help us to define the main spaces in the building. We will do this by using the **Wall** tool, but before we start modeling, some of the current default settings for this tool must be changed.

Double-click the **Wall** tool in the Toolbox or, if the **Wall** tool is already activated, use the icon in the Info Box (*Cmd + T*/*Ctrl + T*):

1. Click on the current **Generic Wall/Shell** setting, and from the **Composite** selection popup, navigate to **Composites** > **Interior** and pick **100 Block Double Plastered** from that folder.

2. Click on **Structural – Bearing** at the bottom of the dialog, and from the **Layer** selection popup, navigate to **Layers** > **Design** > **Architectural**, where you can pick the **Interior - Partition** layer instead:

Figure 3.48: Wall default settings for our interior walls on the ground floor

> **Note for users of Archicad 25 and older versions**
>
> Since Archicad 26, the popup for layers, composites, and other Archicad attributes are shown as a folder structure. Older versions display them as a flat list only, but their functionality is the same. There is a legacy option available if you would like to display them as a flat list again, but that will eventually disappear, so we advise you to get used to the new style.

Just as with the exterior walls, we set the **Reference Line Location** setting of the interior walls to **Outside Face**. This way, their position is not influenced by the plaster layer thickness:

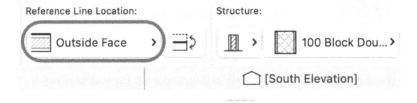

Figure 3.49: Reference Line Location set to Outside Face

Now, you can draw the necessary interior walls, using the **Wall** tool and the appropriate geometry method (**Straight**), referring to the dimensions shown in *Figure 3.50* and the thick blue line indicating the position of their reference line. You don't have to draw the dimensions themselves.

Always try to use exact numeric input via the **Tracker**. This will take a few tries, and you may have to use a few intermediate steps while practicing the various Tracker and snapping techniques we explained earlier:

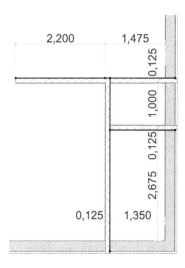

Figure 3.50: Adding interior walls using the Tracker

For the upper-left wall that is designed as a "countertop" in the kitchen with a length of 2200 mm, its height should not be linked to the story height. This wall has a fixed height. Select the wall and go to its wall settings (*Cmd + T*/*Ctrl + T*):

1. Set **Top Link** to **Not Linked**.

2. Enter a height of 1200 from the home story:

Figure 3.51: Unlinking the wall height from a reference story

3.   If necessary, you can "flip" any wall along its reference line:

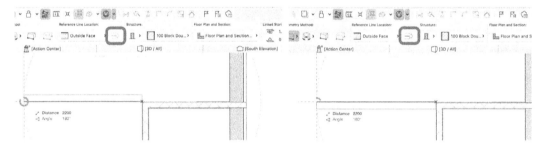

Figure 3.52: Flipping a wall along its reference line

Notice that the wall is still shown as "cut". This is because the default cutting height for floor plans is set at 1100 mm above story elevation—this will be changed later.

Also, notice that the interior walls are not properly connecting to the exterior walls: they join to the outside face. This is due to their material settings. We will adjust this in a later section.

Using the Eyedropper tool to create similar interior walls on the first floor

On the first floor, we must create some interior separating walls as well. These should again be linked to the next level, but otherwise have the same definition as the walls on the ground floor. Instead of adjusting the wall default settings once more, we can use the **Pipet Tool** (or **Eyedropper Tool**) to pick up the parameters from the interior walls on the ground floor:

1. Activate the **Eyedropper Tool** in the Toolbar (or *Alt/Option + C*) to **Pick Up Parameters**.

2. With the **Eyedropper Tool** active, *hover* over the correct wall. Watch the onscreen feedback: the element that is highlighted in blue is preselected, to indicate what you are about to click. The popup near your cursor shows information about this preselection. Click on that wall to have its settings applied as default:

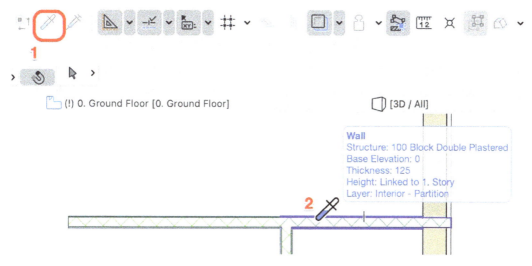

Figure 3.53: Picking up parameters from an existing wall

3. Navigate to the first floor (through **Navigator** or *Cmd + up arrow/Ctrl + up arrow*).

4. Model all the walls on the first floor, according to the dimensions shown in the following screenshot, and using techniques learned up until now. Use the **Stud Partition** composite from the **Composites** > **Interior** folder for the two bathroom walls. Align the wall to the right with the inside face of the exterior wall on the ground floor using **Trace & Reference**:

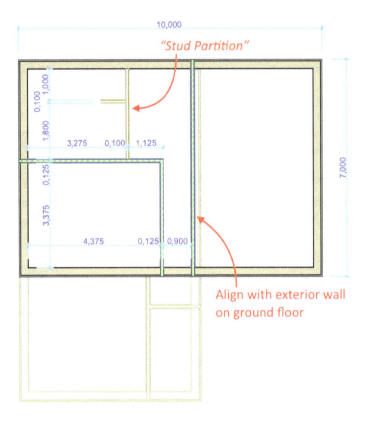

Figure 3.54: Finished first-floor interior walls

We can again use the **Eyedropper Tool** in combination with the **Syringe Tool** to pick up parameters from one element and inject them into other elements. This works best to transfer parameters from walls to walls, slabs to slabs, and so on:

1. If you press the *Opt*/*Alt* key shortcut while left-clicking the element we want to pick up parameters from, the intelligent cursor will change to indicate you are in "eyedropper" mode. Once parameters have been picked up from the bathroom walls (**Stud Partition**), you may release *Opt*/*Alt*.

2. Hover over the two brick walls below the bathroom. Hold down *Cmd + Opt*/*Ctrl + Alt* to get a syringe as a cursor. Click the two walls (one by one) while holding down the shortcut keys to inject the picked-up parameters into these walls, transforming them into **Stud Partition** walls:

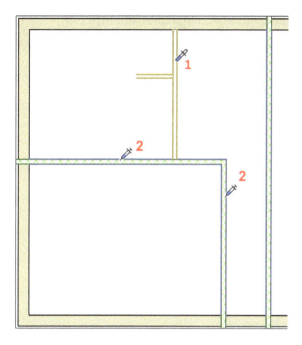

Figure 3.55: Using the Eyedropper to transfer settings between walls

---

Optimizing the use of the Eyedropper

Although there are specific icons in the Toolbar for **Pick up/Inject** parameters and shortcut keys are available for activating these tools, the easiest and quickest way is using *Cmd + Opt/ Ctrl + Alt* in combination with clicking the desired elements. You may have noticed that not all parameters are being transferred in this operation—**Home Story**, for example, changes according to the view/floor plan that we are working in. Though this default setting is best for most scenarios, the parameters that are being picked up can be completely customized through **Edit** > **Element Settings** > **Element Transfer Settings...**.

Use the **Eyedropper** and Pet Palette to create a floor slab on the first floor on your own, creating an opening as shown in the following screenshot by using **Subtract from polygon** and the walls that are already drawn. Please note that we are working in a preliminary design stage and the model may look unfinished. Do not worry, though—further on in this book, we will gradually adjust and finish our model:

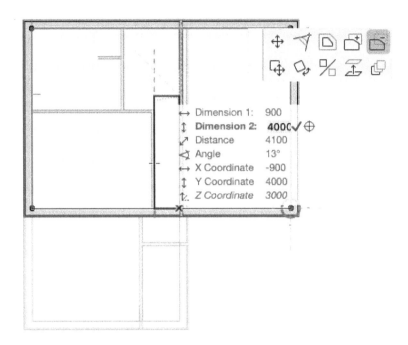

Figure 3.56: Adding an opening into the slab on the first floor

This completes the first part of our small building project. Save your project, as we will continue with it in the next chapter.

Summary

This chapter was a deep dive into Archicad. You have learned a lot about navigation in 2D and 3D, some of the basic modeling tools, and a wide variety of modeling aids, such as Guide Lines, snapping, the Tracker and Pet Palette, **Trace & Reference**, settings, and transferring settings between elements.

The next chapter will introduce more modeling tools that will allow us to complete the basic structure and the overall building envelope of our first small project.

4

Building a Basic Residential Model: SPACEAdding Roofs, Zones, Beams, and Columns

In this chapter, we will continue modeling our small-scale residential project. We will be adding several roofs, some structural elements, as well as the rooms themselves ("space" can be modeled using Archicad). This way, we will keep practicing the basic modeling tools we learned about in the previous chapter while learning new concepts needed to utilize more complex tools.

The following topics will be covered in this chapter:

- Modeling sloped roofs
- Tidying up the model – connecting walls and adjusting walls to roofs
- Defining rooms and spaces by modeling zones
- Adding some structure by adding beams and columns

Technical requirements

For this chapter, you will need the provisional model that you created throughout *Chapter 3*. The same result can also be downloaded from GitHub `https://github.com/PacktPublishing/A-BIM-Professionals-Guide-to-Learning-Archicad`. Save it to a local folder on your system.

Flat and sloped roof elements

Any building needs a roof. In Archicad, there are a few approaches to model roofs. For flat roofs, it is often more convenient to use the **Slab** tool, if you ignore the slight slope that an actual roof requires. In that case, you simply draw the contour and set an appropriate **Composite Structure**. For sloped roofs, this doesn't work. In that case, you can use the dedicated **Roof** tool, which is also based on a contour but provides an additional slope angle.

Modeling a simple flat roof using a Slab

To create the flat roof above the ground floor, we can simply draw a **Slab**. As shown in the settings in the following figure, (*1*) we select the default **Flat Roof** Composite from the **Composites** > **Exterior** folder in the popup. Note that (*2*) the **Reference Plane** property of the roof is set at the top of the Composite, with a negative offset (-150) relative to the **Home Story** property (*3*) to comply with common detailing for roof levels in coordination with interior floor finishes. And finally, at the bottom of the dialog (*4*), select the **Shell – Roof** layer from the **Layers** > **Design** > **Architectural** folder in the **Layers** popup:

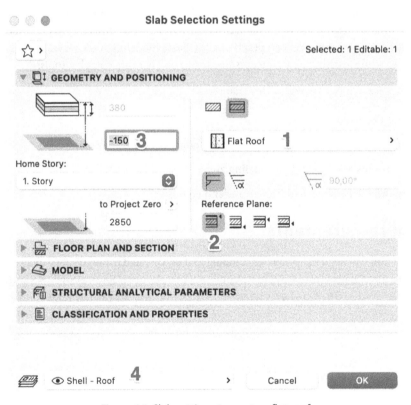

Figure 4.1: Slab settings to create a flat roof

Logically, the **Home Story** property for this roof is the first floor, so open **1. Story** from the **Navigator** menu and use **Trace & Reference** to help you align it accurately with the underlying walls. Here, we will pick the inside corners of the walls (see *Figure 4.2*):

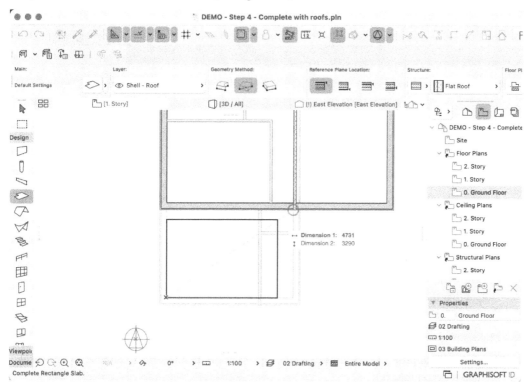

Figure 4.2: Drawing a flat roof contour using Trace & Reference

And with this, we have created the first roof for our project. On the first floor, we will need to add a sloping roof to finish the building envelope – let's see how this works in the next section.

Modeling single-plane roofs

For the sloped roof above the first story, we'll use the **Roof** tool. Since there is only a slope in one direction, we will be using a **Single-plane Roof**.

To define the slope angle and the direction of the slope of a *Single Plane Roof*, Archicad uses the concept of the **Pivot Line**. Like **Reference Lines** and **Planes**, this line anchors the roof, although the roof contour is defined completely independent of the direction and shape of the Pivot Line.

The Pivot Line should be seen as an imaginary axis around which the roof plane rotates. It is the fixed line of a roof and it indicates the **Roof Pivot Height** property in the **Roof Settings** area:

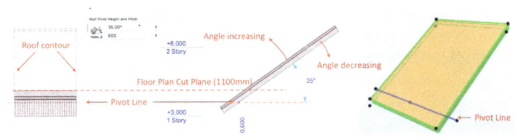

Figure 4.3: Floor plan, section, and 3D view of a single plane roof, to illustrate the Pivot Line

To model the sloped roof, navigate to the first floor (**1. Story**) and work as follows:

1. Activate the **Roof** tool and open the **Default Settings** area in one action by double-clicking the tool in the Toolbox.

2. Set the **Pivot Line Offset to Home Story** property to 2500mm. Notice how the **Pivot Line Elevation to Project Zero** property (**a**) changes accordingly.

3. Choose **Composite Structure** and set it to **Generic Roof/Shell** (from **Composites > Exterior**). Notice that the **Roof Thickness** property (**b**) is grayed out since it is fixed by the Composite's thickness (as we will explain in a later chapter).

4. Make sure you set the **Roof** tool to **Single-plane**!

5. The **Roof Pitch** property should be set to 12°. Notice that you can also choose to define the pitch as a percentage (**c**).

6. Leave **Edge Angle** set to **Set Perpendicular angle for the Roof's Edges**, which automatically sets the cut angle to 90°:

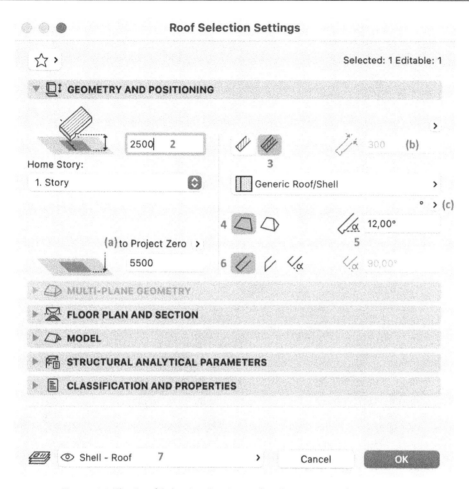

Figure 4.4: The Roof Selection Settings tab – Geometry and Positioning

7. Verify that the roof is assigned to the **Shell – Roof** layer (**Layers** > **Design** > **Architectural** > **Shell – Roof**). Close the dialog by clicking **OK**.

8. Start modeling the roof by defining the Pivot Line first. Place this fixed axis (with an elevation of 2500mm above the current story) at the inside face of the left exterior wall **(a)** and **(b)**. After defining the line with two clicks the **Eye Cursor** will appear, asking us to **Click for Upward Pitch of Roof** in the Status Bar. Click to the right-hand side in our case **(c)**:

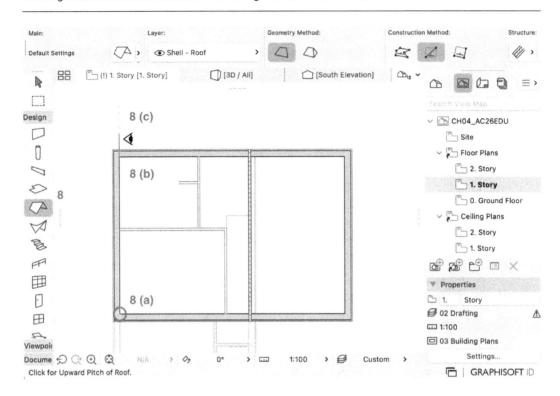

Figure 4.5: Using the Eye Cursor to choose the upward pitch direction

> **Tip**
>
> The Pivot Line is deliberately modeled "longer" than the roof so that it sticks out of the contour. This way, you will always be able to select it easily afterward using the appropriate options from the Pet Palette to edit it to your needs, even if it would be positioned exactly at a roof's edge...

9. Draw the roof contour over the outside faces of the first-floor outer exterior walls using the **Construction Method: Rectangular**.

10. Press *esc* once to stop adding new roofs and reselect the roof: either click on the Pivot Line or hover over a roof edge, press *Tab* until the roof gets highlighted, and select it with the left mouse button. Give all edges an offset of 3 0 0mm using the Pet Palette by clicking on the roof contour:

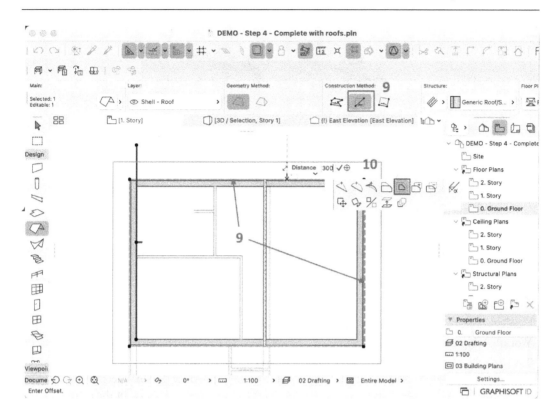

Figure 4.6: Offsetting the edges with the Pet Palette

Turn to the 3D view to check if the result corresponds with the image. Don't worry about the connections – we will get to that soon enough. Got a different result? You can download this stage of the model from GitHub (`https://github.com/PacktPublishing/A-BIM-Professionals-Guide-to-Learning-Archicad/blob/main/CH04_Part1_Roofs.pln`):

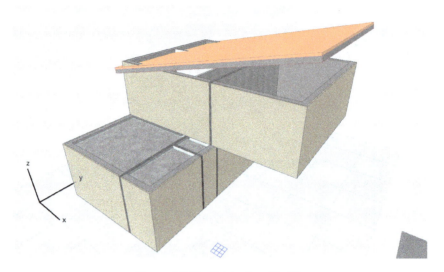

Figure 4.7: The finished roof in 3D

> **Workflow logic**
>
> The preceding workflow covers a lot of steps, and you might get lost along the way. If needed, undo the last few steps, and try again. Don't forget to read as much of the on-screen feedback as possible (including the Status Bar at the bottom left) and follow the "logic" of the software – do not expect the software to follow yours…

Getting to know Multi-plane Roofs

Although we could create complex roof shapes using only **Single-plane Roofs**, Archicad provides **Multi-plane Roofs** as well. This roof type not only covers many common roof shapes in a variety of architectural styles, but it is also a quick way to create a basic sketch for a complex roof. There is no need for a Multi-plane Roof in our basic project, but we will go over the settings that differ from Single Plane Roofs and provide you with some examples.

Multi-plane Roof settings

The basic settings (*Geometry*, *Pivot Line*, *Structure*, and so on) work the same way as for a Single Plane Roof. A big difference for the Pivot Line is that it is more of a *Pivot Polygon* instead of a line. When switching to the multi-plane type (*(a)*), the **Multi-Plane Geometry** tab in the **Roof Settings** area also becomes available:

1. An **Eaves Overhang** can be set for all edges at once, rather than using the Pet Palette as we did for the Single Plane Roof. You can also set this to **Manual**.

2. The cross-section of the roof can have multiple pitches, which can be defined by **Pitch** angle (*(b)*) and **Elevation** (*(c)*), starting from Pivot Line height. The diagram gives a graphical preview of the shape you are defining.

3. Since the Pivot Line may have a circular shape as well, we can use **Curve Resolution** to define the subdivision of circular pivot lines, leading to regular polygons (hexagon, octagon, and others). We will show an example soon:

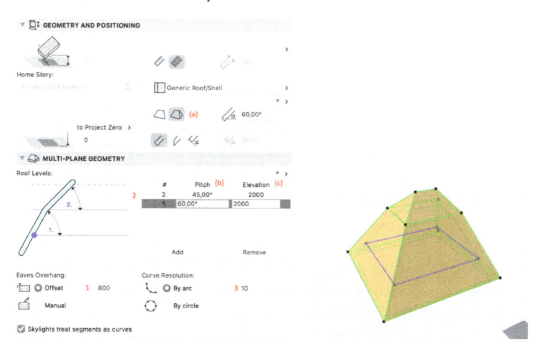

Figure 4.8: 3D results of the settings shown. Notice the rectangular pivot (poly)line

Multi-plane Roof examples

In this section, we will cover some common and less common examples to show the different possibilities of the **Multi-plane Roof**. More examples can be found at the *Graphisoft Helpdesk* (https://help. graphisoft.com/AC/26/int/index.htm#t=_AC26_Help%2F040_ElementsVB%2F040_ ElementsVB-64.htm).

Choose the appropriate **Construction Method** from the **Toolbar**, in the drop-down menu (press *C* to switch between the available options):

- *(a)*: This is the easiest one, the **Rectangular Hip/Gable** roof.
- *(b)*: This shows the use of the **Complex Roof** Construction Method, which is used for an irregular contour.
- *(c)*: From the Pet Palette, select **Stretch Horizontal Ridge** after clicking the intersection node of three ridges.
- *(d)*: Watch the Tracker. When the **Roof Pitch** crosses 90°, you can turn a gable roof into a hip roof (or vice versa):

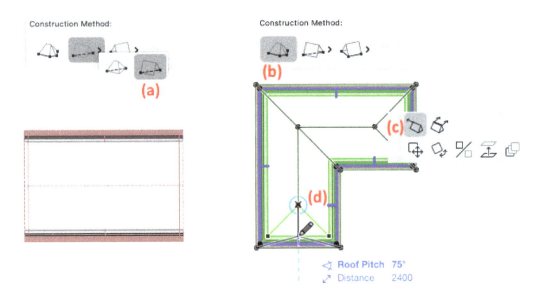

Figure 4.9: Modeling a regular hip roof Construction method and 2D Pet Palette)

Note that there are additional options available in the Pet Palette within a 3D window – for example, you can also adjust the elevation of a horizontal ridge. Otherwise, the geometry settings for both roofs are the same. You can further edit the roof from its **Settings** dialog, where the **MULTI-PLANE GEOMETRY** panel presents the configuration of the roof from one or more pitches and elevations:

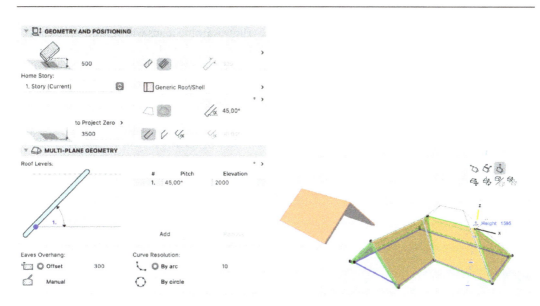

Figure 4.10: Roof settings and options of the 3D Pet Palette

The (originally French) *mansard* roof uses multiple roof levels with appropriate pitches and elevations. Modeling the shape is done in the same way as the first two examples. The Pet Palette contains options for editing its shape the way we would expect for this kind of roof. If you click on one of the intermediate horizontal ridges in a 3D view, the option to **Elevate Roof Level** becomes available:

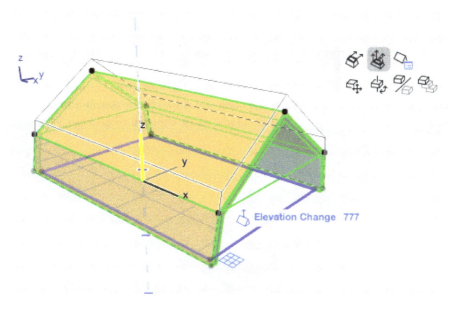

Figure 4.11: Changing the roof elevation from the Pet Palette

When you open the **Roof Selection Settings** area, a few yellow warning icons will become visible. By default, a Multi-plane Roof has the same pitch for every side, edge, or plane, but using the **Rectangular Hip** Construction Method assigns two of the four planes a *custom pitch that differs from the overall pitch setting*. Changing the overall pitch of this kind of roof activates a toggle with a red warning icon, allowing you to apply the change to all other custom planes or edges:

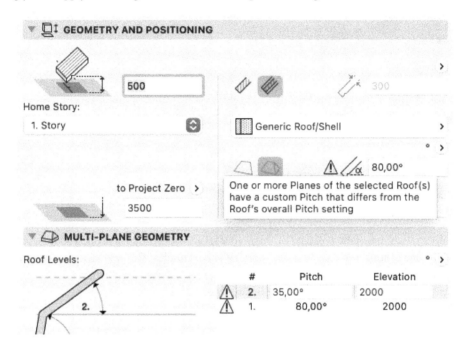

Figure 4.12: Mansard roof Multi-Plane Geometry settings and adapted 3D Pet Palette

Using the **Complex Roof** Construction Method allows us to draw any shape for the Pivot Line. In the following example, we are creating a circular Pivot Line. By setting **Curve Resolution** to 8 segments, **By Circle** results in a polygonal roof. Note that the pitches are constrained between 1° and 89°:

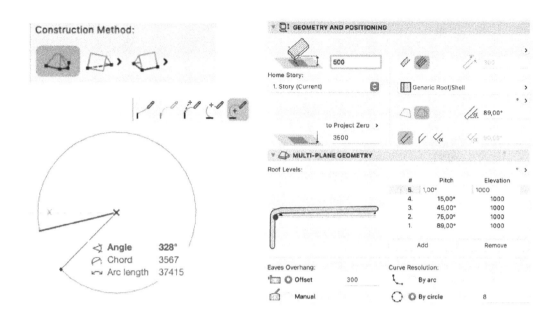

Figure 4.13: Circular roof from an arc and setting Curve Resolution

The resulting circular roof has become a dome with eight segments:

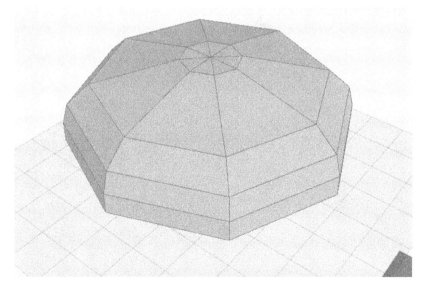

Figure 4.14: Resulting "circular" segmented roof

Now that we've modeled a Single-Plane Roof for our project and have taken a closer look at how to model various Multi-plane Roof types, we need to learn how to properly connect our walls, not only to the modeled roofs but also to each other.

Tidying up the model

Let's return to our house project. We already mentioned we will need proper connections between the walls, and we also need to close the gap between some of the walls and the roof. This can be done in a variety of ways, and in this section, we will introduce common practices for this step.

Changing wall composites for better connections

The use of the **Generic Wall/Shell** Composite has a major drawback: connections with the chosen interior walls do not look correct. How connections between building materials work and which attributes and modeling concepts are involved will be discussed in *Part 2 – Chapter 8*. For now, we will simply choose "matching" Composites:

1.   Open the 3D view.

2.   Make sure **Groups** are active (not suspended).

3.   Click an exterior wall on the ground floor. All four walls should be selected.

4.   *Shift + click* on one of the exterior walls on the first floor to add all four to the selection.

5.   Choose **215 Block Insulated Cavity Plastered** from the **Composites** > **Exterior** folder as a Composite in the **Wall Selection Settings** tab or by using the drop-down list in the Info Box area:

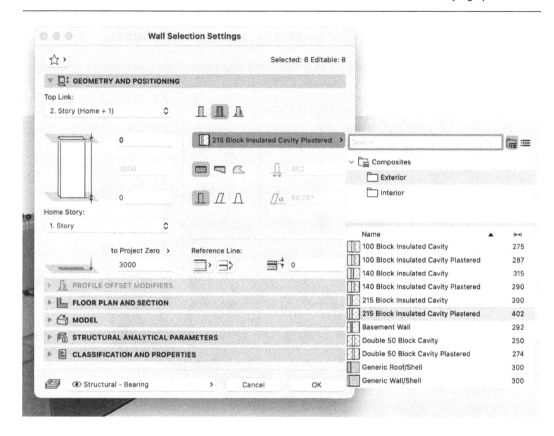

Figure 4.15: Changing the Composite used for the exterior walls of the project

6.   Deselect everything by pressing *esc*.

You just took the first step in refining a model, by choosing the most appropriate Composite for your elements. Don't worry – later in this book, we will learn how to make our own as well.

Adjusting Wall heights to Slabs and trimming with Roofs

To further develop our design, we need to adjust the height of most of our walls. Follow the steps described here:

1. Open the 3D view and select the group of exterior walls on the first floor. **Suspend Groups** (*Option/Alt + G*) and click on one of the bottom wall nodes and select **Stretch Height** in the Pet Palette. Use a node at the bottom of the first-floor slab for reference:

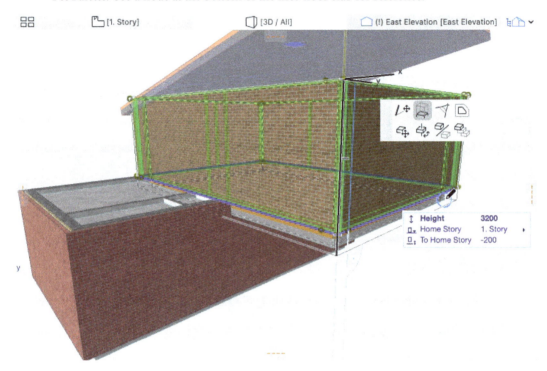

Figure 4.16: Stretching the exterior walls on the first floor downward

2. Press *esc* to deselect the walls. Disable **Suspend Groups** so that you can select all the exterior walls on the ground floor. Using a similar method as before, reactivate **Suspend Groups** and stretch the height of the walls down from a top node to the bottom of the exterior walls on the first floor:

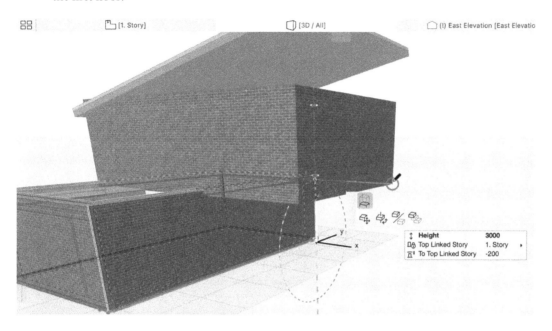

Figure 4.17: Aligning the top of the ground-floor exterior walls with the bottom of the first-floor walls

3. Jump to the floor plan view of the ground floor. Select the three rightmost interior walls that are linked to the story height. Skip the countertop wall on the left. Open their settings and change the **Top Offset to Top Linked Story** property to a value of -530:

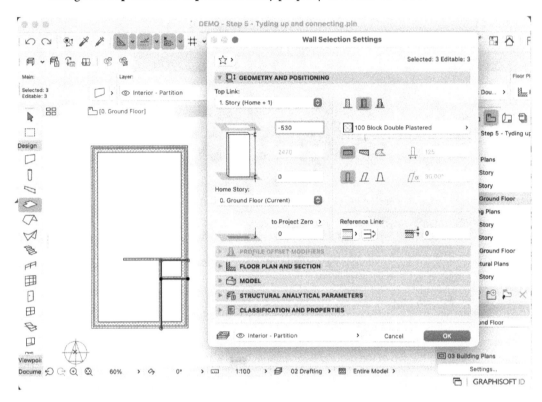

Figure 4.18: Changing the Top Offset property of the interior walls on the ground floor

4.   Change the **Bottom Offset to Home Story** property for the exterior walls on this story to `-200`:

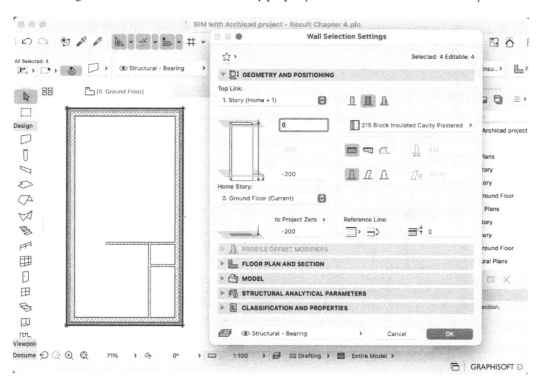

Figure 4.19: Changing the bottom offset property of the exterior walls on the ground floor

5.   Go up one story to **1. Story**. Activate the **Wall** tool in the Tool Box and press *Cmd + A/Ctrl + A*. This selects all the elements in the current view for the active tool – in this case, all the walls on the first floor. Open their settings and change **Top Offset to Top Linked Story** to 3000 – confirm this with **OK**. This way, all the walls will stick through the roof, but that's intentional, as we'll see in the second to last step. Deselect after completing this step:

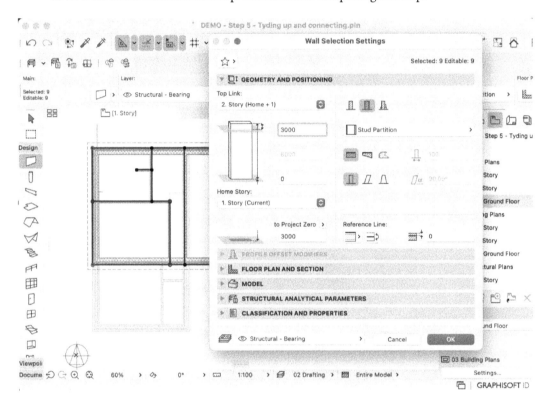

Figure 4.20: Extending all the walls on the first floor to beyond
the roof by changing the Top Offset property

> **Note**
>
> In some designs, a horizontal ceiling (in gypsum board, wood, or similar) is added in the rooms on the upper floor. You could model such a ceiling using the **Slab** tool and then adjust the heights of the inner walls to this slab in a similar way as shown in *Figure 4.20*, but with a different (lower) height of course.

6.   To prepare for lower walls around the flat roof, first, pick up the parameters of the exterior walls on this story using the **Eyedropper** tool. Open the **Settings** dialog and set **Top Link** to **Not Linked0**, and change **Wall Height** to 450:

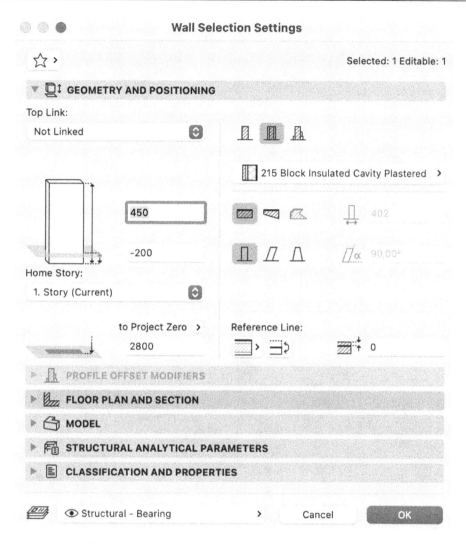

Figure 4.21: Settings for the lower walls around the flat roof

7. Model three new walls around the flat roof using the **Polygonal Geometry Method** option. Using **Trace & Reference**, align to the outside face of the walls from the ground floor and start counterclockwise from the left corner. Click twice at the end point of the third part to stop modeling:

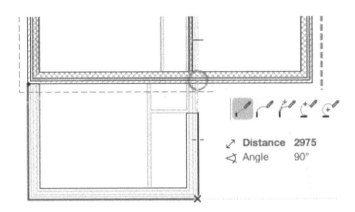

Figure 4.22: Drawing three walls around the flat roof using Trace & Reference

8.   Press *esc* to stop adding more walls or press *W* to switch to the **Selection** tool as an alternative. Open the 3D view and, using a *Cross Selection* by dragging from right to left, select all the walls *and* the sloped roof on the first floor. Do not pick any of the ground floor elements (adjust your camera if needed by orbiting and/or panning):

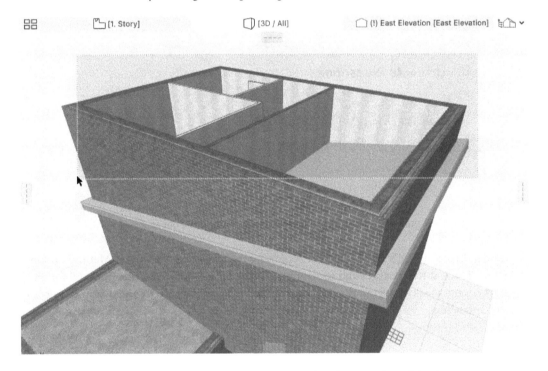

Figure 4.23: Selecting the first-floor walls and roof using a Cross Selection

9. Activate the **Trim to Roof** command in the Toolbar (*(a)*) or right-click and choose **Connect** > **Trim Elements to Roof/Shell** and confirm the settings in the pop-up dialog (*(b)*):

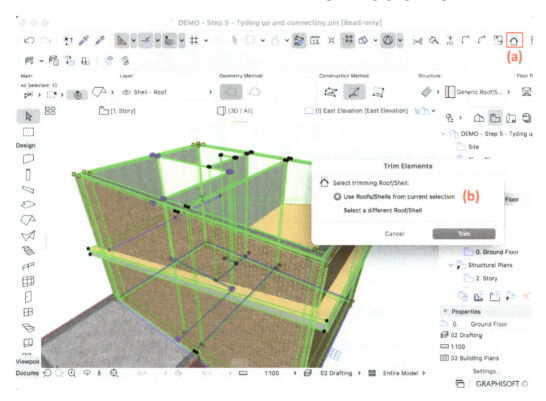

Figure 4.24: Confirming that the selected roof is to be used in the Trim command

10. Look at the result in 3D and save your project:

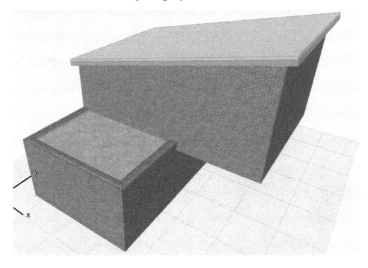

Figure 4.25: The result of cleaning up the wall and roof connections – a tidy model

---

Trim-to-Roof and other Solid Operations

The **Trim-to-Roof** command is the first command we use where we combine several elements in an editing operation. The roof is cutting off the walls at the desired height, at the same time retaining their relationship. There are various connection methods available in Archicad and we will learn how to use and understand them completely in *Part 2 – Chapter 7*.

Now that we have created a nice, closed building envelope (available on GitHub at `https://github.com/PacktPublishing/A-BIM-Professionals-Guide-to-Learning-Archicad/blob/main/CH04_Part2_Walls.pln`), it is time to look at the inside of the envelope and model the spaces inside the building.

Defining rooms and spaces using the Zone tool

One of the major design elements, and the element that defines your design, is the space or room. In Archicad, you can model spaces using the **Zone** tool. In contrast with the other tools we've discussed so far, this is an intangible object. You don't build a space; instead, you create the elements that surround it, such as walls, slabs, or ceilings.

A space is mainly defined by its contour and height. This creates a volume to which information can be attached. When you activate the Archicad **Zone** tool (**Design** > **Architectural Tools** > **Zone**), you must decide between the manual and automatic **Zone** tool **Construction Method** (press *C* to toggle between the available methods).

Figure 4.26: Zone tool Construction Methods in Info Box

The following methods are available for the Zone Tool:

- The **Polyline** method gives you full control over its contour. You manually draw the Zone contour just like you did when creating a slab. The same methods, guides, and Pet Palette can be used.

- The **Rectangular** and **Rotated Rectangular** methods are available as the second manual option. In buildings, many spaces will be rectangular.

  However, we advise that you select one of the two automatic Construction Methods since they allow you to click somewhere in the middle of a series of walls. Archicad traces the contour by looking at the reference lines of the surrounding walls, provided all reference lines are connected properly. This requires a cleaned-up model.

- The **Inner Edge** method is the most common since it looks at the reference lines of the surrounding walls but crops the zone at the nearest side, even if the reference line was placed on the outside. This defines the so-called **Net Room Area** (**NRA**), which is a common method for expressing building areas for clients and facility managers.

- The **Reference Line** method strictly looks at the reference line, so it can lead to Zones overlapping the wall area if the reference line is on the exterior side.

Adding zones using the Inner Edge method

Before we can start adding zones to our project, we have to make sure that they will be fully visible. We have already learned a little bit about **layers** and that we should place each element and object on an appropriate layer (see *Chapter 3*, in the *Modeling tools in general* section). We will go back to layers and how they control visibility in *Chapter 14*, in the *Understanding Quick Options/View Filters* section.

> **Note:**
> Labels will be explained further in *Chapter 7*, in the *Exploring labels* section.

To add zones to the ground floor of our project, we rely on the **Inner Edge** automatic method. Follow these steps:

1. Navigate to the ground floor view and activate the **Zone** tool.

2. Press *C* a few times until the **Inner Edge** Construction Method is active.

3. Click once inside the open space and let Archicad calculate the boundary.

4.   Click a second time inside the same space to place a **Zone Label**, which automatically displays the name, number, area, and height (by default).

5.   Repeat this for the two other spaces (click a second time to place the label).

6.   You should now have **Zone** elements and labels in each space:

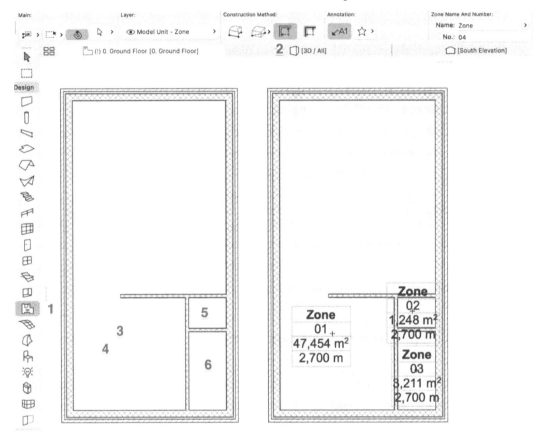

Figure 4.27: Adding zones using the Inner Edge method

Zone Labels

Immediately after you place a zone, Archicad will ask you to indicate the position of the **Zone Label**. This is a symbol that displays information about the zone in the plan view. The **Zone Label** controls how the information it receives from the Zone is displayed.

To select a zone, first, hover your mouse over the zone. Archicad highlights the next element that will be selected. If there are other elements, such as slabs, press *Tab* until the zone itself is highlighted. Click once to select it.

Inside the Info Box or from the **Settings** dialog, adjust the main properties of the zone.

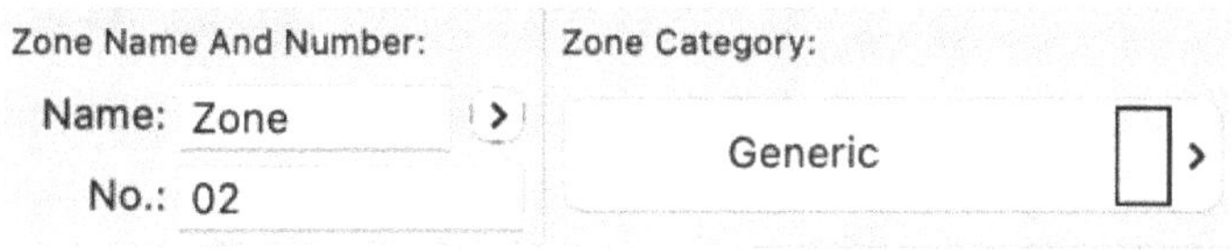

Figure 4.28: Zone Settings in the Info Box

The following properties are shown in the Info Box:

- **Name** is a free text field that you can use to give your zone any name that suits the project. There is a popup list where you can pick a name from a series of preset zone names, such as **Living Room** or **WC**. This indicates the main function of the zone.

- **No.** is also a free text field but is commonly used to write the unique zone number. When you use a string that represents a number, Archicad will automatically increase the number after each new zone you create.

- **Zone Category** gives you a selection of zone categories, which contain both a name and an associated color. This list may seem limited for our house project but can be adjusted (**Options > Element Attributes > Zone Categories...**):

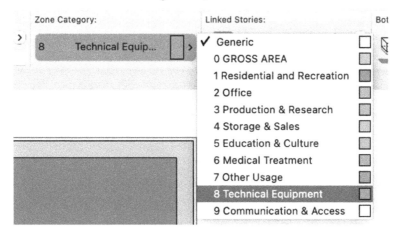

Figure 4.29: Zone Category popup

When you select the label, you will see that it has a series of properties of its own. When you open the **Label Selection Settings** dialog for the label, you can adjust what will be shown for the label. By default, the Zone Label shows a **Zone Name** and **Zone Number**, but you can also display a **Zone Category** and manage other fields to display:

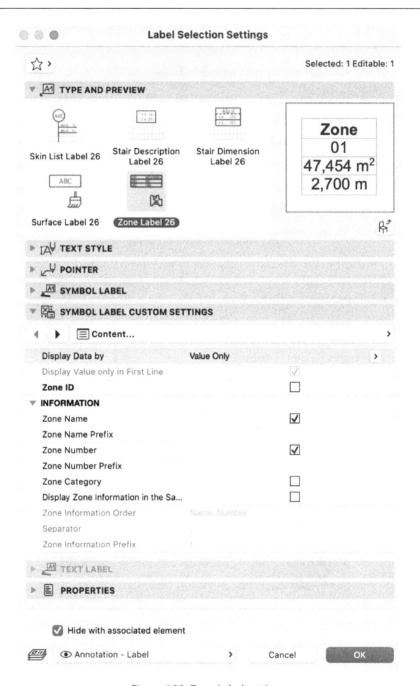

Figure 4.30: Zone Label settings

However, we cannot modify the values displayed inside the label! We must do that via the zone properties and not the label.

This may seem a bit indirect but it is an advantage: there is only one place to manage information and that is within the **Zone** element. The Zone Label is just a way to display that information on a floor plan.

> **The power of BIM**
>
> The power of BIM is that information is at the core of everything you do. You manage the information via the element and all different views and other ways to display that information are updated automatically. You should only worry about the placement of the label and not its content, as that is filled in automatically from the **Zone** properties.

Adding boundaries

The zone may not find a closed contour if there are no walls in certain areas. In that case, you can help Archicad by adding so-called **Zone Boundary** lines. In our example, this is required to distinguish between the kitchen and the living room, as there is not supposed to be a full wall separating them. Let's take a look:

1. Activate the **Line** tool inside the **Document** part of the Tool Box (or go to **Document** > **Documenting Tools** > **Line**).

2. Go to the **Line Default Settings** dialog (*Cmd/Ctrl + T*).

3. Activate the **Zone Boundary** toggle.

4. Click **OK** to close the dialog.

5. Start placing the line from the top-left corner of the countertop wall to ensure it touches the wall's reference line.

6. Pick the end point, which Archicad will find easily using Guide Lines and Snapping. Just ensure the line touches the wall on the left.

7. Press *esc* to return to the **Arrow** tool:

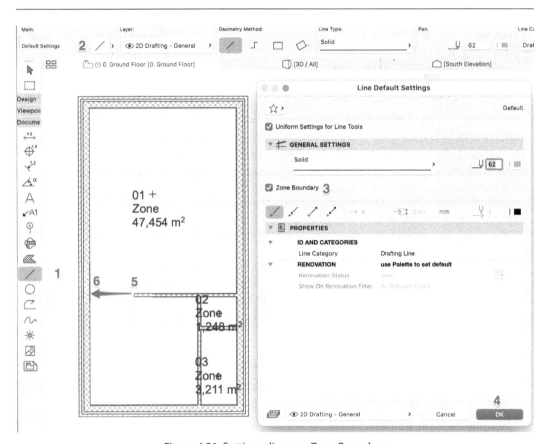

Figure 4.31: Setting a line as a Zone Boundary

But wait, what is going on? Our first zone is still not split across this Zone Boundary. Let's fix this.

Zone updates

One important point of attention with Archicad is that zones with an automatic boundary based on reference lines are not updated automatically! When you make changes to a connected wall, at first, the zone gets left where it was. This may introduce errors and contradictions in the model:

1. To resolve this, invoke the palette where you can update zones (**Design** > **Update Zones...**).

2. Click **Update All Zones** at the bottom. All the zones on the current story will be evaluated again and adjusted if needed. The palette gives you an overview of these changes. If you want to make extensive changes, you can leave this open, since palettes don't block input as dialogs do.

3.   Next, we add a fourth zone in the open area above the Zone Boundary:

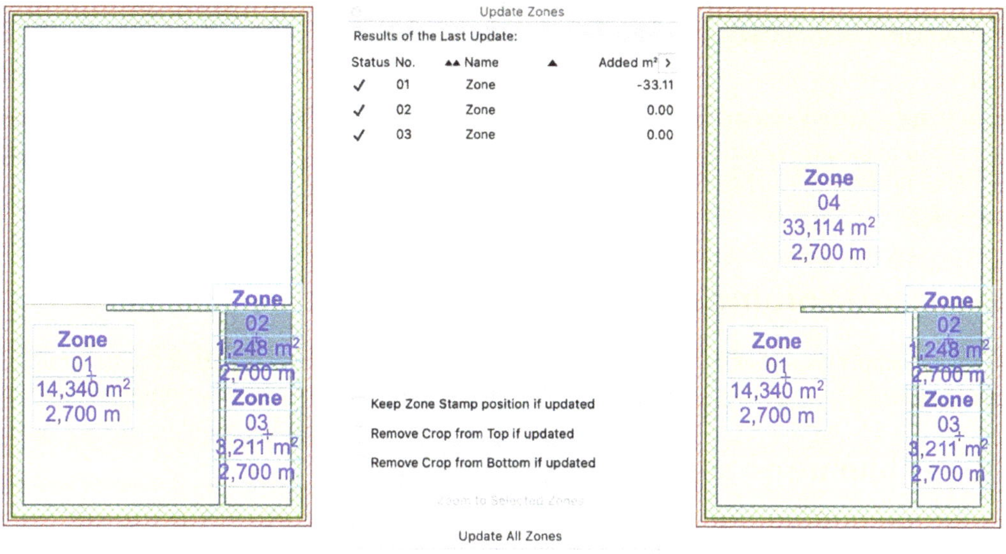

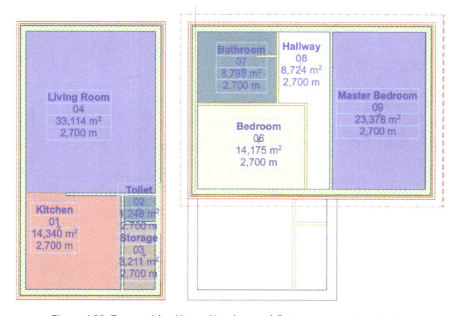

Figure 4.32: The Update Zones palette and updated zones

4.   Now, you can continue adding zones to the first floor as well. Give each zone an appropriate **Name**, **Number**, and **Category**:

Figure 4.33: Zones with a Name, Number, and Category across two stories

> **Tip**
>
> Don't forget to invoke **Update Zones** on each building story before you prepare a submission to your client or the building authorities.

If the automatic zone detection doesn't work or there are simply no enclosing elements available, you can revert to manually creating zones by drawing their contour using the **Polyline** or **Rectangular** Construction Method. Later in this book, you'll learn how to use zones to collect all the space areas in a schedule.

Beams and columns – adding some structure

Our small example project will eventually need a few columns and beams as part of the structure of the building.

Adding columns to support the overhang

We need to add supporting columns to the corners below our first story:

1. Open the **0. Ground Floor** 2D view and use the **Trace & Reference** palette to assign **Above Current Story** as a trace reference.

2. Select the **Column** tool and leave the Geometry Method set to **Simple**.

3. For **Structure**, choose **Circular**:

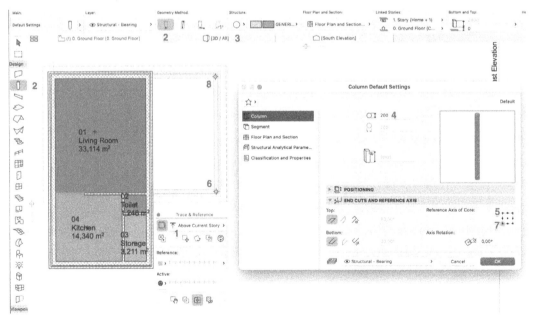

Figure 4.34: Placing columns step by step

4. Set **Cross-Section Size** to 200 to set the diameter from the Info Box or the **Column Settings** dialog:

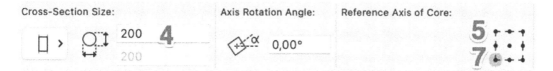

Figure 4.35: Setting the column size and Reference Axis

5. Before placement, set **Reference Axis of Core** to the top-left corner.

6. Place the first column underneath the bottom right corner of the overhang by picking the inner corner of the walls from the story above.

7. Set **Reference Axis of Core** to the lower left corner for our next column.

8. Place the second column underneath the top-right corner of the overhang.

9. Press *esc* to finish adding columns and return to the **Arrow** tool.

Adding beams from walls to columns

To structurally support the first story, we must add beams between the walls and columns:

1. Activate the **Beam** tool.

2. Keep **Geometry Method** set to **Single**.

3. Keep a **Rectangular** section and default material.

4. Set **Elevation** to 2800, which represents the position of the top of the beam above the current story.

5. Adjust **Cross-Section Size** to 350 and 200 in the Info Box or **Settings** dialog.

6. Draw the beam, starting from the center of the column.

7. Complete the beam by clicking perpendicular to the left. Ensure that you click on the inside of the exterior wall so that there's support for the beam.

8. Continue drawing a new beam between both columns (axis to axis).

9. Add a third beam, again starting from the column center, and go to the left.

10. Press *esc* to return to the **Arrow** tool:

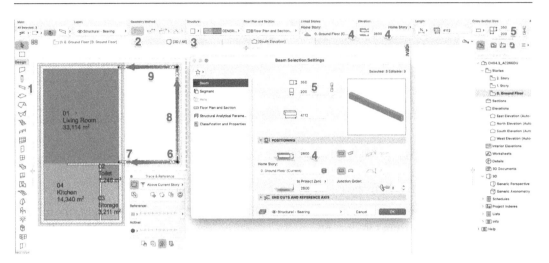

Figure 4.36: Placing three beams between columns and walls

You now have three beams. They may not be completely developed in their final position or with their final dimensions, but this is sufficient at this stage of the design.

Check the result in the 3D window. Good work!

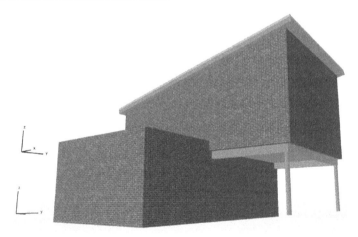

Figure 4.37: The resulting columns and beams in 3D

Summary

This chapter further developed our modeling skills. You have gained experience in modeling and using modeling aids and new tools have been added to help you fully develop a building envelope and its contained spaces. The next chapter will introduce the Parametric Object Library, enabling us to finish this project by adding windows, doors, furniture, and other objects.

5

Building a Basic Residential Model: Modeling Openings, Stairs, and Objects

This chapter continues from the previous one, where we will add a few new tools, including **Doors**, **Windows**, **Stairs**, and the **Morph** tool. We focus mostly on the building envelope and the main circulation throughout the building but will also touch upon the use of parametric library objects and the creation of arbitrary shapes.

The following are the key topics covered in this chapter:

- Modeling joinery (using Doors and Windows) on the outside and inside of the project

- Learning to use the Object Library

- Adding an interior Staircase

- Custom shapes with the **Morph** tool

By the end of this chapter, you will be able to model a complete basic residential project using various tools and techniques.

Adding Doors and Windows to the project

- Now that we have modeled the basic structure for our first project, we need to add some joinery to our Walls to let the light in and allow access to all rooms. Before we can start using tools for Windows and Doors, we should have a clear idea about the key entities in any **Building Information Modeling** (**BIM**) created inside Archicad.

Knowing the key BIM entities and their relationships

Archicad, as a BIM authoring tool, has three key entities that can be modeled:

- **Elements**: This type of entity has a clear definition but no real limits to its dimensions. For example, a Wall is built up a certain way (layers of materials), but neither its length, width, height, nor actual shape are predefined. We have already learned how to model crucial elements for any building with the **Wall**, **Slab**, and **Roof** tools. In this chapter, we will gain a degree of design freedom by modeling with the **Morph** tool.

- **Objects**: These have a more *fixed* definition, although parameters will allow a certain degree of variation for many objects. While some objects can *exist* independently, other types (such as Windows and Doors) commonly depend on a host (e.g., a Wall) and cannot be placed freely. For example, a table has a fixed shape (rectangular), but its size can vary from small to medium to large within certain ergonomic, structural, and/or economic boundaries. The same goes for Doors and Windows: they fit into Wall elements but are rather limited in variations as opposed to a Wall (note that Curtain Walls are to be considered an element rather than an object). This chapter is focused on learning how to use the **Object** tools available in Archicad.

- **Spaces**: Spaces can also be modeled in any professional BIM authoring tool, although – in contrast with elements and objects – this type of entity has no tangible real-life counterpart. This type is (mostly) placed in between other elements acting as its boundaries, and a space can contain other elements and objects. For example, a space could be a single room in a building, but it could also contain multiple smaller rooms (e.g., a fire compartment), or it could even be placed outside the building for example, marking a clearance. We have already modeled spaces in our project using the **Zone** tool from Archicad.

Up until now, we have only been modeling *elements* (Walls, Slabs, Roofs, Columns, and Beams) along with some *spaces* (Zones for the rooms of the project). In this chapter, we will talk a lot about the Archicad **Object Library** and its **Library Parts**, starting in this section with the use of Windows and Doors. Although they appear in the Toolbox as tools on their own, they are actually an integrated part of the Archicad Object Library and in general, *behave* like other objects (for example, a chair or table) in the sense that they are highly parametric in nature and are set up using quite an extensive collection of structured tabs in a similar dialog.

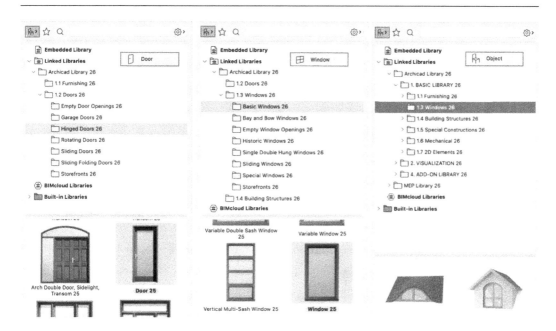

Figure 5.1: The Door, Window, and Object tools next to each other

> **One central Library**
>
> Notice the Object icon in the top left corner and the similar way things are organized and visualized for these three tools, indicating that these are just filtered views from one big central *Archicad Library*.

The Parametric Object Library will be further explained in detail in the *Introducing the Object Library* section later in this chapter, but in the next part, we will see how this works specifically for basic Doors and Windows.

Placing Doors into the interior and exterior Walls

Although Doors and Windows are close relatives within the joinery family, there are some distinct differences. Archicad, therefore, treats them as different tools. The most obvious differences are that Doors lack a frame member at their *bottom* edge, are typically placed directly on the adjacent interior floor (without a bottom offset), and that Doors, in contrast to Windows, always need a moving part to provide access to another space. What they have in common is that they can *only* be placed *into a Wall*. Windows and Doors are dependent on the Wall and are therefore called **child objects**, while the Wall is the **parent element**.

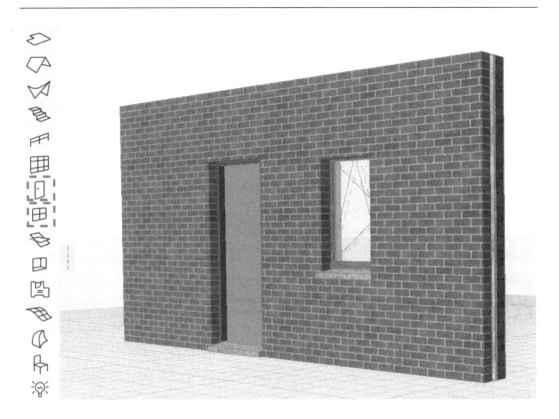

Figure 5.2: Doors and Windows are always placed in a Wall and have other similarities, but are placed using distinct tools from the Toolbox

Placing a Door in an interior Wall

When placing an interior Door, follow these steps:

1. Activate the **Door** tool.

2. Choose the **Door** object you want to use by opening the **Door** default settings and navigating to the **Object** Library. The dialog that opens looks completely different from the ones we have seen up until now. There are two "vertical" columns:

   A. The *left column* (**2a**) shows the available libraries and their objects, arranged in a folder tree in the top half, and displays the contents of this structure's selected subfolder in the bottom half.

B.   The *right column* (**2b**) shows the available settings for the selected Library Part. These settings are mostly arranged similarly, depending on the source of the object. For many objects, a dedicated multi-page user interface is included, defined by the object's creator:

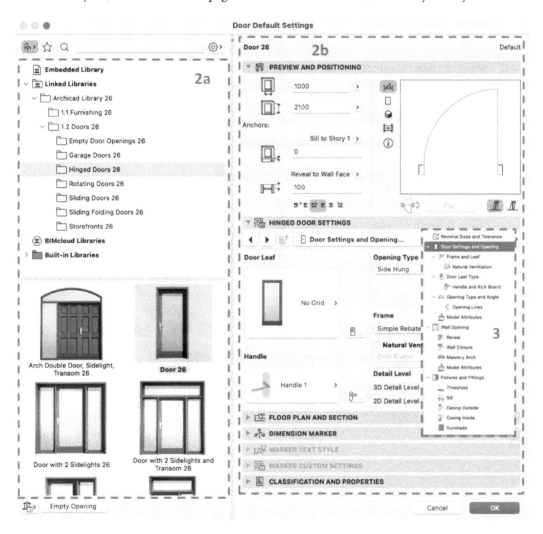

Figure 5.3: The Door Default Settings dialog – quite different from Walls or Slabs

3.   You can adjust the settings of the selected **Door** object in the right half of the **Door Default Settings** dialog and confirm/close the dialog by clicking **OK**.

4.   Here's how to place the Door into a **Wall** element:

A.   First, select the **Anchor Point** option for the Door – this is a point in the visual representation of the Door object, which you use to place the Door. Press *G* to toggle between **Center**, **Side 1**, or **Side 2** (**4a**).

B.   Move the cursor to the Wall where the Door should be inserted (**4b**).

C.   Move the **Sun** cursor icon to indicate the *outside* of the Wall. This marks the **reveal side** for the Door. While interior Walls don't have a real outside, we suggest considering the hallway as outside in relation to the room. This makes the model more coherent for later scheduling. Click once to confirm the position. Use numeric input and drawing aids as needed (**4c**).

D.   Pick any of the four *arrow symbols* to define the Door swing direction (**4d**):

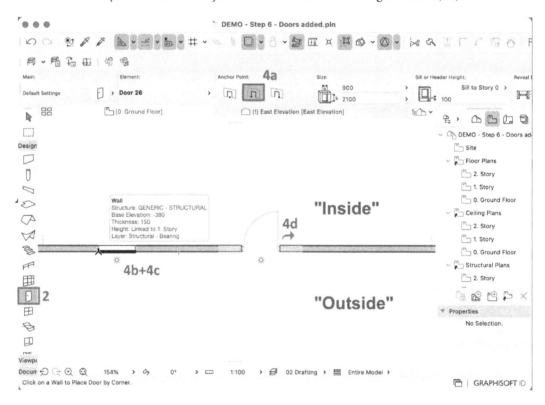

Figure 5.4: Setting the Door swing direction during placement

Let's place some Doors in our project. Use the result you achieved at the end of *Chapter 4* (or download them from GitHub at `https://github.com/PacktPublishing/A-BIM-Professionals-Guide-to-Learning-Archicad/blob/main/CH04_Result.pln`). In the floor plan view, our newly placed Zones are still visible, and they may clutter the view. To solve this, we are going to choose a different Layer Combination. As briefly explained in *Chapter 3*, in the *Modeling tools in general* section, hiding a Layer hides the elements on this Layer. A Layer Combination is a stored configuration of the different Layers (hidden or visible). Layers and Layer Combinations are explained in depth in *Chapter 14*, in the *Understanding quick options/view filters* section. For now, just choose **03 Plans - Preliminary** from the drop-down menu in the **Quick Options** bar, so the Zones will be hidden:

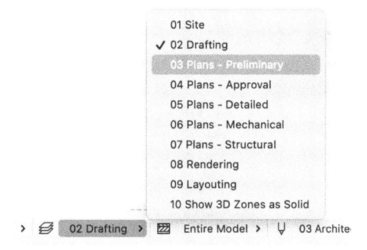

Figure 5.5: Choosing a different Layer Combination

To place the interior Doors in our project on the ground floor and the first floor, choose **Door 26** from the `Hinged Doors 26` folder in **Library**. You can keep its size, but you should change the value for **Sill to Story 0** to 0, as you can see in *Figure 5.3*. You should also adjust some of the settings mentioned under *step 3*:

1. From **Door Settings and Opening...**, choose **Style 1** as **Door Leaf**:

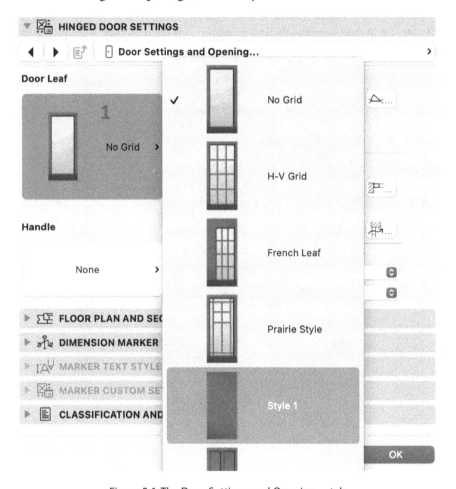

Figure 5.6: The Door Settings and Opening… tab

2. In the **Frame and Leaf...** tab, choose **Block** as **Frame Style** and tick the box for **Frame Thickness=Wall Thickness**:

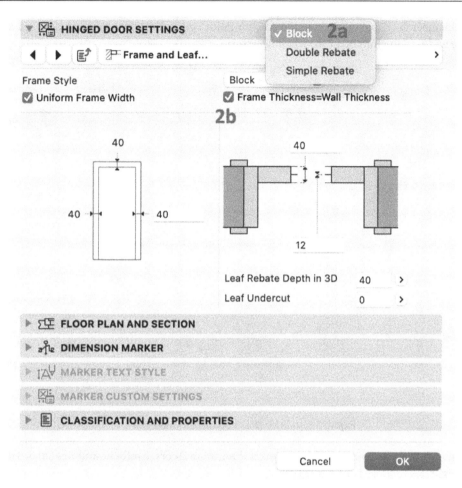

Figure 5.7: The Frame and Leaf tab

Notice how the preview images in the user interface panel follow the settings you apply.

3. Place the interior Doors for the project following *Figure 5.8*, using the drawing aids and numeric input (a tip is to place a Door at a certain distance from a corner or perpendicular Wall using the methods we learned previously for defining the first node relative to a reference point). We place doors one floor at a time – doing this in a 2D Floor Plan View is easier than doing this in 3D. Use the settings in **Info Box** for efficiency.

Remember to choose the most useful **Anchor Point** option (**3a**) when placing Doors and adjust the **width** in the Info Box (**3b**) when necessary:

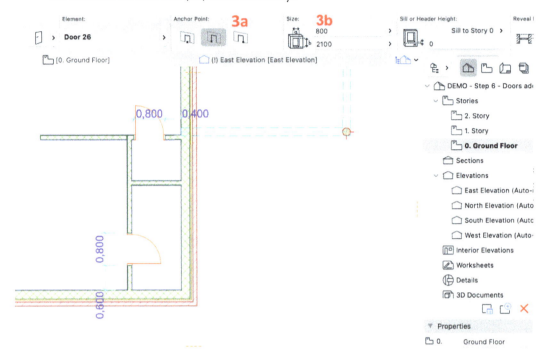

Figure 5.8: Interior Doors on the Ground Floor

Navigate to the **1. Story** first floor to insert three more Doors. Notice how under Sill or Header Height, Sill to Story is 1:

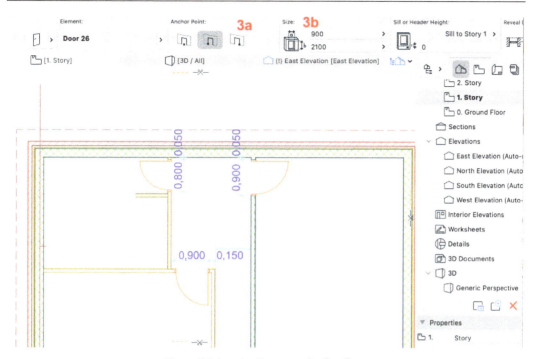

Figure 5.9: Interior Doors on the first floor

Note

For now, these few Door settings will suffice. You can, of course, freely explore other settings and experiment if you want, but keep in mind that Archicad *remembers* the last used settings as the default for any tool. So, start by placing a Door that was correctly set up, as you can always reset all settings by using the **Pick Up Parameters** tool, as explained in *Chapter 3* .

Placing a Door in an exterior Wall

Although all rooms are now accessible through interior Doors, our virtual building is still a closed envelope. Adding a front Door will provide an appropriate entrance. Before we can place this Door, we should change a few settings starting from the interior Doors we have created in the previous subsection:

1. Navigate to **0. Ground Floor**.

2. Pick up parameters (*Option/Alt* + click) from any interior Door.

3. Open the settings (*Cmd/Ctrl* + *T* or through the **Info Box** option).

4. Navigate to the first tab (**Nominal Sizes and Tolerance...**) using the arrows (**4a**). Set the **Wallhole Dimensions** options (**4b**) to 1000 by 2100 (**Width** by **Height**). Notice how the different available dimensions (**4c**) in the first settings tab change accordingly.

Note

Notice how the different Dimensions' correlation is determined by settings in other tabs within **Door Default Settings** or **Door Selection Settings** (when one or more Doors are selected).

Archicad allows any of these Dimensions to be **Set As Nominal**, meaning the chosen size is shown in the Info Box when you select and edit a modeled Door (or use **Info Box** to change the default settings to model a new Door). Toggle through **Set As Nominal** for **Leaf Dimensions** and confirm by clicking **OK**. Now take a look at the **Size** section in the Info Box but change the settings back to **Wallhole Dimensions** afterward, as this works better for this project.

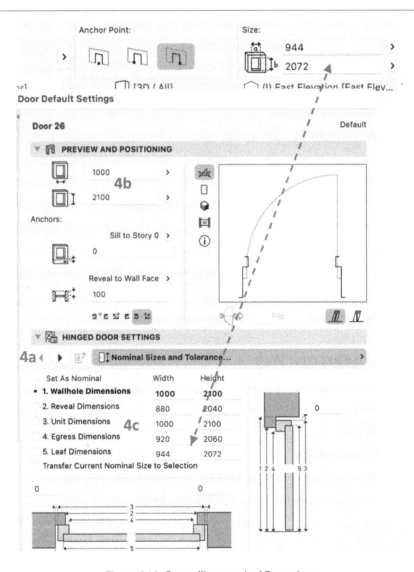

Figure 5.10: Controlling nominal Door sizes

> **Pro tip**
>
> We advise you to choose one of the provided **Dimensions as Nominal** options for all interior Doors (or at least per type) and one for exterior Doors. Your choice can be influenced by local legislation or industry standards (e.g., in Belgium, **Leaf Dimensions** of 830-880-930-980-. . . are standard for interior Doors and correlate with **Wallhole Dimensions** of 900-950-1000-1050-. . .). Should you change this per element; editing your Doors using **Info Box** will become very confusing, especially in larger projects.

5.   In the **Door Settings and Opening...** tab, select **No Grid** as the **Door Leaf** option (**5a**), and for **Handle** (**5b**), pick **Handle 1**. Notice the small buttons next to the different settings, including **Opening Type and Angle** (**5c**), **Frame and Leaf** (**5d**), and **Natural Ventilation** (**5e**). These are shortcuts to other tabs. If you use the drop-down menu (**5f**) at the top, just below the **HINGED DOOR SETTINGS** title, you will see the same items as part of the hierarchy of the dialog tabs, just like we saw earlier in *step 3*. The **Door Settings and Opening...** tab gives an *overview*, while the five available buttons lead you to *detailed settings* for each item. We suggest starting with the drop-down option first, as navigating using the arrows gives you less insight into the structure of these complex joinery objects:

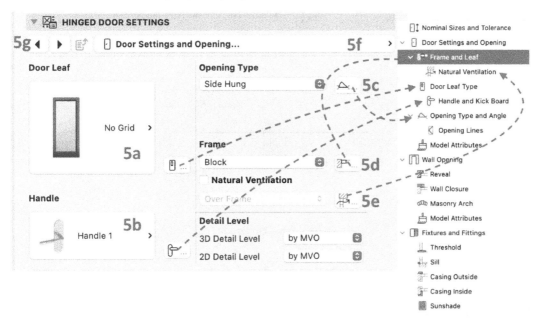

Figure 5.11: Navigating the Frame and Leaf settings with shortcuts or through the drop-down menu

> **Note**
>
> Although we chose to model the Handle in this example, this is not common practice: a Door Handle is a detailed part of the 3D design, and its representation in 2D documents is not always desired. Furthermore, the complexity of these 3D components may negatively affect Archicad's performance, especially in larger projects with hundreds of Doors. Fortunately, there is a solution for each of these concerns – via **View Settings**, **Element Properties**, and the like – each of which will be covered in later chapters!

6. Navigate to **Frame and Leaf...** in whichever way you prefer and change both **Frame Width** (**6a**) and **Leaf Thickness** (**6b**) to 60 and **Glass Thickness** (**6c**) to 20, as we have chosen a glazed **Door Leaf** in *step 5*. Uncheck **Frame Thickness=Wall Thickness** (**6d**) and set the **Frame Style setting** to **Simple Rebate** (**6e**):

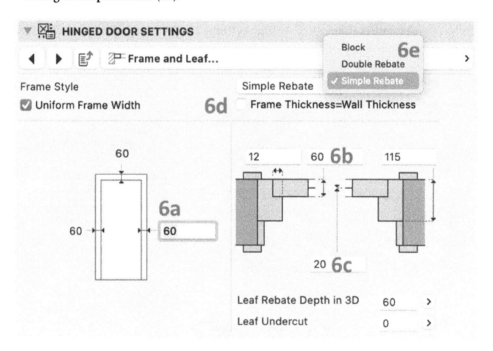

Figure 5.12: The Frame and Leaf settings

7. Next are the settings under **Wall Opening....** Under **Reveal**, pick Reveal from the dropdown for the **Reveal** option (**7a**) and set **Reveal** Depth (**7b**) to 100, which corresponds with the thickness of the face brick of the exterior Walls. For **Jamb Depth**, use the shortcut button (**7c**) to navigate to its detailed settings.

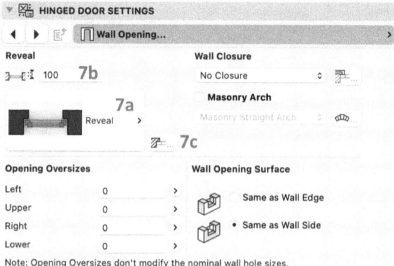

Figure 5.13: The Wall Opening settings

8. Set **Jamb Depth** (**8a**) to 40 so the Door leaf isn't hitting the Wall when opening. Notice that **Reveal Depth** (**8b**) is also visible here. The value was already set to 100 in *step 7*, as well as the chosen **Reveal Type** (**8c**) in the top row of options being set to Reveal. You can change them in either panel as they refer to the same parameter:

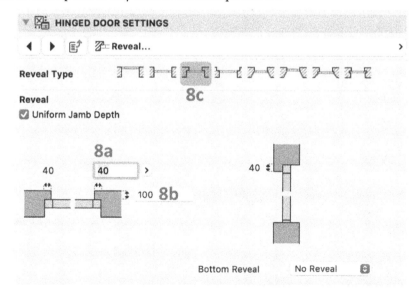

Figure 5.14: The Reveal settings

9. We also need a sill on the outside. Using the drop-down menu, navigate to the appropriate tab in the **Fixtures and Fittings** section. Tick the box next to **Sill – Door (9a)** to add a sill. For **Sill Type**, choose **Stone Sill (9b)**:

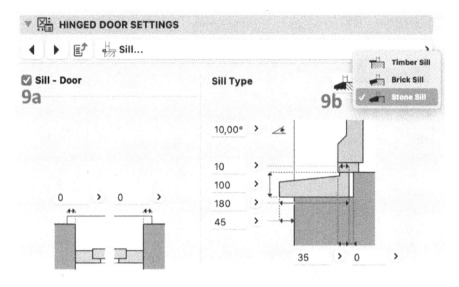

Figure 5.15: The Sill settings

10. On the overview tab for **Fixtures and Fittings...**, untick the **Inside** and **Outside** boxes under **Casing**:

Figure 5.16: The Fixtures and Fittings settings

11. Navigate to **Wall Closure...** and set **Turn Plaster** to **Automatically** for **Outside Face** and **Inside Face**:

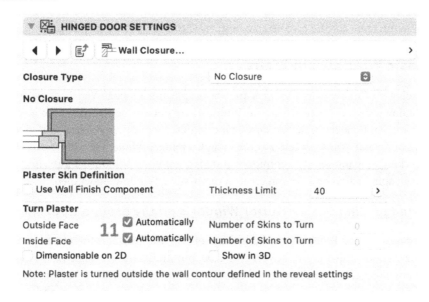

Figure 5.17: The Wall Closure settings

12. Confirm by clicking **OK** and place the Door as you did for the interior Doors.

Pay attention that the Sun cursor properly indicates the outside, as this will determine the reveal side and, thus, the correct connection between the **Door** object and the **Wall** element:

Figure 5.18: Placing the entrance Door with the correct orientation and swing direction

That was an extensive deep-dive into the countless settings for our Door. But we are not done yet, as we also need to look at Windows. So, now is a good time to save your project. Our version can be found on GitHub: `https://github.com/PacktPublishing/A-BIM-Professionals-Guide-to-Learning-Archicad/blob/main/CH05_Part1_Doors.pln`

Placing basic Windows in Walls

In general, placing a Window works in the same way as placing a Door. We will show some extra settings for Windows, such as the *Sill Height*, as Windows are not always placed at floor level. We will also explain how to combine multiple Windows, for example, for a corner Window with two parts or a configuration of several Windows into a more complex joinery whole. We will frequently use the technique of picking up parameters to configure and place different Windows more effectively and, finally, explain an approach to how we can create a multi-story Window. Stay tuned.

Setting up and placing basic exterior Windows on the ground floor

The settings for our *exterior Windows* resemble those of the *exterior Doors* for a large part. As with Doors, exterior Windows also need a reveal and a sill, but unlike the entrance Door, most of our Windows will not be placed at an elevation of 0 above the finished floor level. Furthermore, we will learn to use different opening types, such as fixed glass, turning/tumbling, sliding, and so on. Some of these options are also available for Doors and can be used in the same way.

Let's go ahead and place some Windows on the ground floor. Navigate to the correct **0. Ground Floor** floor plan view, double-click the **Window** tool to activate it, and open **Default Window Settings**. The default Window from this template is **Window 26**, which can be found in the `Basic Windows 26` folder.

> **Template default settings**
>
> As previously mentioned, Archicad remembers the last used settings for any tool as the default setting the next time that tool is used. For example, when starting a new project based on a template, all the default settings for every tool are the settings last used when creating the template.

We will use the default Window to quickly design the remaining openings for our building envelope. Our only concerns, for now, are setting the correct sizes, sill height, and Wall opening settings. All other settings are the same for all Windows in this project, such as the material or sill. After placing any Window, you can still edit individual instances and change their **Opening Type** options or even replace a placed object with another one from the Library. Let's navigate through the general settings for our first Window, just like we did for the Door settings:

1. Pick the **Wallhole Dimensions** option for **Set As Nominal** and enter 2000 by 2100 (**Width** by **Height**, aligning the top of the Window with our entrance) (**1a**) and set **Sill to Story 0** (**1b**) to 0. Change **Reveal Type** to **Reveal to Wall Face** and set it to 100 (**1c**). Notice how this Window dialog looks exactly like the settings for the **Door** tool:

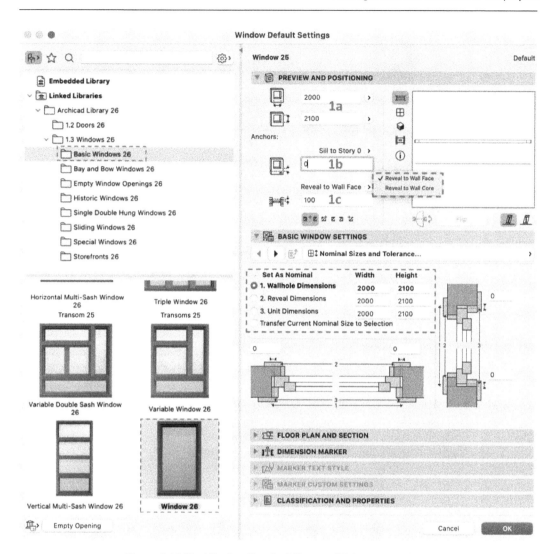

Figure 5.19: The Window Nominal Sizes and Tolerance settings

2. The **Opening Type** option of **Main Sash** should be set to **Fixed Glass**.

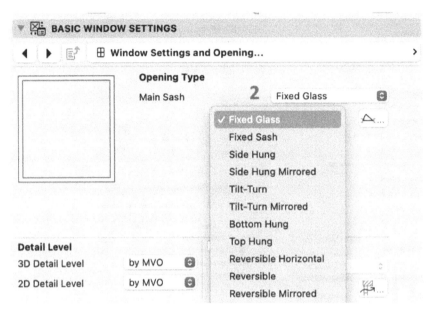

Figure 5.20: The Window Settings and Opening… settings

3. **Reveal Type (3a)**, **Jamb Depth (3b)**, and **Reveal Width (3c)** are set just like the entrance Door: **Reveal**, 100, and 40:

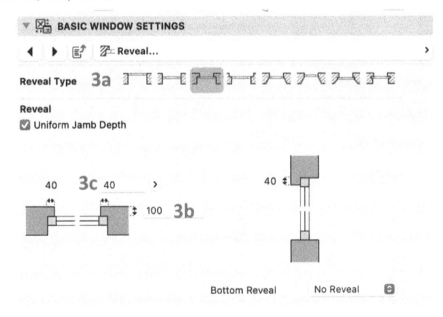

Figure 5.21: Reveal settings

4. On the **Frame and Sash…** tab, change the **Frame Thickness** and **Frame Width** settings (**4a**) to 60 by 60, check **Uniform Frame Width** (**4b**), and choose **Mitered Joint** (**4c**) instead of **Butt Joint**. Notice that we don't have to go through the settings in the hierarchical order of the drop-down menu. You are free to follow your own logic!

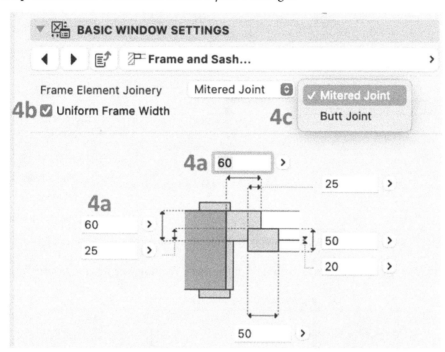

Figure 5.22: The Frame and Sash… settings

5. Activate **Sill** and set the type to **Stone Sill** like our exterior Door.

6. Confirm by clicking **OK** and then placing the two big Windows in our project's kitchen and living room area, using the dimensions given in *Figure 5.24* and appropriate Anchor Points and modeling techniques, aligning these Windows with other elements from the project.

7. Change the **Nominal Size** settings of the Window to 900 by 5100 for the Window located near the Staircase, making it stretch along the height of two building stories:

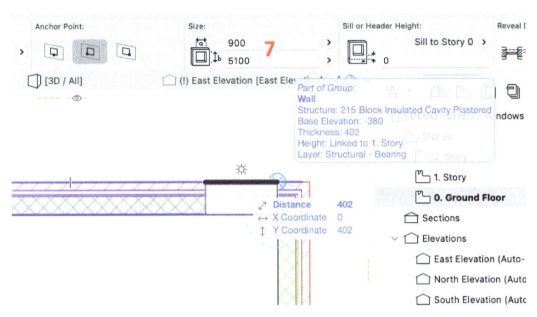

Figure 5.23: Position the Window in the corner

The following diagram gives an overview of the three Windows we need to insert:

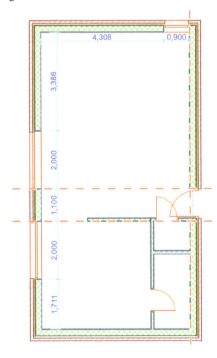

Figure 5.24: The Windows on the ground floor and their position

> **Note**
>
> Aligning with other elements using snaps and guides generally works more easily than using dimensions. The "odd" numbers are caused by the fact that exterior dimensions were originally used for the basic shape of the project, but we have since detailed these Walls into a Composite containing plasterwork skins on the inside, which is a common practice.

At this point, we need to take a closer look at some of the things we have done so far in this chapter. First, open a 3D View of your project and orbit until you see your newly placed Windows. Surprisingly, our Staircase Window is only visible on the ground floor story, both in a 3D View and on the 2D Floor Plan View. Switch between **0. Ground Floor** and **1. Story** to confirm that the Window is not shown on the first floor:

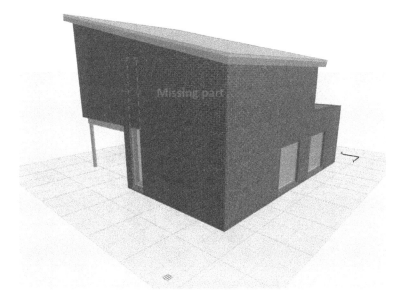

Figure 5.25: The upper half of the Window on the Staircase does not show in 3D

As described in the introduction for placing Doors into interior Walls, Windows and Doors can *only* be placed *into a Wall* – and we can *only* add them *to a single Wall*. Since this specific Window is about two stories high and is placed on the ground-floor Wall, it will not show on the first-floor Wall as that Wall is not its parent element. However, there are several solutions for this challenge:

- **Option 1 – model a Wall two stories high**: This solution seems the easiest choice. Beware that the first-floor exterior Wall must be split up since the *overhang* remains separate, only placed on the first floor. This method creates one parent element into which the (child) Window object fits perfectly. The Window also appears in 3D and 2D as expected, although there are some imperfections. For example, the plasterwork has *disappeared* on the ground floor in 3D. You will learn how to solve these imperfections later, but for now, this is an acceptable solution. To create this solution, use the Pet Palette to adjust the length of the upper Wall (only the overhang). Adjust the height of the lower Wall so that it can be trimmed by the Roof, and use this command to connect the Roof and Wall correctly. The Window needs no further adjustments:

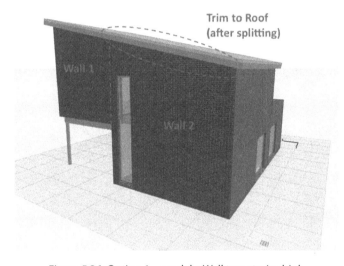

Figure 5.26: Option 1 – model a Wall two stories high

- **Option 2 – model two Windows, one above the other**: From a design point of view, this seems more realistic. The glass in front of the concrete floor slab of the first floor isn't the greatest idea. Of course, we could choose a different Window type, with a panel halfway, to improve the first solution. Another reason why this second solution could be more appealing is that modeling elements and objects "per story" as much as possible is a good practice, often required in *Modeling Agreements* in the **BIM Protocol** and/or the **BIM Execution Plan**. The rest of this solution is straightforward: you just must pay attention to the **Ganging** settings for both Windows, as this makes it possible to cleanly combine multiple Window objects: tick which sides of your Windows will be connected to another one – as seen from the outside. Using this setting, you will have to uncheck **Uniform Frame Width** for both Windows and set each **Frame Width** option as shown. Two Windows suffice, of which the upper one should have a **Lower Transom** or Lower Sash set up as a `Solid Panel` (as shown in *Figure 5.29*). Adjust the dimensions of each part to make them fit your Story Settings:

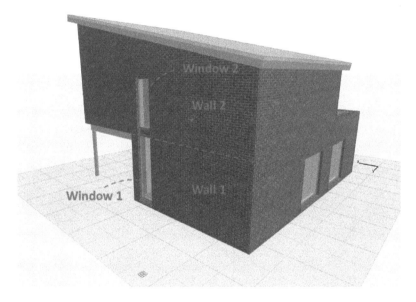

Figure 5.27: Option 2 – model the Window in two parts

Here are the settings for **Window 1** on the ground floor:

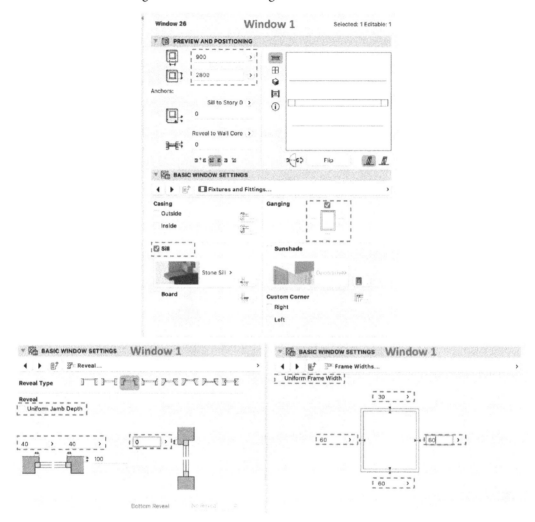

Figure 5.28: Settings for Window 1

Next, we show the settings for **Window 2** on the first floor:

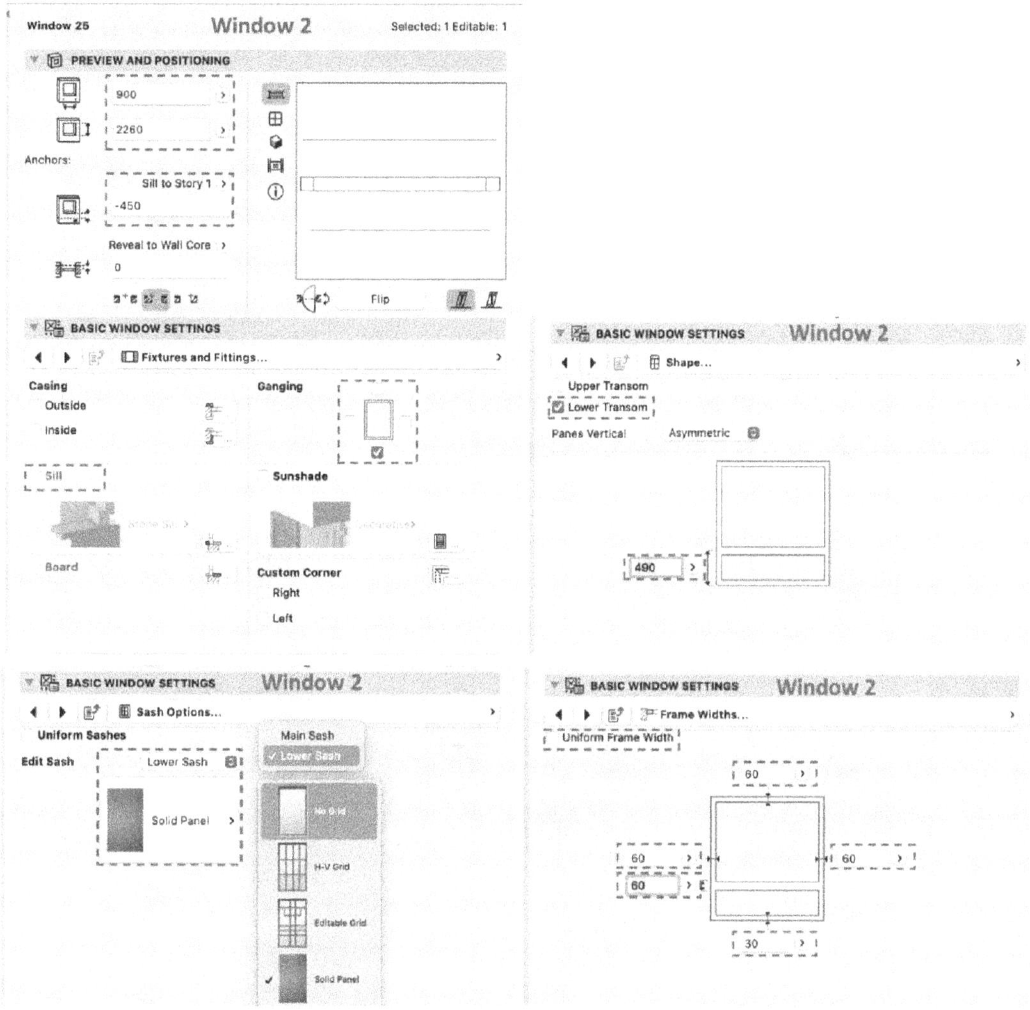

Figure 5.29: Settings for Window 2

Now that this is solved, you are ready to finish the Windows by modeling the remaining ones on the first floor.

Placing the Windows on the first floor

Placing the two solitary Windows is quite easy. Pick up parameters from one of the two large Windows on the ground floor, navigate to the first floor, and adjust **Sill to Story 1** and the **Height** setting in **Info Box**, so the top height matches with the two-story Window we just created. Place these solitary Windows at the positions indicated in *Figure 5.30*:

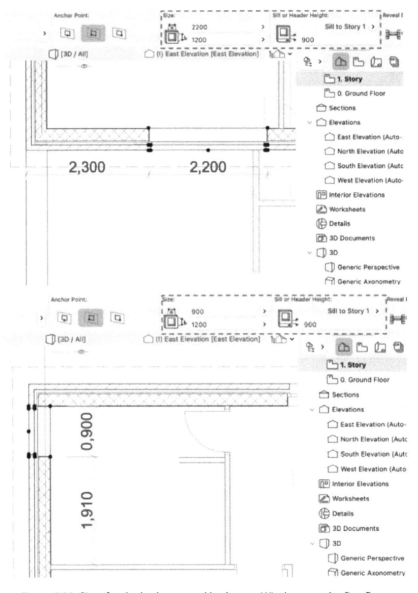

Figure 5.30: Sizes for the bedroom and bathroom Windows on the first floor

Modeling a corner Window requires you to model two Windows and connect them using the **Custom Corner** setting. Setting **Corner 1** or **Corner 2** can be confusing, as this is influenced by the *swing direction* selected when placing your Window: **Custom Corner 1** is placed at the side of the chosen swing direction and **Custom Corner 2** is placed at the opposite side of the chosen swing direction. Placing the two corner Windows requires a few steps and adjustments with the Pet Palette. The angled corner connection makes precise placement in one click quite a challenge.

For the Window in the northern Wall, we picked the swing direction to point at the corner, so **Custom Corner 1** is the corner to be set. For an outside corner, we set 90°:

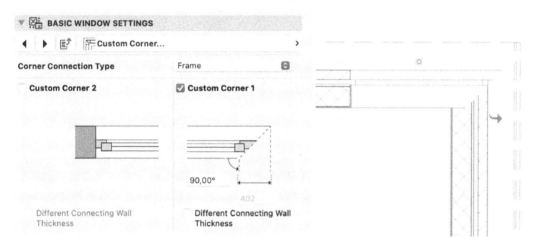

Figure 5.31: Setting Custom Corner 1 and the effect of the swing direction at placement

The second Window in the eastern Wall was picked with its swing direction away from the Wall, so here, we enable **Custom Corner 2** and again use the 90° angle:

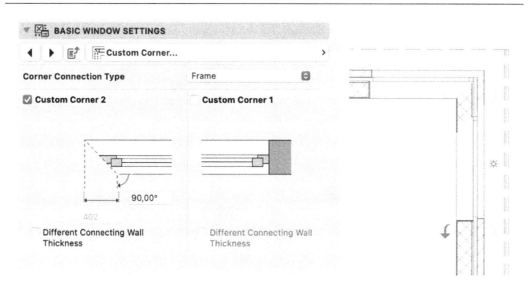

Figure 5.32: Setting Custom Corner 2 and the effect of the swing direction at placement

Afterward, we set the Windows' nominal sizes: set their height to 1200 and width to 1500 and 2200 respectively, and use the **Move** command (*Cmd/Ctrl + D* or through the Pet Palette) to shift them right into the corner to get a nice and clean fit. Use the node at the interior corner as the point of reference. The **Sill to Story 1** setting can be set to 900:

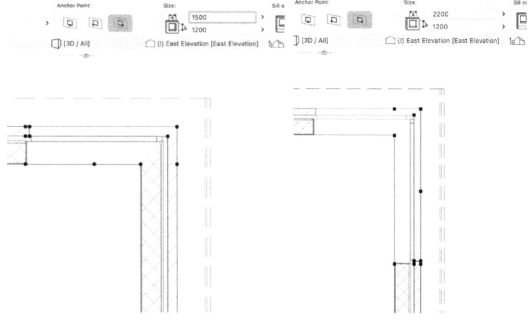

Figure 5.33: Dimensions for the corner Window in Info Box

> **Note**
>
> Archicad also provides a **Corner Window tool**, which you can find at the bottom of the Tool Box. This actually brings you back to the **Window** tool. It is a legacy solution, replacing the former Corner Window object, which was a somewhat limited option compared to the Windows we have been using. The **Corner Window** tool lets you set up a Window like you are used to, after which you can place it at a corner, automatically creating two Windows with the appropriate Custom Corner settings – this can speed things up!

Open the 3D View and admire the result. Don't forget to save your project (our version can be found on GitHub at `https://github.com/PacktPublishing/A-BIM-Professionals-Guide-to-Learning-Archicad/blob/main/CH05_Part2_Windows.pln`). Now that we have placed all our Windows with the correct sizes and sill heights, let's change how they open!

Changing the Window type without changing other parameters

It may seem odd that we have modeled all our Windows as the **Fixed Glass** type. Although we could change the type or size settings every time we create a new Window, it is a good practice to first use this "generic" type to make sure settings such as material, sill, or frame size are set correctly and all Windows are in the right place. Afterward, we can easily change some of the settings of our Windows without losing what we already set up!

To change the **Opening Type** option, follow these steps:

1. Select the Window (for example, the one in the bathroom).

2. Open **Window Selection Settings** (*Cmd/Ctrl + T*).

3. Navigate to **Opening Type and Angle...** in the **BASIC WINDOW SETTINGS** tab and choose a different setting for **Main Sash** (e.g., **Tilt-Turn**):

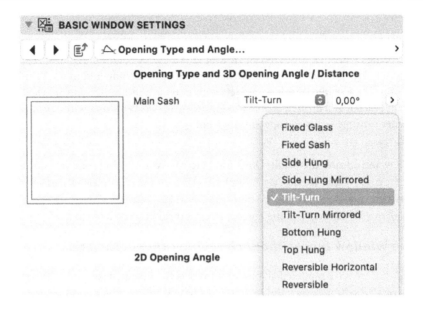

Figure 5.34: Setting Main Sash to Tilt-Turn

To replace a Window with another one from the library without losing settings, follow these steps:

1. Select one or more Windows you want to change into another type from the library. In this example, we will select the two large Windows on **0. Ground Floor** to change them into sliding Windows.

2. Open their selection settings (*Cmd/Ctrl + T*).

3. Navigate to the folder of the desired Window type (`Sliding Windows 26`) in the Library, but *don't click any Window yet!*

4. While holding down the shortcut key for injecting parameters (*Cmd + Option/Ctrl + Alt*), click the `2-Sash Sliding Window 26` folder in the library. Voilà, you have successfully changed your basic Window into a sliding one while retaining all (compatible) parameters already set for the newly chosen type.

> **Note**
>
> This works because Graphisoft ensures all these Windows follow the same structure and use the same parameter names. Otherwise, you'd have to set them up from scratch, and you remember that quite a few parameters had to be adjusted previously.

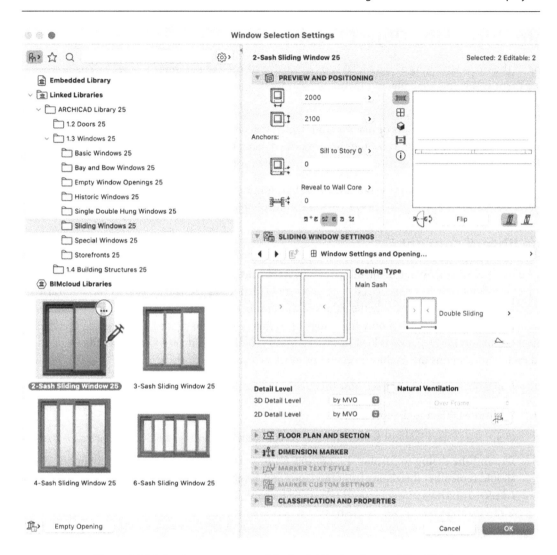

Figure 5.35: Injecting parameters of the basic Window into a sliding Window

By now, you should feel quite comfortable using the **Window** and **Door** tools for a variety of joinery elements. Of course, there is still a lot to explore in this unique section of the Object Library, but at least now you understand how dialogs work and what settings are used to configure a Door or Window correctly. Apart from joinery elements, the Object Library contains many more elements, and it is time to take a look at how these can be used in a basic residential project.

Introducing the Object Library

One of the main strengths of Archicad is its extensive **library** of parametric objects. You already used them when you were placing Doors or Windows, and the concept is applied to many other commands as well.

The library is a series of scripts or macros from which objects are generated. Rather than drawing or modeling such objects graphically, they are described using the **Geometric Description Language** (**GDL**). This language is at the core of Archicad, even when you, as a user, are not writing these scripts.

While this introductory course will not cover programming objects ourselves, we will look at GDL objects from a user's perspective: how to use and control objects!

The library concept

As we mentioned briefly in the introduction, an **Archicad project file** (*.pln) contains all your project information and references to one or more loaded libraries. The objects in these libraries are not stored in your project; only the values of their properties are. So, you may have thousands of parametric chairs included, yet only the values to configure each chair need to be stored. The chair **object definition file** (*.gsm) is kept in either a library folder or a **Library Container file** (*.lcf) and each item contains the Archicad version number in its name (e.g., 26 or 25).

Let's set up a bookshelf object to see how this works in practice:

1. Select the **Object** tool, represented by a chair icon.

2. Open the **Object Default Settings** dialog to configure the settings before we place an object.

3. Expand the dialog to display the *Library browser* at the left if it is not already open. You should remember this dialog from the **Door** and **Window** tools. Navigate to `Navigate to Linked Libraries > ACHICAD Library 26 > 1. BASIC LIBRARY 26 > 1.1 Furnishing 26 > Cabinets and Shelves 26` and select the **Bookshelf 01 26** object icon.

4. The panel to the right has the same subdivisions as any other Archicad element but with an added *Library panel*, where you can configure the object, typically in great detail. Here, the **Bookshelf 01 26** object has a dedicated user interface (see *Figure 5.36*) titled **BOOKSHELF SETTINGS** to make its configuration self-explanatory: each property or parameter has a tooltip, a clear name, and is often documented with an icon or figure. The figures even adapt when other values are selected.

5. You can access multiple pages and subpages using the arrows or the pop-up button, just like with Door and Window objects.

6. The last page, called **Description**, is a text-based table of values. When an object lacks a user interface, this table is all you have. There was a time when this was the only way to configure objects.

> **Note**
>
> Revit users may find this tabular display of parameters familiar, as that is the only interface they get with Families. However, we personally prefer the icons and preview pages.

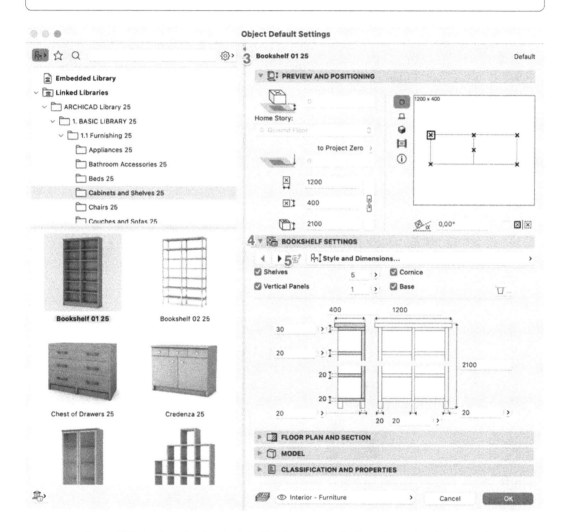

Figure 5.36: Settings for the Bookshelf object, with a dedicated user interface panel

Adding some objects

We can place a few furniture objects in our project using the **Object** tool. Look for tables, chairs, seats, and beds. They are straightforward to configure and can be adjusted in detail, so we invite you to experiment a bit. We'll wait until you return.

Kitchen Cabinets and equipment

Some objects behave like super-objects, controlling a whole group or setup of related objects. A good example is the **Kitchen Layout 25** object, found in `1. BASIC LIBRARY >1.1 Furnishing 25 > Furniture Layouts 25`.

With this object, you can configure a complete kitchen via the user interface and have detailed control over the cabinet styling, the position of the equipment, and the dimensions of cabinets, shelves, countertops, and so much more.

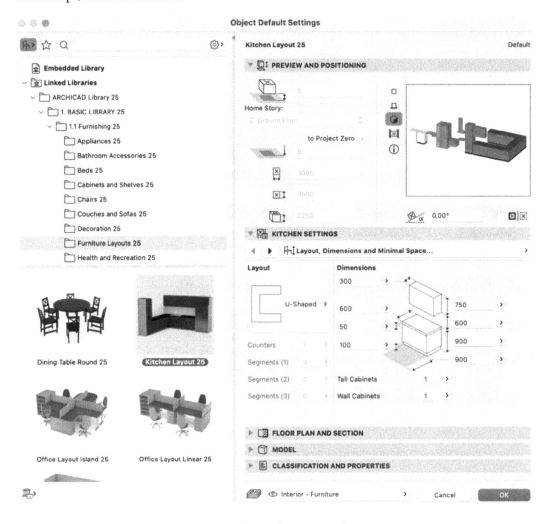

Figure 5.37: The Kitchen Layout object

After you place the object in the floor plan, you have **editable hotspots**. These are *magenta-colored reference points* from which you can start shaping the model interactively. You are provided with a familiar Pet Palette with the last command selected, so you know you can use Snap Guides, Guide Lines, and the Tracker to position the hotspots. In this example, we are moving the Refrigerator, but you can see multiple hotspots for this object, including the corners of the Kitchen itself, to ensure it fits into the room.

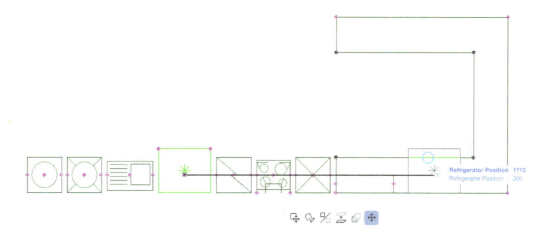

Figure 5.38: Configuring a Kitchen Layout object using editable hotspots

Fitting the kitchen into our floor plan

The default placed kitchen doesn't really fit into our design, so let's redo it:

1. Erase our first attempt and reopen the (default) settings dialog.
2. In the first **KITCHEN SETTINGS** panel, set **Layout** to **L-Shaped**.
3. In the **Accessories…** panel, deactivate **Hood**.
4. Pick the **5-burner Cooktop** option.
5. Deactivate **Refrigerators**.
6. Select the **Double with Drain Board Sink** option.

7.   Deactivate **Washers**:

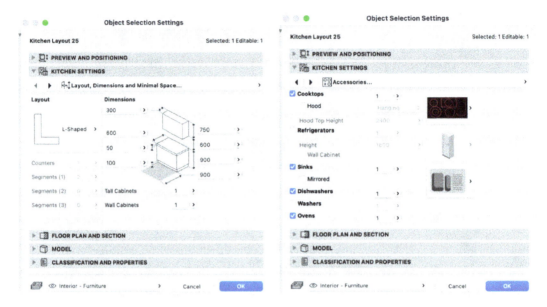

Figure 5.39: Adjusting the Kitchen Layout 25 settings

Rotate and resize the L-shaped kitchen layout to match two kitchen Walls. Play around with positioning the equipment and the top shelves, using the (magenta) diamond-shaped editable hotspots to ensure they all fit into the kitchen corner. Use *Figure 5.40* as a reference and look at the Tracker to understand what each hotspot is controlling:

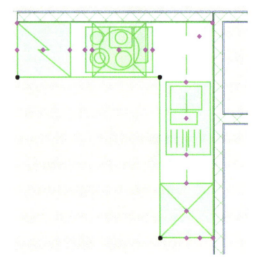

Figure 5.40: Kitchen Layout configured into an L-shape

For example, we found a layout that works well with **Sink Position** set to 900 and 300, **Oven Position** set to 2700 and 1900, **Cooktop Position** and **Dishwasher Position** both set to 2700 and 970, and **Tall Cabinet Position** set to 1000 and 300. Also, adjust the top shelf to your liking, such as aligning it with the eastern kitchen Wall.

Pro tip – when to use objects and when to use tools?

When you have looked through the Library browser, you may have discovered that some of the objects closely resemble tools we already have available, such as Beams and Columns. Why is that?

The answer is *flexibility*. Sometimes, it is more straightforward to model a Beam or Column using the dedicated tool, but these commands don't cover every possible type of Beam or Column, so you can fall back on one of the library objects.

Another example could be standalone Doors or Windows. As an object, they don't have to be inserted into Walls anymore, which may be something you might need for a project.

Beware, though, that you may arrive at a mixture of objects and dedicated commands and thus must take them into account when extracting information from the model.

The two floors in our basic residential project have been modeled, and they now have Windows and furniture. So almost every tool we need for small-sized residential models is known by now. Only the vertical connection between the stories is lacking! Time to learn how to model a Stair.

Basic Stairs between the building stories

The Stair systems in Archicad is a **System** tool: it is a hybrid mixture of a library object to manage components and profiles and a modeling tool to allow broad geometric flexibility. You define a Stair using its main reference line (called the **Baseline**), just as with Walls or Beams, and can link the bottom and top to stories, but you also have a contour to control the outline, just as with Slabs or Roofs. The Baseline contains multiple segments, which can each be set to become a **Flight**, **Landing**, or **Winder part**, and you have deep control over how the Stair, including Landings and Treads and Railings, will be generated from its Baseline and contour.

Let's dive in with a basic setup for a straight Stair:

1. Activate the **Stair** command from the Tool Box.

2. Set the **Baseline** position to the **Right** side of the Stair, but keep **Offset** at 0.

3. Pick the first Baseline node in the 2D Window and start moving to the right.

4. Archicad shows a preview of the threads and shows a familiar Tracker to help you enter precise dimensions. You can use all Snap Guides, Guide Lines, and other methods as usual.

5. The Pet Palette also shows the different available modes for the current Baseline segment: **Flight**, **Landing**, **Winder with Equal Angles**, and **Winder with Equal Goings**. Keep it at Flight (first icon on the top row).

6. Continue to move the cursor all the way to the highest possible thread to maintain a straight flight:

Figure 5.41: The Stair creation command, with the preview, Tracker, and pet palette

We have not configured all the settings yet, but rest assured, you have control over the Riser height and length, Stair width, and even the full setup of all the Stair components:

Figure 5.42: The Stair display in the 2D Window

You can use the same approach to create a Stair with an *L-shape* by picking additional in-between points when defining the Baseline. For a *U-shape*, it works best when you draw the reference line against the inside of the Walls where the Stair should be placed:

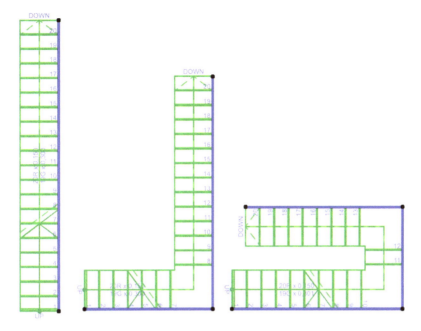

Figure 5.43: Three Stair types (straight, L-shaped, and U-shaped), with the Baseline highlighted

Don't worry about perfecting your Stair on the first go since you can fully edit the Stair at a later stage. When you select a Stair, you can dive into the editing mode using the **Edit...** button, which pops up:

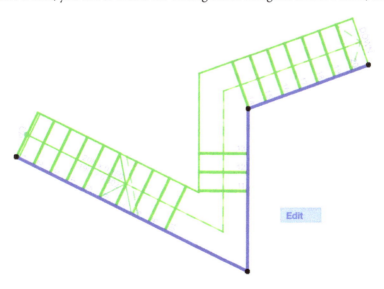

Figure 5.44: Stair with the Edit pop-up button

In this **Edit** mode, the Window switches to a more focused mode, and you have a small dialog at the top left of the view where you can toggle the display to different structures of the Stair system. In this mode, you can access individual threads and start shaping them with the familiar tools you already used to shape a Slab or Roof contour. Press **Exit Edit Mode** to return to the regular display:

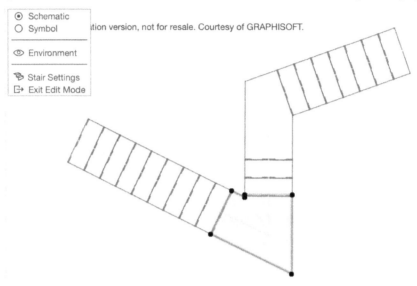

Figure 5.45: The Edit mode for Stair

Fitting the straight Stair into our design

When you start from a default Stair and try to place it into our design, you may notice that it is too long to fit. We can adjust the Risers and Goings to make the Stair a little shorter while still ensuring that the Stair adheres to comfortable and safe Stair calculation rules. Let's get started:

1. Activate the **Stair** tool.

2. Open the **Stair Settings** dialog (*Cmd/Ctrl + T*).

3. You'll notice that it uses a different user interface, which displays a hierarchical structure at the left. Select the **Stair** entry at the top of the tree.

4. The right side of the interface is more familiar. We only need the **GEOMETRY AND POSITIONING** panel.

5. We can confirm that **Top offset to Top Linked Story** and **Bottom offset to Home Story** of the Stair are set with an offset of 100 (millimeters (mm)) to take the floor finishing into account, but as we modeled our floor Slabs with a default offset of 0 (see *Chapter 3*, in the *Adding a Slab* section), you should set both offsets to 0 accordingly.

6. Set **Stair Width** to 798 (mm).

7. Set **Number of Risers** to 17.

8. As a result, **Riser Height** is automatically adjusted to 176 (mm). If you select another **Riser Height** value from the dropdown, Archicad will adjust the **Number of Risers** value accordingly.

9. This is the result of the following **RULES & STANDARDS** setting, where we enforce that **2 Riser + 1 Going (2*R + G)** can only be allowed between 600 and 650 (mm) – which is a rule of thumb for ergonomic Staircases. You can adjust this, but only do this when you understand the consequences for your design and for the accessibility of the Stair.

10. Set **Baseline** to **Right**, so that you can model along the inside face of the exterior Wall.

11. You can close the dialog by pressing **OK**.

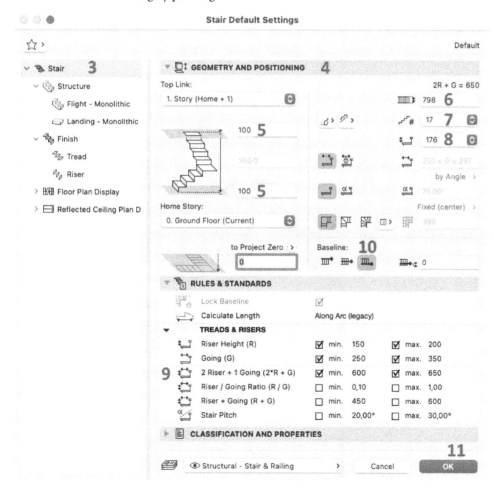

Figure 5.46: Stair Selection Settings to define the Width, Riser, and Going dimensions

Your Stair is ready to be placed in the design now. Follow these steps to do so:

1. With the **Stair** command active, hover over the bottom-right corner of the main open room on **0. Ground Floor** until a snap point appears.

2. Move the cursor in the upper direction and press *D* to set a **Distance** value of 1000 (mm) in the Tracker.

3. Orient the cursor upwards to get the Stair preview and pick the furthest point when the last thread is displayed to indicate the endpoint of the Baseline.

4. The Stair is created, so you can press *esc* to finish modeling it:

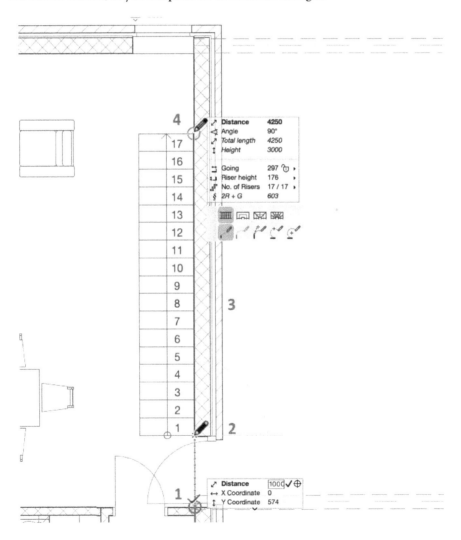

Figure 5.47: A straight Stair in four clicks

To confirm that the Stair looks right, switch to the 3D Window (*Fn + F3/F3*) and use your navigation skills to look at the Stair inside the room. The **3D Explore** navigation mode is suitable for moving into this viewpoint:

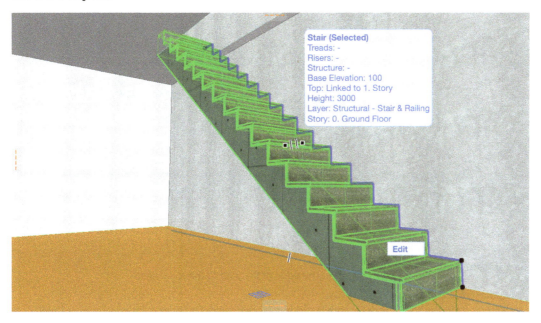

Figure 5.48: Looking at the Stair in 3D from inside the room

The finishing touch is to ensure that the contour of the Slab on **1. Story** is adjusted so it closes cleanly against the highest thread of the Stair. This is easy to do in a 2D Window.

Let's get started:

1. Navigate to **1. Story**.

2. Select the Slab with the **Arrow** tool.

3. Use the **Offset Edge** command in the pet palette (the fourth icon in the top row) to move the edge of the slab and snap it to the last thread of the Stair:

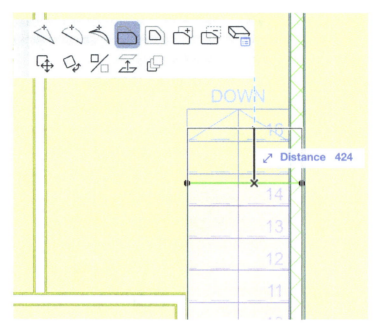

Figure 5.49: Offsetting the edge of the Slab to the top thread

As an Archicad System tool, the **Stair** command is very extensive. You have deep control over the Stair and all its components: the walking/reference line, the outline or contour, the individual threads, the thread generation rules, the materials and finishing, the Railings, and the Stringers. There is no end to it!

You are advised not to try to make the most complex Stair and expect it to be perfect from the first attempt. Instead, start from the Baseline and begin with the overall sizes (length, width, and height). Having a few Walls or Slabs as a reference makes it easier to align the Stair.

Be prepared to switch from 2D to 3D and back to check the Stair from different viewpoints.

> **Pro tip**
>
> Remember that a Stair is also a very important design tool: it is one of the few objects in a building that crosses building stories.
>
> In 2D CAD drafting, we often keep it very simple, but even then, there is a huge chance of making mistakes. With 3D parametric Stairs, Archicad gives you a very powerful tool to figure out the feasibility of the Stair, the required free space and head height, the Going length, and Riser heights. Use it to your advantage: let the software work for you to guarantee that you design a Stair that works well.

By now, every "normal" object in any small residential project is covered. Sometimes, you need special or peculiar shapes, though, which are not easily found in the Library or for which no particular tool is located within the Toolbox. For these elements, the versatile **Morph** tool comes in handy!

Custom shapes with the Morph tool

Despite the wide variety of tools and the plethora of parametric library objects, you may need an object with a particular shape that is not available out of the box. Designers and manufacturers make countless designs in all sizes and shapes.

While you could try to look online for that perfect library object or use other 3D software to create the geometry, Archicad provides a very flexible modeling command: the **Morph** tool. This is a single tool with a very powerful pet palette providing a flexible set of geometric operations: extruding, filleting, adding edges, moving vertices, and so on. If you have some experience with *SketchUp*, you'll feel at home.

And the biggest advantage? It's right there, inside Archicad, where it behaves as a proper, native BIM object with all the power that Archicad brings to it. You can measure them, include them in 2D and 3D views, or extract information.

Let's start with the basics.

Basic Morph shapes

Let's see how we can create a basic shape with the **Morph** tool:

1. From **Toolbox**, activate the **Morph** tool.

2. Select **Box** Geometry Method (by pressing *G* a few times).

3. Outline the box in the 2D Window and use the Tracker to make it 1000 by 800 (mm).

4. When the **Enter Extrusion Vector Length** pop-up dialog appears, enter 1200 (mm) for **Extrusion Length** and press **OK**:

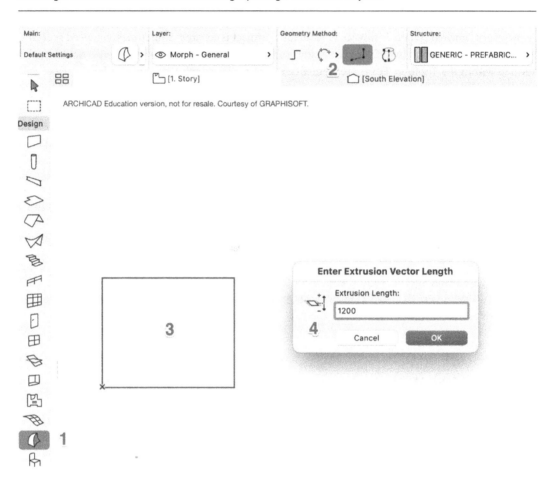

Figure 5.50: Working with the Morph tool

5. Press *ESC* to return to the **Arrow** tool.

The result is a custom box with the given dimensions:

Figure 5.51: Using the Morph tool to create a simple box

This resulted in a 3D object, but you can also start from just edges:

1. Reactivate the **Morph Tool**.

2. Select **Polygonal** Geometry Method (press *G*).

3. Use the well-known drawing methods to draw an arbitrary polygonal curve in the 2D Window using the Tracker and pet palette tools.

4. Press *Enter* to complete the command. You have created a curve, but this is not all.

5. Press *Fn + F3/F3* to open the 3D Window. The curve is also visible as a 3D object in its own right: visible and editable with the rest of the model.

> **Note**
>
> It is not possible to display regular annotation entities, such as lines or fills, in 3D.

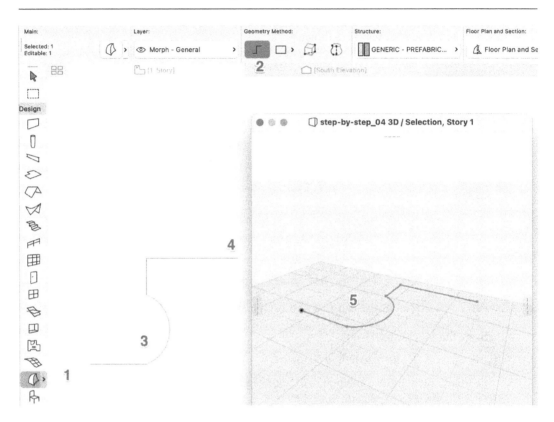

Figure 5.52: Morph to create a polygonal curve

Using the pet palette to edit the shape

The real power of the **Morph** tool is in its extensive editing using the pet palette. We'll illustrate this by further modifying the object we just created.

Freeform sculpting

When we switch to our box in the 3D Window, we can start sculpting from a face:

1. Ensure that the Morph box is selected.

2. Click on its *top face* and select the **Push/Pull** command (*aptly named after the tool made famous by SketchUp*) (the first on the top row).

3. You can read on the Status Bar that this performs a perpendicular extrusion of the face: move it outside to enlarge or inside to reduce the extrusion.

4. Push the top face upward by 500 (mm) using the Tracker.

5.   Press *Enter* to confirm.

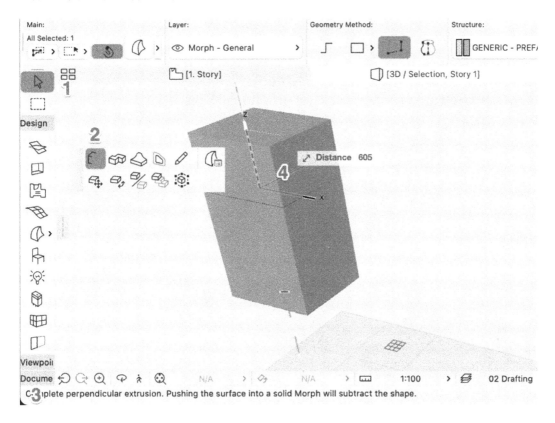

Figure 5.53: Morph using the Push/Pull command

Using the **Offset** command, we can further edit the shape, but this time from an edge:

1.   Ensure the object is still selected.

2.   Click on the bottom *edge* of the box.

3.   Use the **Offset** command (the third icon in the top row of the pet palette) to stretch the edge horizontally.

4.   The Status Bar explains what will happen.

5.   Use the Tracker to indicate a displacement **Distance** of 1000 (mm).

6. Press *Enter* to confirm.

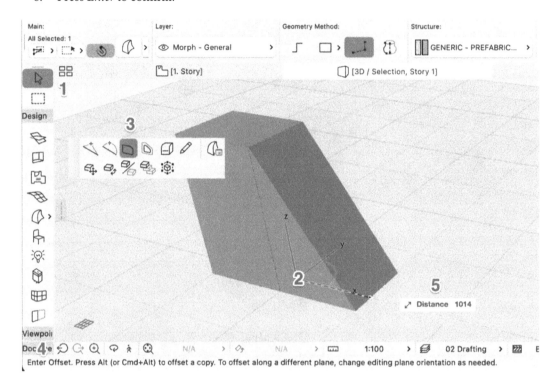

Figure 5.54: Morph to offset an edge

We can also split faces using the Pencil from the **Morph** tool:

1. Ensure the object is still selected.

2. Hover over the left vertical edge and click on it.

3. Select the **Add Polyline/Rectangle/Box/Revolved Morph** command – what's in a name? – to open a special sub-palette (the sixth icon in the top row, indicated by a Pencil in *Figure 5.54*).

4. Switch to **Polygonal** Geometry Method (press *G*).

5. Draw a horizontal edge line all the way to the diagonal edge on the right to split the face into two.

6. Click on the lower front face of the newly split face.

7. Click the last icon on the sub-palette, to go back to the main palette.

8. Use the **Push/Pull** command to extrude the bottom part, using the Tracker to indicate a perpendicular extrusion of 750 (mm).

9. Press *Enter* to confirm.

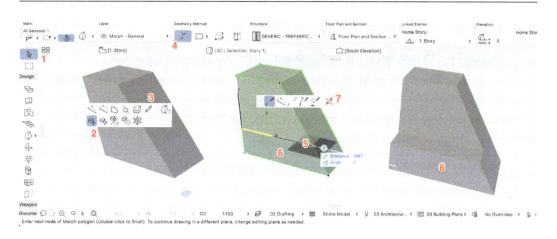

Figure 5.55: Splitting a face to Push/Pull a part of it

By now, you should feel ready to start digging deeper into exploring the various modeling methods of the **Morph** tool.

Here are a few suggestions:

1. **Offset all edges**: Do this by pressing *Alt* (or *Cmd + Alt*) for an offset copy to create an inset face. Use the Tracker with a negative distance if needed. After that, use **Push/Pull** to make an indent.

2. Add a **Rectangular** series of edges on top of a face using the Pencil and use **Push/Pull** to punch the face through the Morph to create an opening.

3. Pick an edge and use **Curve Edge** (the second icon at the top) to turn it into an arc and create a more organic bulge.

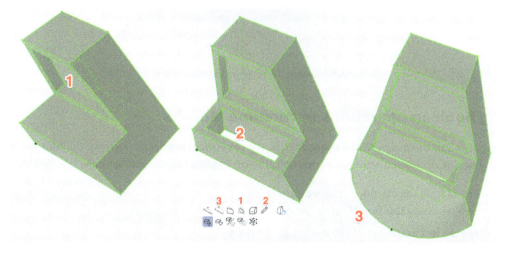

Figure 5.56: Morph step-by-step shape sculpting

Follow these steps to create a cone shape, starting from an arc:

1. Using the **Morph** tool, select the **Arc** Geometry Method option.

2. Draw a complete arc by picking a center, setting the radius, and moving around the cursor until the arc goes all the way around.

3. Click on the closed circle to call up the pet palette.

4. Use **Push/Pull** to extrude the circle into a cylinder.

5. Activate the **Offset All Edges** command as in the previous example.

6. Offset to the inside to shrink the top face.

7. Click to confirm and press *Enter* to finish.

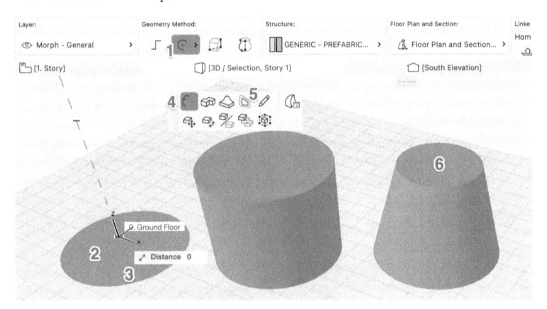

Figure 5.57: Morph to create an extruded cone

Using Boolean operations to combine Morphs

Some shapes can be easily created using a sequence of geometric operations from the Morph pet palette. But sometimes you want to use one object to carve out a hole in another object:

1. Starting from the extruded cone we already have, add a simple extruded box using the **Morph** tool – try doing this in a 3D view (or switch to 3D after creating the box in 2D).

2. Drag the box around in 3D so it overlaps the cone.

3. Select the box and call the **Subtract…** command from the context menu (**Boolean Operations > Subtract…**).

4. When the box is highlighted in red, you can select the cone as the object to subtract from, which will highlight in blue upon hovering.

5. The box will be subtracted from the cone, and you will arrive at our combined Morph:

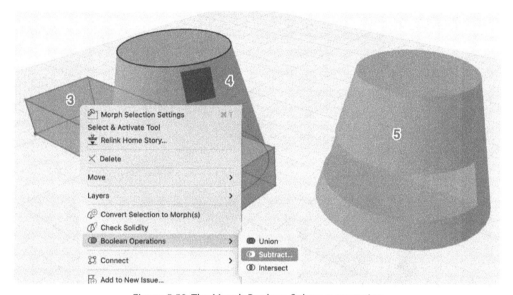

Figure 5.58: The Morph Boolean Subtract operation

This type of shape would be hard to create in any other way, yet using **Boolean Operations** makes it straightforward.

Other **Boolean Operations** options include **Union**, to merge two or more Morphs, and **Intersect**, to only keep the part that overlaps both shapes:

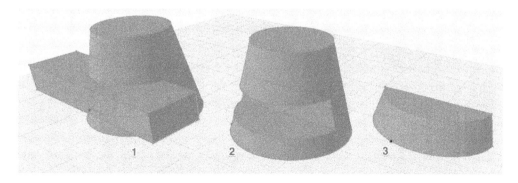

Figure 5.59: The Morph Boolean results: (1) Union, (2) Subtract, and (3) Intersect

Such operations are very popular, as they are easy to understand, yet they bring a lot of modeling power to the user.

Turning any object into a Morph

The modeling freedom of a Morph is unique, but you have the option to turn any object in Archicad into a Morph. Follow these steps to do so:

1. Draw a Wall with a Door somewhere outside of the building (as we don't want to mess up our design).

2. While the Wall is selected, open the context menu to **Convert Selection to Morph(s)**. Archicad presents an important warning to ensure you know what you are doing (see *Figure 5.60*): the new Morph will completely replace the Wall and Door.

3. The Wall and Door are now converted into a series of Morphs:

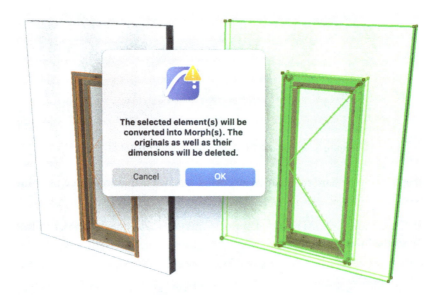

Figure 5.60: The Archicad warning appears while converting a regular object into a Morph

If you select them, they have become multiple independent Morph objects, and if you want, you can start pushing and pulling and carving them to your liking. The pet palette and the Settings dialog of the Wall or Door are no longer available.

> **Caution**
>
> This brings all the modeling freedom of the Morph tool, but the original object loses its parametric behavior. A Wall turned into a Morph is no longer a Wall. You can no longer insert a Door or Window, nor does it join with other Walls to clean up the junctions.

So, when would you use a Morph tool, and when would you stick with the regular commands? It depends on what you want to do with the object:

- If one of the regular commands or library objects does not provide the exact shape you require, you can create it with a Morph

- If you need the flexibility of the regular command, such as making automatic connections or inserting Windows into Walls, don't turn the object into a Morph

In both cases, try to be as organized as you can when defining dimensions or selecting the Home Story and Layer, so that you won't notice anything different in your output documents. People sometimes maintain a copy of the original objects in a hidden layer or in a separate project file.

Summary

This chapter completes our trilogy of chapters on the preliminary design by introducing the Library concept, which is applied to Doors, Windows, and other parametric objects, but also pops up in many other tools as well. This is the core power of Archicad and its GDL. We also introduced the nifty and versatile Morph tool to make almost any shape you want.

While we weren't able to show you every possibility that these tools bring, you should have gained a good understanding of how Archicad works, how it expects you to enter accurate dimensions and sizes, and how you can combine all of these to make a basic model of a building, including Walls, Slabs, Roofs, Beams, Columns, Stairs, Zones, Doors, Windows. And lastly, we have also explored the Morph tool for those rare situations where other tools do not suffice in terms of geometry.

In the next chapter, we will compile our results into a drawing set, ready for printing.

6
Basic Drafting and 2D Views

By now, you may have the impression that Archicad is all about 3D and virtual building. That is true to a certain extent, but rest assured that the software has everything you need, including all the necessary drawings, to complete a project – such as sections, elevations, or details. We have already worked on the main 2D plan, where the majority of our modeling takes place, together with the 3D window, to help us position everything and inspect the design from all sides.

The majority of the building elements will be created as objects with both a 3D and 2D representation, but there is still a need to draft dimensions or labels or place 2D symbols. This will be performed on plan views and within sections or elevation windows. There is also the possibility of using existing 2D CAD drawings as underlays or even using independent 2D drawings for design options or sketch studies. We don't need external CAD software to finish our documents or produce traditional drawings.

In this chapter, the following topics are covered:

- Creating traditional 2D views for sections and elevations, based on the 3D model as well as independent 2D documents (for sketching or as an underlay, for example)
- Learning how to use basic 2D drafting tools and techniques and 2D objects and learning about attributes and how they define elements and objects in Archicad

In this chapter, we will use the result of the project that we worked on at the end of *Chapter 5*. You can download this from GitHub: `https://github.com/PacktPublishing/A-BIM-Professionals-Guide-to-Learning-Archicad/blob/main/CH05_Result.pln`.

Using basic Section, Elevation, and Independent Viewpoints

In traditional 2D drafting, we draw sections or elevations alongside the floor plans. By aligning them and using auxiliary lines, we are able to check how the façade aligns with the floor plan. But here, we are working in dedicated BIM software. In Archicad, you get dedicated **Viewpoints** using the **Section** tool and the **Elevation** tool. These tools are almost the same. Once you know how a section works, you understand the **Elevation** tool as well.

While Sections and Elevations are derived from the 3D model, in a design process, we sometimes need independent views. In Archicad, we use **Worksheets** to create such a non-linked view. These can be used to sketch out alternative elevation designs or add external content to use as an underlay. In *Chapter 12*, in the *Creating a detail viewpoint* section, we will also see how we can use Worksheets for detailing purposes.

Creating a Section view

Creating any view derived from the 3D model is done by inserting a special object (elevation or section) into a floor plan view. For a section, follow these steps:

1. Select the **Section** tool in the **Viewpoint** sub-group of the **Toolbox**.

2. For **Geometry Method**, we'll stick to **Single** and not **Staggered**.

3. Pick the starting node of the section line, somewhere to the left of the building.

4. Using the well-known Snaps, Guide Lines, and Tracker, pick the end node of your section by picking a point outside of the building to the right.

5. Finally, you get an eye cursor, which you use to pick the view direction of the section, hinted by a *black triangular arrow*. Let's pick a point above the line.

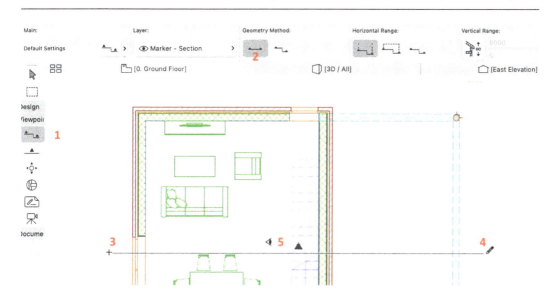

Figure 6.1: Adding a section to a floor plan

You have a section symbol on the floor plan now.

6. With the section selected, open its **Settings** dialog (*Cmd/Ctrl + T*) to find extensive options with many panels to configure.

7. For now, we only need to worry about **Reference ID** to set the short code and **Name** to add a descriptive name to the section.

8. You can also limit the reach of the section from **Zero Depth** to **Infinite** for **Horizontal Range** and **Infinite** or **Limited**, based on elevation heights, for **Vertical Range**.

9. There is a whole range of other panels to control how the 3D geometry is visualized, the style of the **MARKER** symbol, and the way **STORY LEVELS** are displayed. We'll leave everything as their defaults for now.

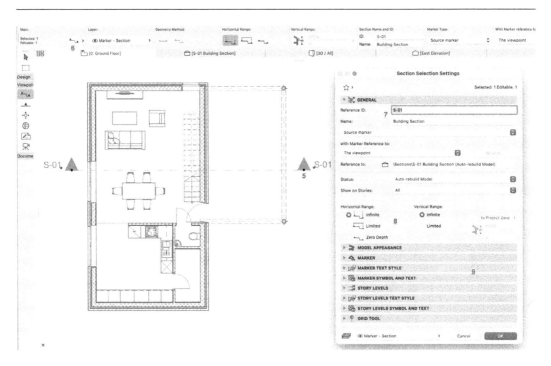

Figure 6.2: Section Selection Settings dialog

The symbol you see is in fact a visual representation of the section viewpoint. You can open the attached viewpoint in different ways:

- From the context menu when you right-click the section symbol (**Open Section in New Tab**)
- By double-clicking the name of the section in the **Navigator**

In both cases, the section viewpoint opens as a separate tab displaying the section Archicad created from the model. It appears very similar to a floor plan window, with the same navigation methods as the 2D Floor Plan.

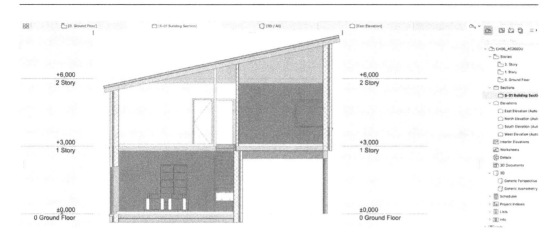

Figure 6.3: Section viewpoint opened in a separate tab

At first glance, this looks like a 2D drawing, albeit generated by Archicad and not drawn by us. But the main difference is that everything you see here is *extracted* from the model: you get a clipped projection of the 3D geometry, with the clipping plane located at the section line and pointing into the direction you just indicated when creating the section. This means that the contents of the section will be updated if you change something in a floor plan or 3D view.

Be aware that this is not simply a 2D drawing but a new viewpoint into the underlying virtual building. When you select an object in the section, you get exactly the same Info Box as you would in the 2D Floor plan or the 3D window. It is the same object!

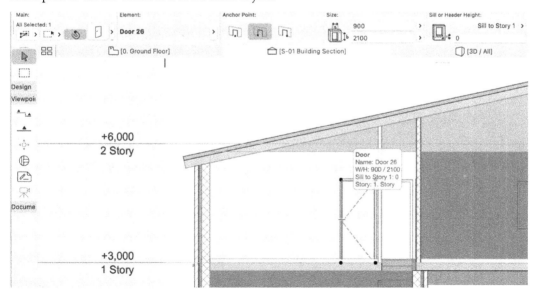

Figure 6.4: Selected door in a section viewpoint

When you drag an object in a section, it is moved in the model and thus in all the viewpoints where the object is visible.

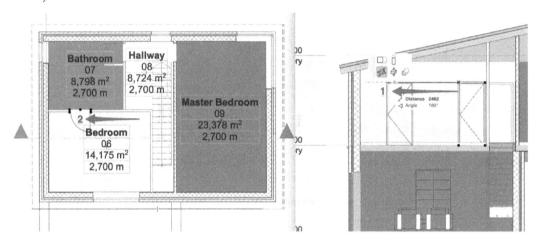

Figure 6.5: Drag a door in the section – the floor plan follows

To keep Archicad responsive, viewpoints are not refreshed in real time, so you have to switch to the other viewpoint before you see the effect of your changes. That said, Archicad employs multi-core processing to keep all viewpoints in sync, so the refresh happens quickly.

Since a section is also a regular 2D window, you can draw on top of the section to add an annotation, such as text or dimensions, which will be discussed shortly. The annotation will only be visible in the Section itself. It won't become part of the 3D model or any other viewpoint.

Please undo the door move (*Cmd/Ctrl + Z*), as the bedroom needs to remain accessible.

Adjusting the Section

If you select the section in the floor plan, you can adjust its settings from the Info Box. Let's focus on the most important ones:

1. Select the section and look at the Info Box to follow along.

2. We have set the section **ID** to S-02 and **Name** to East-West Section. **ID** should be a short code or number, while **Name** can be more descriptive text.

    Notice how the symbol is adjusted in the floor plan and the name is also updated in the **Navigator** under the **Sections** branch.

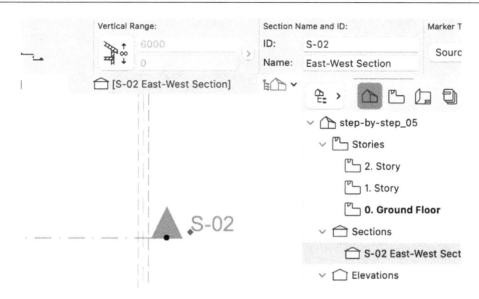

Figure 6.6: Section in the Navigator and on the floor plan

In *Chapter 12*, we will go into more detail on configuring a section and how you can alter the model projection. For now, the configuration we get out of the box from our template is already very usable.

Creating Elevations

From what you have learned so far in the chapter, you should now be able to manage an **Elevation**, as it behaves exactly the same as a section. You'll see a visual difference in the symbol and icon used, but it has similar settings to a section. In fact, in our default template, four elevations have already been added to the project and they can also be found in the **Elevations** branch of the **Navigator**. Before using these, you should check that they are positioned correctly in relation to the designed building. You can simply move **Elevation Marker** in the floor plan views if you would like to adjust the position. To open an elevation view, work as follows:

1. Open **East Elevation** by double-clicking on its name in the **Navigator**.

2. Since these views have been provided from a template, you have to review their position and extent for the current design. You can directly drag the small black **Horizontal Boundary** line in the section view. You can also drag the endpoints of the section marker in a floor plan, and the section view will be cropped accordingly.

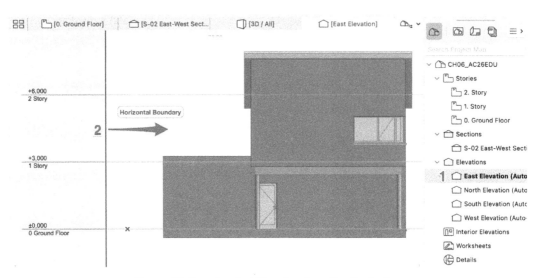

Figure 6.7: Opening East Elevation from the Navigator

At this point, this suffices to understand Sections and Elevations. In *Chapter 12*, we will explore more options and possibilities.

Using Independent Worksheets

The third kind of window we introduce here is also a 2D viewpoint, but this one is not related to the model: the **Independent Worksheet**. You can use it for any purpose you see fit: sketching, documentation, experimentation, and so on. The content of this worksheet will not be visible anywhere else, so it doesn't affect the model.

To create a new independent worksheet, follow these steps:

1. From the **Navigator**, select the **Worksheets** branch and click the **New Viewpoint** button.

2. In the **New Independent Worksheet** dialog, confirm or set **Reference ID** (a number or short name) and **Name**, which is more descriptive.

3. Then, click **Create**, after which Archicad adds an entirely new worksheet to the Navigator and opens a new tab containing an empty viewpoint.

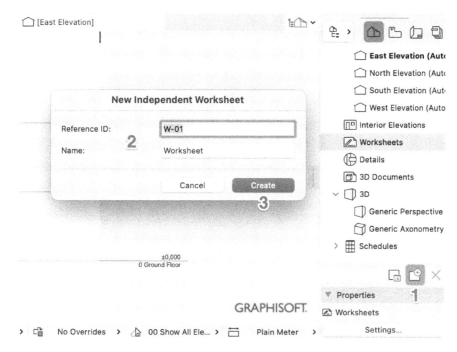

Figure 6.8: New Independent Worksheet

Think of this worksheet as an empty 2D environment, where you can work in isolation if needed, but know that you can reference the sheet as an underlay when working in any other 2D window, using Trace and Reference.

Just like Sections and Elevations, only a limited set of tools is available, relevant to 2D Viewpoints. So, let us introduce you to the drafting tools next.

Getting started with drafting tools

Although 2D drawings and other outputs are derived from the central 3D model of the virtual building in a BIM workflow, sometimes we still need 2D drafting tools in the design and construction process. 2D elements are added for various reasons: annotations, markings, visualizing legislation (e.g., plot boundaries), quick elevation sketches, or simply enhancing the presentation of the design.

Archicad has very good 2D drafting tools that allow the user to do all this and more. Most of these tools have a similar workflow to their 3D counterparts, so learning how to use them should not be that hard. Keep in mind, though, that the resulting 2D elements are less "intelligent" than 3D objects, elements, or spaces. They can have multiple properties, which we can use for visualization purposes (using Graphic Overrides – discussed later in *Part 2*, *Chapter 13*), but we will not be able to add them to any Schedule (also explained in *Chapter 13*) and thus we are not able to "measure" 2D elements, although there are some exceptions.

Now that you know these general principles, let's have a look at how the most commonly used 2D tools work.

Drawing and annotating with 2D linework

Drawing and drafting 2D lines, arcs, circles, ellipses, rectangles, polylines, and splines is very common in any **Computer-Aided Design** (**CAD**) process. Obviously, Archicad provides ample tools to do this. Since you've already learned about several 3D modeling tools by now, applying them will be quite easy.

The **Line** tool is a good place to start exploring these tools. How you draw a line, the Geometry Methods that are available, and the options the Pet Palette presents you are very similar to how the **Wall** tool (covered in *Chapter 3*) works.

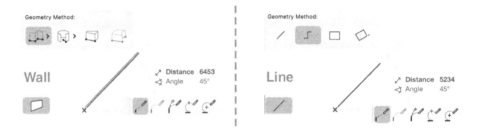

Figure 6.9: Similar Pet Palettes for the Wall and Line tools

Drawing a line is pretty straightforward, and extra options are limited. The only "option" you have is adding an **arrowhead** at the first and/or last node. In a **Chained** line (the second Geometry Method), this is the same – an optional arrow at the first and last node, but not at every node.

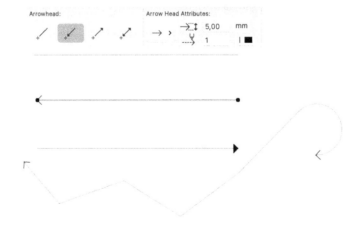

Figure 6.10: Arrowheads and their attributes – options for 2D lines are limited

Although the Pet Palette provides us with options for drawing non-linear 2D elements, such as arcs and circles, we still need separate tools for the full range of 2D drafting. The **Circle** tool is evidently meant for circles but is also the only option you have to create an ellipse in Archicad. Again, the Pet Palette and Geometry Methods allow us to do this in a way that best fits the needs of the project.

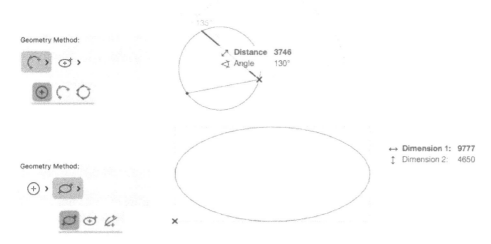

Figure 6.11: Geometry Method for circle and ellipse shape types

Using the **Line** tool with the **Chained** Geometry Method results in grouped lines. It looks like what we call a *polyline*, but has the advantage of allowing the easy editing of individual parts. There is also a **Polyline** tool in Archicad. The only difference between using this tool as opposed to the **Chained** line lies in the result: the Polyline tool creates a single element, not a group of lines. Explore the difference for yourself!

The last distinct type of 2D line is **Spline**. A spline can be defined as a smooth parametric curve, controlled by nodes on the curve and control points outside of it. It is a relatively easy way of drawing curved and complex 2D shapes.

Archicad provides three Geometry Methods for drawing a spline. They influence the way you place nodes and/or how you can edit the spline afterward:

- **Natural Spline:** The resulting curve is defined by the placement of every next node. Every next node thus influences the way the spline curves. Using Natural Spline is very easy and results in a good approximation when using an organic shape as a reference, for example. Editing Natural Splines is limited to moving its nodes or adding new ones if needed.

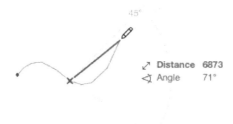

Figure 6.12: A fluently shaped Natural Spline is immediately created while nodes are added

- **Bézier Spline**: Named after French engineer *Pierre Bézier*, who introduced these curves into the car manufacturing industry at *Renault* in the 1960s, Bézier curves allow for more control. They are defined by the nodes you place, but they also have editable handles at each node. These handles are represented by tangent lines in both directions. The angle and length of each handle define the shape of the Bézier Spline. Their placement and editing are more complex than for Natural Splines, but this type of spline is very common in graphics software. Although you can "just" place the nodes, clicking and dragging allows you to define the initial tangent for each separate node. This results in two handles of equal lengths. Changing lengths (independently) on either side afterward lets you further adjust the smoothness of the curve. To do this, use the **Move Tangent Handle** option in the Pet Palette.

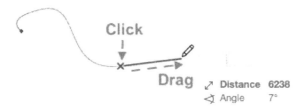

Figure 6.13: Dragging after clicking helps to control the Bézier Spline

- **Freehand Spline**: This Geometry Method can result in a less smooth curve, as it lets you draw the spline without defining exact positions for the nodes. After clicking for the first node, you can release the mouse button and scribble away. Finish by clicking a second time, after which the resulting spline will be calculated. Using the Pet Palette allows you to adjust the (many!) nodes of a Freehand Spline, which makes it harder to adjust afterward, as compared to the other two Geometry Methods for splines.

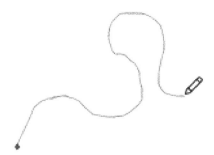

Figure 6.14: After you finish the scribble, you will get a smoothly curved spline

Any "type" of spline can be changed into a Natural Spline or Bézier curve. You can change between these two Geometry Methods without limit. By default, a freehand curve is always "translated" into a Natural Spline after you finish drawing it.

Learning how to use line types as an introduction to Archicad attributes

More than when we modeled walls or slabs, using linework will likely involve using **linetypes** and CAD**pens**. These are two basic concepts in CAD that are implemented in most drawing and modeling software. In Archicad, these two concepts are called **attributes**. **Layers**, **Building Materials**, **Composites**, **Surfaces**, **MEP Systems**, and **Zone Categories** are also examples of attributes.

Attributes are the elementary particles that make Archicad work. They are in essence groups of definitions or configuration settings that are used throughout a project. They show up in numerous dialogs and pop-up menus and their definitions can be accessed through **Options** > **Element Attributes**.

Figure 6.15: All element attributes in the Archicad Options menu

Let's look at some examples of different attributes and their use and relations:

- A line is defined by its **Line Type** (attribute 1) and **Line Pen** (attribute 2) and is placed in a **Layer** (attribute 3).

- A wall with a composite structure is defined by a **Composite** (attribute 1), which consists of several **Building Materials** (attributes 2, 3, etc.), and **Skin Separators** (covered in *Chapter 8*), which are defined as line attributes with the pen and line type as attached (sub-)attributes. Each Building Material is defined by a **Surface** attribute and **Section Fill**, which uses a Foreground and Background **Pen** (two more attributes). And it doesn't stop there, as even the Surface can have a Vectorial Fill in its definition(this option is shown in Figure 6.24 and Figure 6.25). Are you still following?

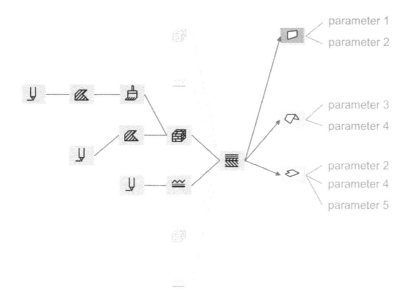

Figure 6.16: Example of how the Composite attribute relies on other attributes

As shown in *Figure 6.16*, the Composite attribute relies (for each skin) on other attributes, such as Lines, pens, and Building Materials (to the left), and is used in tools such as Wall, Roof, and Slab. Attributes are clearly *not* parameters (such as height, element ID, and so on), although they are chosen from the same settings dialog when editing a tool.

We will further expand our understanding of attributes in *Part 2*, *Chapters 8* and *11*, but for now, you should remember that they are the cornerstone of Archicad models, and understanding how they interact is key to robust and efficient modeling! To get to know these essential elements, we will introduce three attributes in this chapter: **Line Types** and **Pens** for linework and **Fills** for covers and hatches.

Exploring Line Types

As well as for 2D linework, we also use line types in settings for Composites (for skin separators), 3D elements such as walls or slabs (for 2D and 3D visualization), parts of 2D annotations (dimension lines), and many other tools in Archicad. Line type definitions are accessible through **Options** > **Element Attributes** > **Lines....** How you can create your own line types is explained in *Part 2, Chapter 12*.

There are three categories of Line Types:

- **Solid Line**: Holds only one Line Type, a continuous, that is, solid line.
- **Dashed Lines**: A pattern of two alternating elements: dashes and gaps. A dashed line can have a maximum of six dashes (and six gaps) and, obviously, has a minimum of one dash (and one gap) – otherwise, it is a solid line. This line type (and the solid line) does not define a width. All of its parts are arranged *exactly on* the same "line" or path.

- **Symbol Lines**: A repetition of drawing fragments, possibly separated by a gap. This line type allows for a lot more complexity than dashed lines do. Depending on the elements they consist of, symbol lines have a visible width defined. Although all of its parts are also arranged *along* the same path, not every element is *exactly on* this path.

Figure 6.17: Examples of a Solid Line (top), a Dashed Line (middle), and
a Symbol Line (some Christmas lights at the bottom)

We will show, in *Chapter 12*, how to create your own Dashed and Symbol Lines (yes, even the Christmas lights) – for now, you just need to know the difference between the types.

Scaling Line Types

There is another important concept to explain when discussing line types, which is linked to the scale of typical construction documents, such as plans and sections. For most dashed lines, we don't want the size of dashes and gaps of a certain line type to be different when printed at a different scale (**Scale Independent**) – they are symbols and should be recognizable as such. On the other hand, lines such as our Christmas lights should have a size that stays the same in relation to other construction elements (Scale with Plan). All 2D annotations, such as text, lines, and fills, provide a scale-related size setting.

Model Size (**Scale with Plan**) means that elements with this setting represent *a real physical object*, the printed size changes of which align with the chosen scale. **Paper Size** (**Scale Independent**) means that those elements will stay the same size, regardless of the scale you choose to print on. The following figure shows several examples.

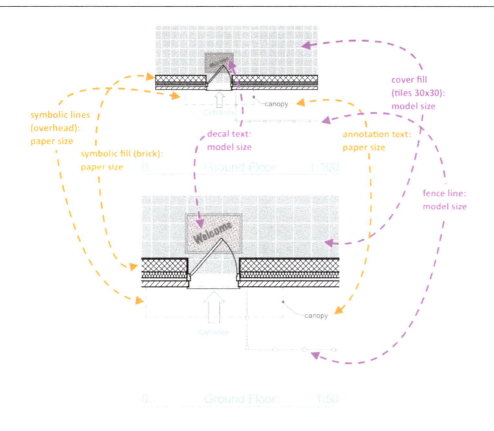

Figure 6.18: Same design (part) on two different (relative) scales, showing the difference between model size and paper size

Now that we have learned how to draw all sorts of lines, it is time to see how Archicad tackles that other basic component of 2D drafting: fills or hatches.

Using Fills for 2D drafting and annotating

Although using linework with different line types can be "enough" to make an architectural or construction drawing, using **Fills** (sometimes called *hatches*) greatly improves the readability of any construction document and expands our means of expression with a whole new vocabulary. The **Fill** tool can be found in the **Document** section of the **Toolbox**. Remember, you can expand and collapse the sections in the Toolbox by clicking the headers of each section.

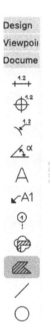

Figure 6.19: The Fill tool in the Toolbox

How you draw a **2D Drafting Fill design**, the Geometry Methods that are available, and the options the Pet Palette presents you are very similar to how the **Slab** tool works, as Fills and Slabs are both defined by their contour.

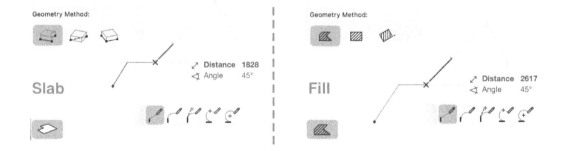

Figure 6.20: Similar Pet Palettes for the Slab and Fill tools

Try drawing and editing some fills to explore the tool and confirm you know how it works.

Getting to know the different Fill Types and how to edit them

A **Fill** can be defined as the repetition of a pattern (the fill definition) within a Fill Outline (a closed polygon as a boundary). Creating a Fill in Archicad requires setting up several attributes: Line Type for **Outline Type**, setting **Fill Type** (a pattern), and three pens that have to be set: the optional **Contour Pen** and the **Foreground** and **Background** Pens, with the former being used for the pattern that is shown in front of a background color, which is defined by the latter.

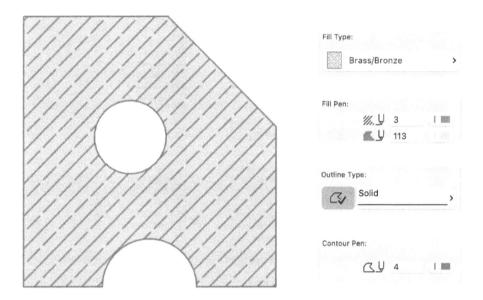

Figure 6.21: A Fill with a clearly visible distinction of the different attributes

Just like the **Line** tool, the **Fill** tool corresponds with the fill attribute, accessible through **Options | Element Attributes > Fills...**. There are four available fill types.

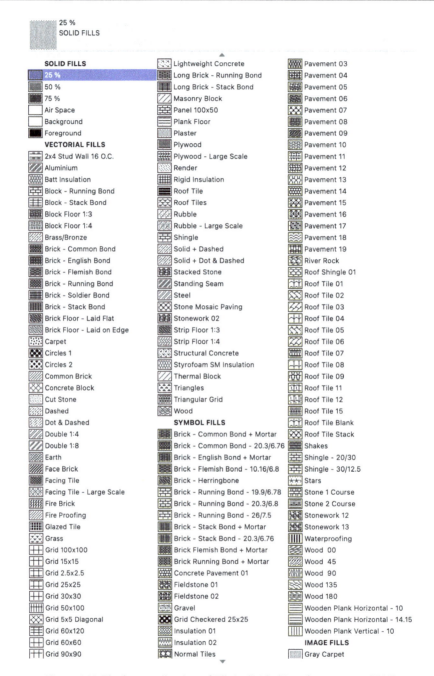

Figure 6.22: The large collection of Fills is divided into four groups: SOLID
FILLS, VECTORIAL FILLS, SYMBOL FILLS, and IMAGE FILLS

The available Fill Types are:

- **SOLID FILLS**

  This type is defined by the "amount" of the foreground pen that is used in the fill – editable by using the **Foreground opacity** slider. A fill of 100% fully covers the background pen (color), while 0% results in only showing the Background Pen. A solid fill of 50% with a yellow foreground pen and a red background pen will result in an orange fill. This category contains one special kind of fill: **Background**. In this definition, the described system is inverted: the background Pen is now in front of the foreground pen. This seems odd, but has a very useful application in combination with certain special pens (and **Graphic Overrides**), as we will learn later on in this section and in *Chapter 13*.

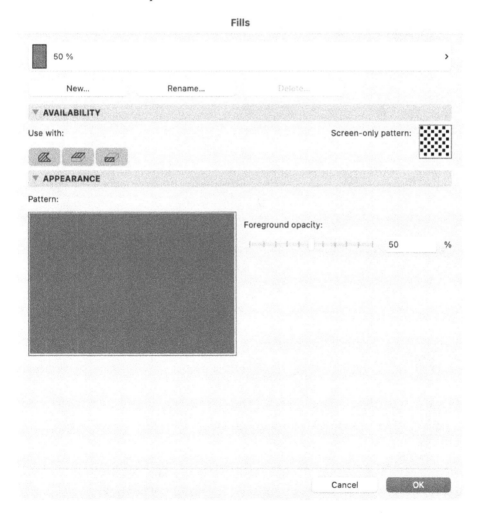

Figure 6.23: Solid fill definition with the opacity set to 50%

- **VECTORIAL FILLS**

  Using parallel or intersecting lines in a pattern is a very common way to create fill patterns in any CAD or BIM application. In the settings, these fills can be edited by changing the **Pattern Unit size** value in any of the two available directions and adjusting the **Rotation** angle of the pattern as a whole. Vectorial fills are defined using GDL and have a clear mathematical definition, but their definition is not directly accessible through the Archicad interface.

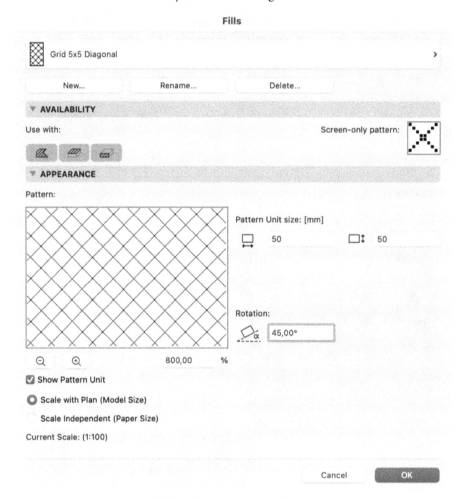

Figure 6.24: Basic vectorial fill with two perpendicular lines forming a "cross-hatch"

More complex vectorial fills follow the same logic but have (many) more lines in multiple directions within **Pattern Unit**, visualized by the yellow marking in the preview window, when activated. This pattern can also be mirrored in the definition of this type, as you can see in the following screenshot (*Figure 6.25*).

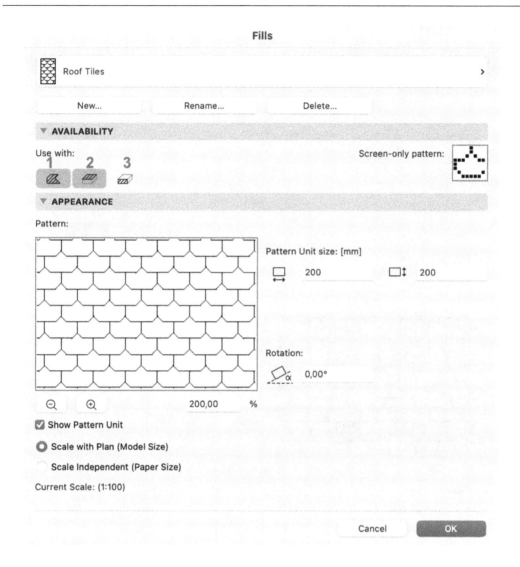

Figure 6.25: The yellow area marks the pattern that is repeated and mirrored in the Fill

Use with

Notice that in every Fill definition, we have the option to choose one or more Fill Categories (**Drafting Fill** (**1**), **Cover Fill** (**2**), and **Cut Fill** (**3**), in the preceding screenshot) for which a definition should be available. The **Roof Tiles** fill in *Figure 6.25* is set to be usable with drafting and cover fills, but not for cut fills – which seems logical, given what the fill looks like. These categories are explained in the next subsection, *Getting to know the different Fill Categories and how to use them with the Fill tool*.

- **SYMBOL FILLS**

  This type appears to be quite similar to the Vectorial Fills, but instead of using an (invisible) GDL definition, they are created from a pattern that has been drawn in 2D in Archicad. Lines, Arcs, and Hotspots are copy-pasted into the Symbol Fill definition.

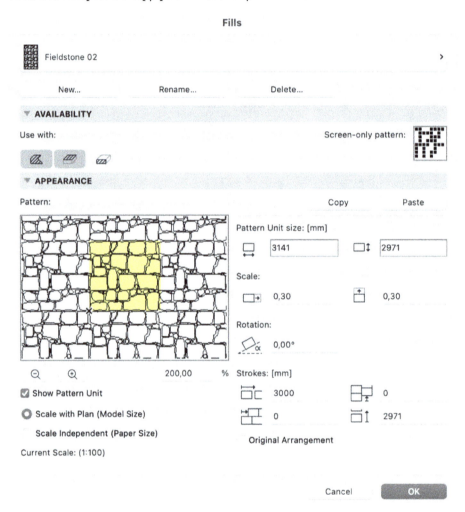

Figure 6.26: A typical vectorial fill showing a clear repetition of a pattern

Due to the nature of this type, there are more controls when editing it. **Scale** and **Rotation** are self-explanatory, but **Strokes** needs some explanation. The upper-left and lower-right settings under **Strokes** control the position of the *first occurrence* of the pattern unit (in relation to its size) in the **X** and **Y** directions, while the lower-left (**dX**) and upper-right (**dY**) settings control the shift of the pattern in the next *row* or *column*, respectively.

In *Figure 6.27*, the **X** and **Y** values have been set to 438 mm and 156 mm, respectively, creating a gap of 40 mm in the X direction and 20 mm in the Y direction of the pattern. The size of the gap is the difference between these values and the corresponding **Pattern Unit size**. The value of **dX** is set to 20 mm and the value of **dY** to 10 mm, shifting the rows and columns each by half of the gap size. Try this for yourself to fully understand these settings!

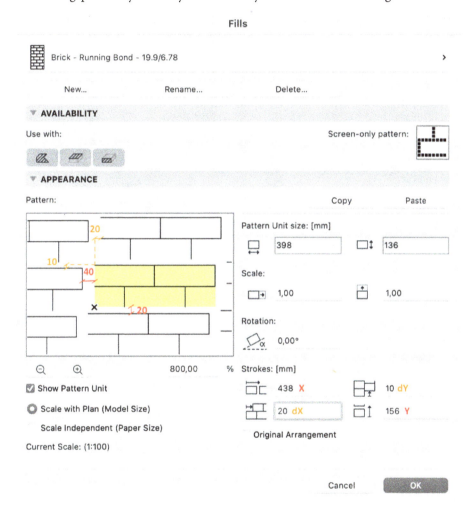

Figure 6.27: Applying strokes distorts this brick pattern

> **Note**
>
> We will see in *Chapter 12* that this type is relatively easy to create, but be aware: this kind of fill uses more resources for visualizing the pattern than vectorial fills. In large projects, this can lead to performance loss!

- **IMAGE FILLS**

  The last type does not use lines, arcs, or points. It creates a pattern using an image that is loaded into its definition. Needless to say, Image Fills use even more resources for visualization than Symbol Fills. They can be used for presentation views, for example, in an elevation showing semi-realistic materials (a technique we will explain in *Part 2, Chapters 12* and *15*).

Figure 6.28: An elevation view entirely made within Archicad, using several
visualization techniques, including the application of Image Fills for the bricks

When setting **Pattern Unit size**, you can choose whether or not the original proportion should be respected. As always, a **Rotation** angle can be set. A specific setting for this type of fill is **Distribution**, providing several mirroring and repetition settings to let the final fill be displayed as seamlessly as possible.

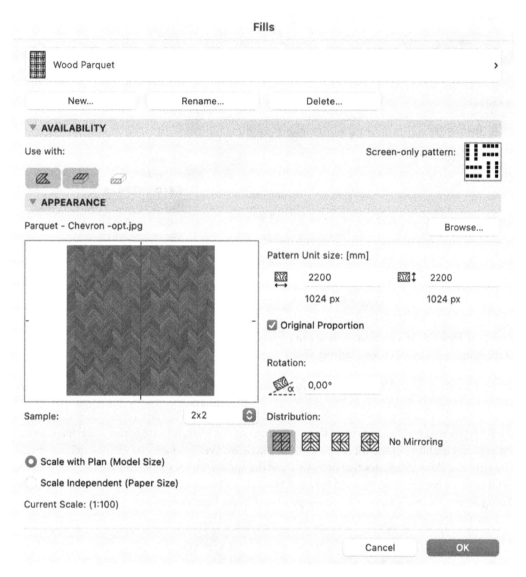

Figure 6.29: A good, clean image with a correct crop does not
need any mirroring to result in a seamless pattern

Pro tip

There is in fact a fifth Fill Type available: **Gradient Fill**. Strangely enough, it is not visible in the attribute dialog. It is only visible when you choose **Drafting Fill** under **Fill Category**. The Fill Categories are covered next!

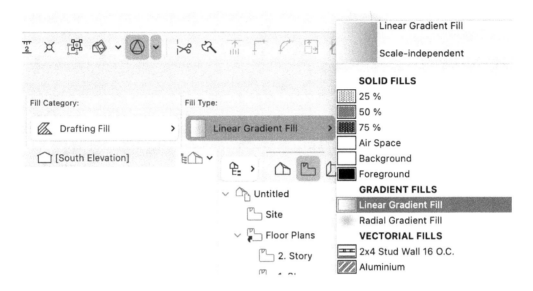

Figure 6.30: Choosing a Gradient Fill as Fill Type

As shown in *Figure 6.30*, Gradient Fills are not found in the attribute settings, but are available as a Fill Type under Fill Category for Drafting Gill.

Getting to know the different Fill Categories and how to use them with the Fill tool

While fill types make a distinction in the definitions of the Fills, fill categories are related to the function or applicability of the fill in the project. There are five categories in total, which we divide into two groups: three categories that directly show the application of the fill (and determine – in the fill definition – what they can be used for) and two categories that actually refer to other attributes (Building Materials and Surfaces), which depend on fills for their definition.

Figure 6.31: Five Fill Categories shown in the Info Box dropdown for the Fill tool

Let's define the Fill Categories.

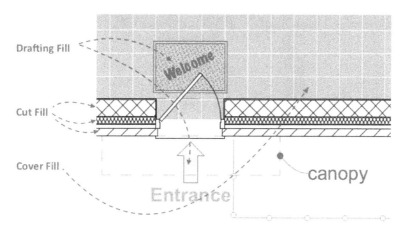

Figure 6.32: Three "main" categories of fills shown in one image

There are three main fill categories in Archicad:

- **Drafting Fill**

  Used for 2D drafting fills, as a 2D markup, or for annotation purposes. This category is only available for the **Fill** tool. In the *Archicad 26 Template*, all Fill Types are available as **Drafting Fill**. Drafting fills only show in 2D Views.

- **Cut Fill**

  Cut Fills are used to represent the composition of elements that are being cut in any View (e.g., horizontally cut Walls in 2D Floor Plans, vertically cut Walls and Slabs in a 2D Section, or anything that is cut in a 3D view using *3D Cutaway* – which is explained later). Cut Fills are symbolic by nature (they use graphics that do not mimic real life appearance), so we need a legend in our documents to explain the meaning of each fill used (e.g., the "double diagonal line" fill represents concrete masonry blocks in a section).

- **Cover Fill**

  This category is shown on elements that are *not* being cut in a view (e.g., a brick pattern in a 2D elevation or a floor tile pattern in a 2D floor plan).

  Cover fills typically show a pattern that is also visible in the real-life counterpart of our model, although often simplified. This makes the 2D representation graphically "lighter" and also saves on resources. In some scenarios, a fully detailed pattern can prove its use though, as the following figure shows.

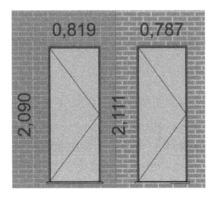

Figure 6.33: Using a brick pattern with mortar lines (right) allows for a more precise size and precise placement of a door as opposed to the simplified pattern on the left

- **Building Material (Cut Fill)**

  Cut fills are applied in the definition of Building Material attributes, as we have seen before (and is further explained in *Part 2*, *Chapter 8*). This is actually how Archicad integrates the concept of a material legend into the definition of the virtual Building Materials used in a project. When you set **Fill Category** for a 2D fill to **Building Material (Cut Fill)**, you cannot choose a fill definition directly. Instead, you are presented with a drop-down menu of the Building Materials in the Info Box (or in the default/selection settings of the **Fill** tool). The pattern shown in the fill is defined by the settings of the chosen material.

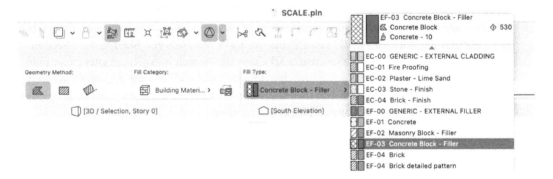

Figure 6.34: Choosing a Building Material to get the corresponding 2D fill

This can be useful in a number of situations. When adding 2D drafting elements to a worksheet, for instance, it is convenient that we can choose a material used in 3D elements of the project. Should the definition of the Building Material change later on, the 2D fills in the worksheet will be updated accordingly!

- **Surface (Cover Fill)**

Using the **Surface** category for a 2D fill has similar advantages. Instead of choosing a Building Material, this selects a Surface in the dropdown in the Info Box. Each Surface has several other attributes and parameters making up its definition. Of all those settings, **Surface - Color Fill**, **Surface – Cover Fill**, and **Surface - Texture Fill** are relevant for our 2D fills, as these are available as sub-categories in the settings for the 2D **Fill** tool.

Figure 6.35: Setting Fill Category to Surface (Cover Fill) has three sub-category options

As for the Building Material category, changing a Surface definition will also affect 2D fills relying on this definition. The sub-categories allow us to choose to show one of the three relevant aspects of the Surface. Using and editing the **Surface** Attribute is further explained in *Part 2, Chapter 15*.

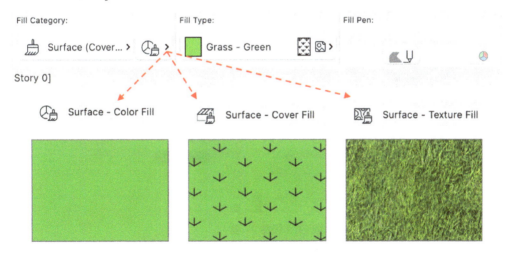

Figure 6.36: Switching the Surface sub-category gives a distinctly different result in 2D

Now that we can use 2D drafting to complete the views of our residential design project, we should also learn how to set the correct width and colors for all the lines and fills in our project.

Using Pens and Pen Sets to change the weight and color of lines and fill patterns

Eventually, any 2D or 3D View – independent worksheet or linked view derived from the model – will be printed in one way or another. Paper prints are still very common, but "digital prints" – mostly in the form of **Portable Document Format (PDF)** files – are also widely used to share output from BIM models. We will learn how to create basic output in the final section of *Chapter 7*, but in order to create clear and readable documents, we have to define the colors and line weights of all the elements shown on these final documents. In the next two subsections, we will learn about Pens and Pen Sets to do so.

Understanding the Pen and Pen Set attributes

The Pen and Pen Set attributes are accessible through **Options** > **Element Attributes** > **Pens & Color (Model Views)....** This dialog can also be opened using the button in the **Quick Options** bar at the bottom of the screen. Both methods bring up a relatively small dialog, showing **AVAILABLE PEN SETS** in the top half and the individual pens of the currently selected set in the bottom half.

Figure 6.37: The Pens & Color (Model Views) dialog, showing the Pens and Pen Sets

To understand the Pen and Pen Set attribute and its (great) powers, we will first focus on explaining the bottom half, containing 255 of a total of 256 available pens, numbered from 1 to 255. There also is a pen with index number 0, which is not shown in this table as it is a system pen. This **Pen Number 0** always results in a transparent color and is available for fill-related settings only. There are two more of these hidden pens, one showing **Window Background** (*index -1*) and one referring to an RGB color, without an index and only applicable in certain cases, for example, for Surface Category Fills.

Let's compare these pen numbers with a physical box containing 255 pens. Each pen has a specific color and draws lines with a specific line weight. There can be multiple pens in this real-life Pen Set with the same color and/or the same pen weight. So far, this seems to be the same as any set of pens, pencils, or crayons. But there are two things that are special about our virtual Pen Set in Archicad:

- Every Pen has a *fixed position within the Pen Set* – imagine this as individual compartments in the box containing the real-life Pen Set. Each of these compartments has a number, called its **Pen Index**.

- Many of the Pens (but not all of them) have *a very distinct function* and are only to be used for that *specific purpose*. For example, one pen is used for drawing the contour lines of brick walls, while another one is used for the hatch lines (fill) of reinforced concrete.

Pens in different Pen Sets that have the same index *should* have the same purpose. Refer to the number and letter labels in *Figure 6.38* to follow the example: the pen at index **#150** *(a)* is a pen with a dark gray color *(b)* with a pen weight *(c)* of 0.25 mm and has `Reinforced concrete - Cut Lines` for **Description** *(d)*. The pen is located in the **03 Architectural 100** Pen Set *(1)* intended for documents on a scale of 1:100. For an **Electrical** plan, we will use a different Pen Set, but the pen at index **#150** will still be used for cut lines of concrete walls. Pen **#150** in the **08 Electrical** Pen Set *(2)*, however, is much thinner (0.15 mm), while still using the same dark gray color. In the **07 Mechanical** Pen Set *(3)*, this pen is black instead of gray and has the same weight as for architectural plans. It is still meant to be used for drawing the contour lines of concrete elements that are cut. These are three examples of a *structured Pen Set*, in which most Pens have a clear description, showing the single purpose of the sets. The **09 Color** Pen Set *(4)* is an example of a *generic, unstructured set*, with no descriptions (and all weights set to 0 mm).

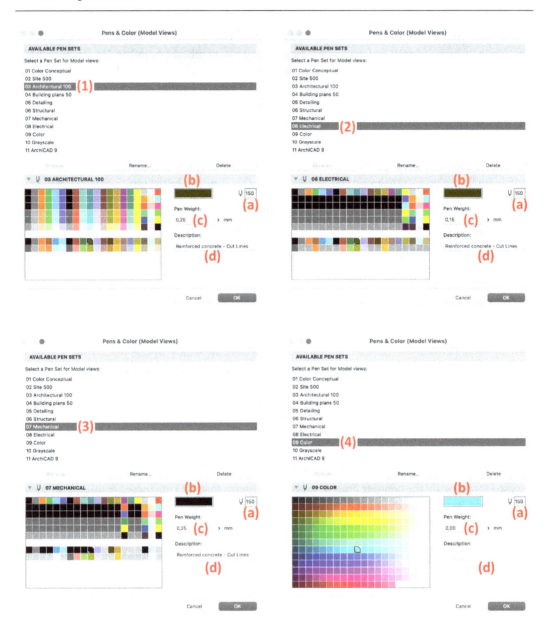

Figure 6.38: Comparing one pen (index #150) across different Pen Sets

In reality, having multiple boxes of Pen Sets, each containing 255 pens, would be overkill, especially because within one set *and* across multiple sets, one specific pen (e.g., 0.18 mm red) would occur multiple times. In Archicad, this system allows us to create views with a fixed Pen Set, which guarantees consistency in the graphics of 2D and 3D documents. The settings in elements and objects in the template are configured according to the functions linked to certain Pen Indexes, so the user does not have to set the correct pen for the outline of every object every time. In other words, this system avoids using the wrong pen for the wrong purpose, but requires a carefully configured template to start from!

Creating a black and white Pen Set for documents at a scale of 1:100

Going through every single Pen in the shown structured sets reveals the logic in the *Archicad 26 Template*. Pens are organized in a grid, in which the columns group *element types* (**General**, **2D Elements**, **Openings**, etc.) and the rows group certain *pen purposes* (**Cut Structural**, **Symbols, and Separators**, etc.). If you follow this logic, it is quite easy to create your own Pen Set.

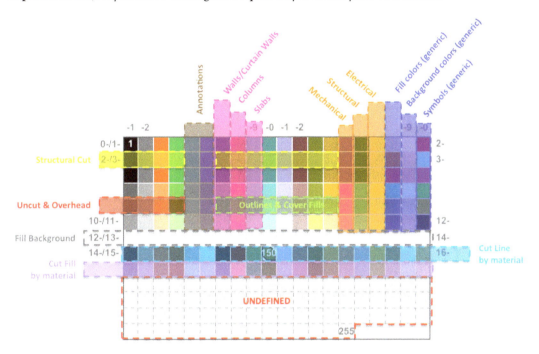

Figure 6.39: Explanation of the Pen Set logic in the INT Archicad 25 template

Let's create a custom Pen Set for black and white 2D documents in our project file, based on the **03 Architectural 100** set included with the INT template. The "100" indicates that it is to be used at a scale of 1:100. Let's get started:

1. Open the **Pens & Color** dialog and select the **03 Architectural 100** set.

2. Select the row of pens from **#141** up to and including **#160**. You can do this by clicking pen **#141** and, while holding down the *Shift* key, clicking pen **#160**.

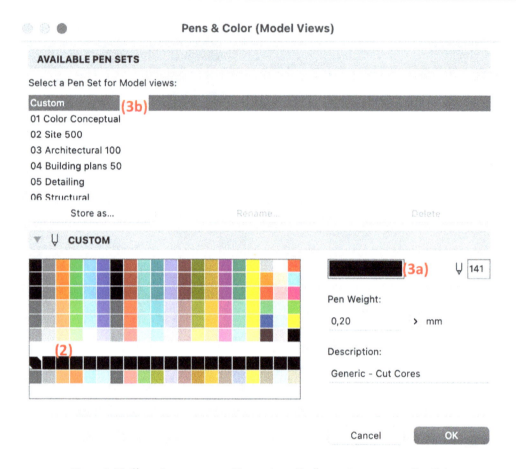

Figure 6.40: Changing any pen setting automatically creates a custom Pen Set

> **Pro tip**
>
> You can select a "block" of pens by clicking the upper-left corner of the target selection first and then pressing *Shift* and clicking the lower-right corner second. This also works for rows and columns of pens. Using *Cmd*/*Ctrl* and clicking can be used to select multiple non-adjacent pens.

3. Click the color *(a)* and set it to black using the color picker dialog method, which is different on macOS and Windows. Notice that a **Custom** *(b)* Pen Set is automatically created.

4. Do the same for the row beneath *(a)* the Cut Line pens (**#161** to **#180**). After setting the color for these "fill pens" to black, set the value for **Pen Weight** *(b)* to 0.05 mm instead of 0.10 mm.

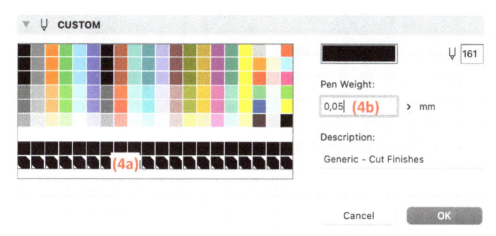

Figure 6.41: Creating fine black hatch lines

5.   The pens used for the cut lines for the objects and elements also have to be set to black. These are pens **#2** to **#4**, **#22** to **#24**, and **#42** to **#44** *(a)* and the pens in the first three rows from column 7 up until and including column 13 *(b)* (see *Figure 6.42*).

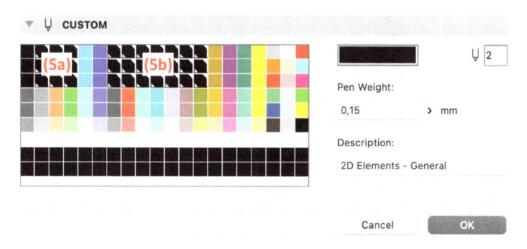

Figure 6.42: Following the logic explained earlier, we continue to set
the pens for all the elements shown in 2D to black

6.   The three rows beneath these two groups of pens contain pens for uncut and overhead lines. For these, we will use the same grayscale color used in pens **#62**, **#82**, and **#102**. From within the color selection dialog of your operating system, you can store these grays locally, so you can easily reuse them afterward. Change the columns for openings (**#3/#23/#43**) and objects (**#4/#24/#44**) to a dark and light gray, respectively, as well – it will give a better result than all black...

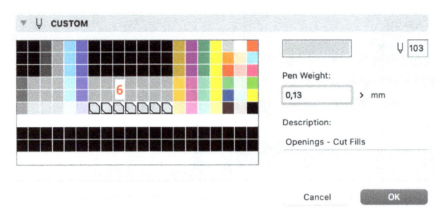

Figure 6.43: Finish the Pen Set by setting overhead lines to a grayscale color. Do the same for openings (windows and doors) and objects – columns 3 and 4

7. Finish by clicking **Store as…** *(a)*, and in the pop-up dialog, select the **Store Custom Pen Set as** option and enter an appropriate name in the text field *(b)*. Do *not* select **Overwrite following Pen Set**, to avoid replacing an existing Pen Set. Confirm by selecting **Store** *(c)*.

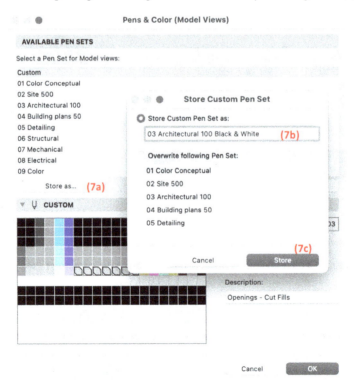

Figure 6.44: Storing a custom Pen Set

8. Look at the result of any 2D plan where this Pen Set is applied – it should look like the following screenshot. You can repeat this process starting from **04 Building plans 50** if you would like a black and white Pen Set at a scale of 1:50, but pay attention to the pen weights as they differ from the **03 Architectural** set! Also, try figuring out what you should change in the Pen Set to make plumbing fixtures no longer look pink (hint: look at the system explained in *Figure 6.39* – plumbing fixtures belong to the **Mechanical** group).

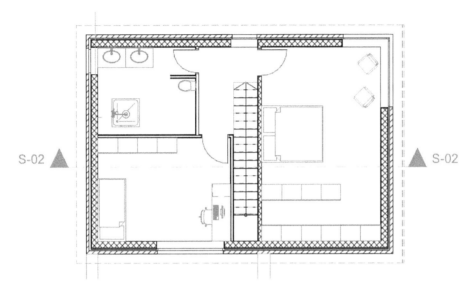

Figure 6.45: Black and white result for the first floor of our project – ready for (digital) printing

This section taught us about some relatively easy drafting tools, such as Lines and Fills, but we also took quite a deep dive into important Archicad concepts, such as attributes and how they influence the appearance of what we model. Later on, we will combine this knowledge with other settings and options (such as **Graphic Overrides**), but we should first further explore some more drafting tools and objects, which we will do in the next section.

Other tools and objects for drafting

We also have multiple objects, symbols, and other tools, such as figures, at our disposal that we can use for 2D drafting.

Adding 2D Objects

We already discussed parametric GDL objects when we introduced the Door, Window, and Object Tools in *Chapter 5*. The library also contains a series of objects that don't have 3D geometry but are still usable as **parametric 2D symbols**.

Typical examples include symbols for trees, furniture, cars, and electric appliances or details. They can be added to every 2D viewpoint: floor plans, sections, elevations, independent worksheets, and also a few we haven't introduced yet. Let's get started by adding the car symbol:

1. Activate the **Object** tool and go to its settings to open the Library panel on the left. There are a few folders with 2D objects within the Library.

2. Open the **1. BASIC LIBRARY 25 > 1.7 2D Elements 25 > Vehicle Symbols 26** folder. There, you have a few typical CAD symbols, each containing multiple viewpoints and even variants.

3. Select **Car Symbol 26**. To get the side view of a classic *Porsche* 911, choose **Style 4** for **Symbol Style** and **Side** for **View Type**.

4. Click **OK** to close the dialog.

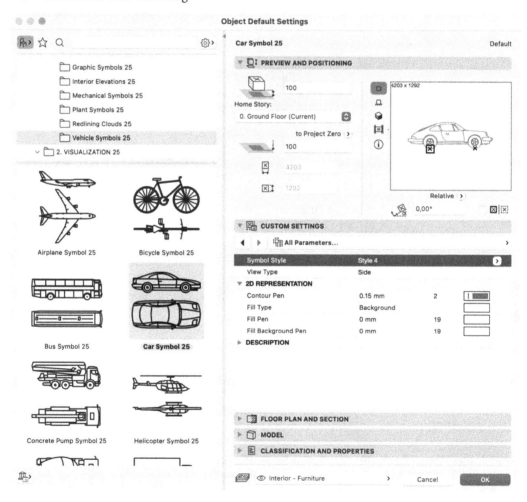

Figure 6.46: 2D symbols in the Archicad Library

This symbol can now be placed somewhere in the model. It is best to open a section or elevation viewpoint for this. Notice that the car symbol is already *properly scaled* for you. It even has two grips, indicated by black dots, to help you align the wheels with a road or parking.

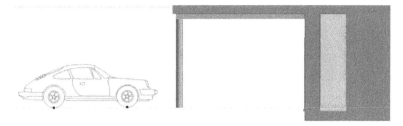

Figure 6.47: Placing a 2D object in a Section Viewpoint

The main use of 2D GDL objects is, of course, to add symbols, or "blocks" as they are sometimes called. But there are also other nice examples of 2D objects:

- **Graphical symbols**, such as legends, floor openings, entrance indicators, or a scale indicator
- **Electrical symbols**, such as outlets, switches, transformers, or light bulbs
- **People contours** (inside the **2. VISUALIZATION** section of the library)

Why would you use 2D symbols in 3D software? Well, they are lightweight, so they don't slow down the creation and regeneration of sections or elevations. Also, they allow you to add a few touches to those sections without cluttering the 3D model.

In other cases, they are also part of common drawing conventions, such as parking or electric symbols, which don't always need 3D geometry.

Also, be aware that many 3D objects actually have a dedicated 2D representation when shown in a floor plan. If you check their settings, some objects allow you to even disable their 3D geometry if you don't need it. That is the flexibility of having GDL-scripted objects in the Archicad library.

Apart from 2D objects used for visualization purposes or as symbols, Archicad also has a "point" tool, with which we can create Hotspots. These are mainly used as a drawing aid and are covered in the next section.

Hotspots

Hotspots are non-printable reference points. You use them for convenience, to add a permanent snapping point where you need it, for example, as a permanent reference or snapping aid. Placing a hotspot works as follows:

1. Activate the **Hotspot** tool from the *Toolbox*, inside the **Document** section.
2. Open the **Hotspot Default Settings** dialog.

3. Notice that you can only set **COLOR** by selecting a pen number and the Layer onto which it will be placed.

4. Confirm the settings configured in the dialog by clicking **OK**, and then place the hotspot using the regular snapping and guides. That's it!

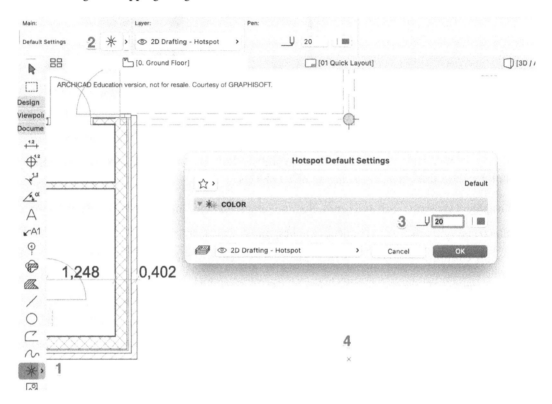

Figure 6.48: Adding a hotspot to have a permanent snapping point

Adding a Figure

You can also include a **Figure** in your drawing, from a variety of file formats. This can be site pictures, renderings, or any other image that you want to include in a printout. Let's add a figure to our drawing now:

1. Activate the **Figure** tool from the **Document** section of the *Toolbox*.

2. From the **Figure Selection Settings** dialog, you can import any figure with the **Open…** button, which gives you a regular open file dialog. In the following screenshot, we've imported the Archicad logo just for fun, but you can choose another one of course.

3. Within the dialog, you can set the real-world figure size of the image. It is best to enable **Keep Proportions**, so the image doesn't distort.

4. If needed, you can also set **Rotation** and **Anchor Point**, which are self-explanatory.

5. Confirm the settings configured in the dialog by clicking **OK**, and then place the Figure in your view.

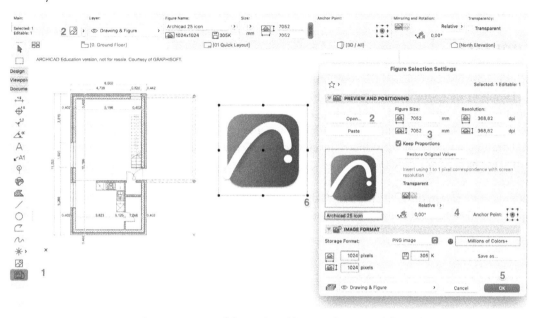

Figure 6.49: Placing a Figure of the Archicad logo in the ground floor plan view

When you select the image, you can still rescale it using the black handles.

A common use of Figures is when you want to trace over a scanned drawing, for example, to roughly set out a site plot from a scanned image.

Pro tip

To resize a Figure to a real-world scale, you can use the **Resize** command (**Edit** > **Reshape** > **Resize…**) with a known distance as a reference. Many scanned drawings have a scaling ruler to help you with that, but any known distance will work. Try to use the longest known distance to avoid rounding errors since picking a point on an image is not that accurate.

Resize your newly placed image (or any other figure) by following these steps:

1. Keep the **Define Graphically** option enabled and click **OK**.

2. Pick one point of a known distance for **Resizing Center Point**, which will stay fixed in place.

3. Pick a second known reference point for **First Vector of Resize Ratio**, which will get stretched.

4. Use the **Distance** field in the Tracker to enter the value of the known distance.

Now that you have learned about a variety of 2D tools and their attributes to add lines, fills, objects, hotspots, and even external figures to your project, the next and final chapter of this first part will explain how to add important information, such as dimensions and labels, to your model using annotation tools in Archicad.

Summary

In this chapter, we introduced the tools required for the creation of 2D content. We added Section and Elevation viewpoints, which help you to reuse the 3D model for creating construction drawings and other 2D documents. We also looked at the independent worksheet, which can be used when you don't want to draw in one of the model Viewpoints. Since Archicad is effectively a complete CAD drafting software, we introduced the creation of 2D drafting elements using lines and fills, and learned the basic concepts of the Archicad attributes – the elementary particles that create the whole Archicad universe. We will get back to the attributes in the first chapter of the second part, but let us first finish this part in the next chapter by adding annotations and dimensions. We will then be ready to print the 2D output of our first project or export the result to PDF.

7

Adding Annotations and Creating 2D Output

Although our first project is completely modeled and we have already created the necessary views based on this 3D model, these views are not finished yet. To explain a design to other parties, the 2D drawings also contain dimensions and other annotations.

In this chapter, we will learn how to annotate a basic project efficiently and correctly. We will learn how to add plain text, use labels to derive data from the building information model, and add the dimensions of the project.

The following topics are covered:

- Understanding and adding annotations to your project
- Linking (external) reference drawings to your project
- Learning a basic way to create the 2D output of your project by exporting and printing

In this chapter, you can use the result achieved at the end of *Chapter 6* or download our version on GitHub: `https://github.com/PacktPublishing/A-BIM-Professionals-Guide-to-Learning-Archicad/blob/main/CH06_Result.pln`

Understanding annotation

Annotations explain drawings further and add non-graphical information to a construction document. Annotations can consist of plain text (e.g., explaining a certain design choice), they can show the dimensions of different elements in the design, or they can simply provide construction information (e.g., about materials). The tools used in Archicad for these main three options are explained in the following subsections.

Adding and adjusting text annotations

Any technical drawing typically has lots of annotation text and dimensions. They are not really part of the 3D model in BIM software, but they are required to clarify and document the design via different drawings. For a typical building permit drawing, you are often required to add at least the main exterior dimensions of the project and a few reference levels in elevation drawings. Text is also widely used to explain the design, in either floor plans, sections, and elevations or in detail drawings, which we haven't discussed yet.

Creating text is straightforward, as it involves most of the formatting you know from any text editor, such as *MS Word* or *Apple Pages*. In Archicad, adding text works as follows:

1. Activate the **Text** tool from the **Documentation** section of **Toolbox**.

2. Drag a rectangular box on your view, which defines the width of the textbox. You can always resize it afterward if needed.

3. Format the text to your liking, using **Text Font** > **Pen** to set the color and the regular **Bold**, **Italic**, and **Underline** formatting options.

4. Adjust the font size of the text in mm (not "points" as you would do in a text editor).

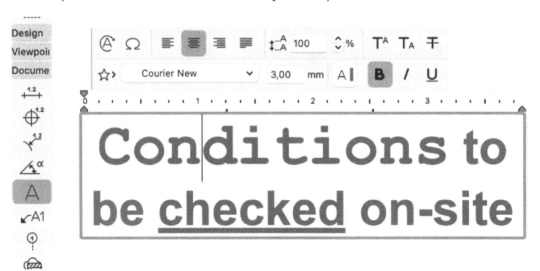

Figure 7.1 – Rich text editing

You also have a settings dialog for text, which presents a few additional options in the **TEXT BLOCK FORMATTING** panel, such as adding a frame, an **Opaque** fill, and the option to rotate the text to be **Always Readable**, even when the drawing is rotated.

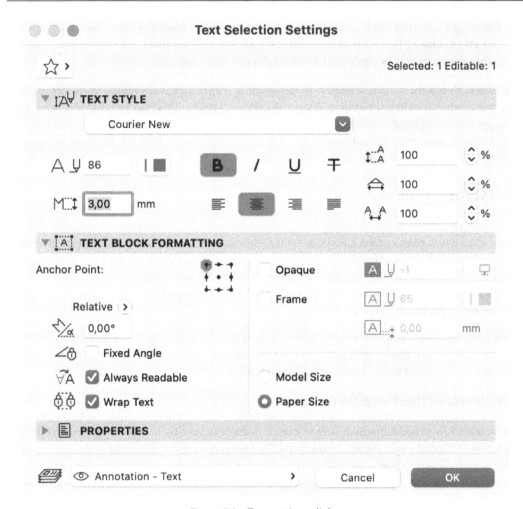

Figure 7.2 – Text settings dialog

However, the font size of text takes a little more explanation. In Archicad, or any CAD system for that matter, you have two scaling contexts (as we already briefly explained in the *Exploring line types* section in *Chapter 6*): the model, which should be understood as being on a 1:1 scale, and the output documents, where you look at a scaled drawing, such as 1:100, 1:50, or 1:20:

- **Model Size** expresses the height of text in the context of the model: how large is the text in the real world? This can be used to add signs, painted characters to indicate a parking lot, the number of a house, or other textual objects.

- **Paper Size** expresses the height of text in the context of the output: what is the height of the text on a printout? This is the default and you express this in the main units (mm in our case), rather than in points as you would do in regular office software.

Model Size is always the same size relative to the building. **Paper Size** is always the same size in the output, but when you switch the scale of a view, the text will resize. This may be what you want, so you can get away with reusing the same piece of text on, for example, 1:100 and 1:50, provided that it doesn't overlap too much. The alternative is to use different layers and have, for example, one layer for 1:50 text and another for 1:100. It's up to you to decide.

Understanding dimensions

Every building has dimensions. Defining these dimensions is one of the most important tasks of any designer when developing a project, and correct dimensions in 2D documents are indispensable for the problem-free realization of the design. It should come as no surprise that within BIM, dimensions are generally a lot more intelligent than in the CAD world. All used elements are always at least geometrically correctly defined, and "measuring a plan" in fact comes down to requesting the right parameter values from the database.

In the following subsections, we will learn how to properly set **Working Units** and **Dimensions** (and what these mean), how to set up and use the **Dimension** tool, and how we can efficiently add dimensions to the project using **Automatic Dimensioning**.

Setting Working Units and Dimensions

Before we learn how to add dimension lines to any 2D document, we have to make sure we understand how units in Archicad work. As you know from *Chapter 1*, our model should be seen as a *virtual building*, a digital equivalent of a real-life project. Keeping this concept in mind, it makes sense that we are always modeling at actual size (as already mentioned in *Chapter 3*), so we don't need to set any drawing units (something you might know from CAD systems). However, we *can* choose which **Working Units** will be displayed when modeling – whatever is convenient for you. By default in the **Archicad 26 Template** included in the INT version, **Working Units** is set to millimeters. This can be adjusted through **Options** > **Project Preferences** > **Working Units** and may be set differently in other projects.

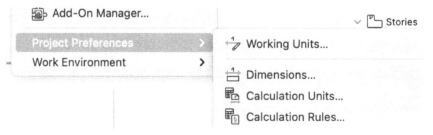

Figure 7.3 – Project Preferences – notice the separation line between
Working Units... and the other menu items

You may set the length or any other measure to the units you like to work with. This will *not* affect the size or scale of your model, but it will "translate" all displayed units, for example, in the Tracker or dialog boxes. Try this by changing the units to meters, drawing a line 1 meter long, then changing it back to millimeters and drawing a line with a length of 1000 mm. You can use the **Measure Tool** (*M*) to check the length of the line you drew.

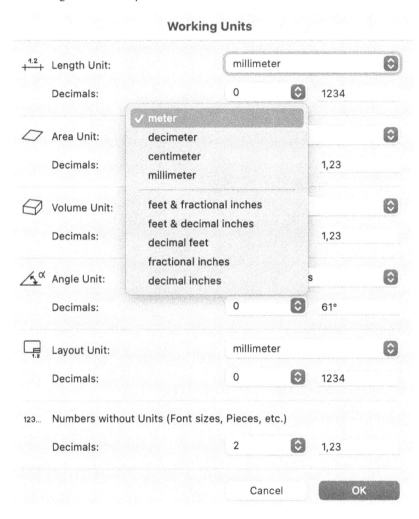

Figure 7.4 – You can choose units from the drop-down menu. Notice that metric and imperial units are available – neatly arranged into two groups

Working Units does not affect the type of unit that is displayed in placed **Dimension Lines** in any way. In fact, we can create more than one "type" of dimension and apply the different types to different views (for example, steel construction drawings in millimeters and site plans in meters). You can set up different dimensions through **Options** > **Project Preferences** > **Dimensions...**.

Let's create a dimension in centimeters as an example:

1. The first step is to open the **Dimensions** preferences inside the **Project Preferences** dialog, which we did a moment ago.

2. Using the icons **2(a)** in the top row, you can navigate to **Linear Dimensions**, **Angular Dimensions**, **Radial Dimensions**, **Level Dimensions**, and **Elevation Dimensions**, but also **Door, Window, and Skylight Dimensions**, and **Sill Height Dimensions** and **Area Calculations**. You can choose any combination of units for all these different measures. To the right of these icons, you can see a **Sample 2(b)** of the dimension you are setting up. Start with **Linear Dimensions**.

Figure 7.5 – Let's create a new linear dimension

3. Click the drop-down next to **Unit** and choose **centimeter** (*3a*). Notice how a **Custom** setting (*b*) is automatically set.

4. Set **Decimals** to **1** using the dropdown.

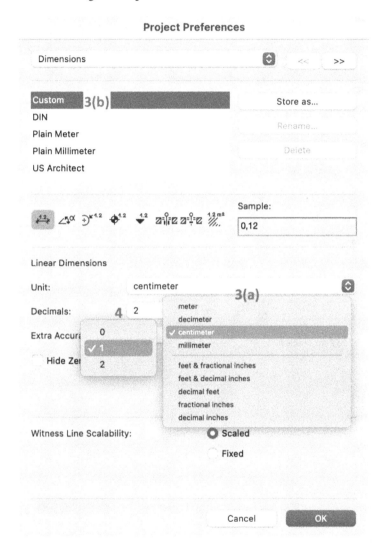

Figure 7.6 – Changing Unit and the number of decimals shown in the dimension

5. **Extra Accuracy** can be left off. This setting allows you to add rounded numbers behind the set number of decimals (**5**, **25**, **1**, or **01**). This extra accuracy is displayed in superscript when active.

6. **Hide Zero Wholes** and **Hide Zero Decimals** should not be checked either, as we want to see every dimension with all of its defined digits (in this case, **1**).

7. **Witness Line Scalability** can have a **Fixed** size, but the **Scaled** setting is often more convenient (as it makes the definition more usable for different scales).

Figure 7.7 – Finalizing settings for our centimeter dimension definition

8. You can continue exploring the settings if you want; for example, choose your own settings for the other measures. Notice that the appropriate units are available in each tab, for example, **radian**, **decimal degrees**, and so on for **Angular Dimensions**.

9. After finishing, click **Store as…**, save your settings as `Plain Centimeter`, and confirm with **OK**.

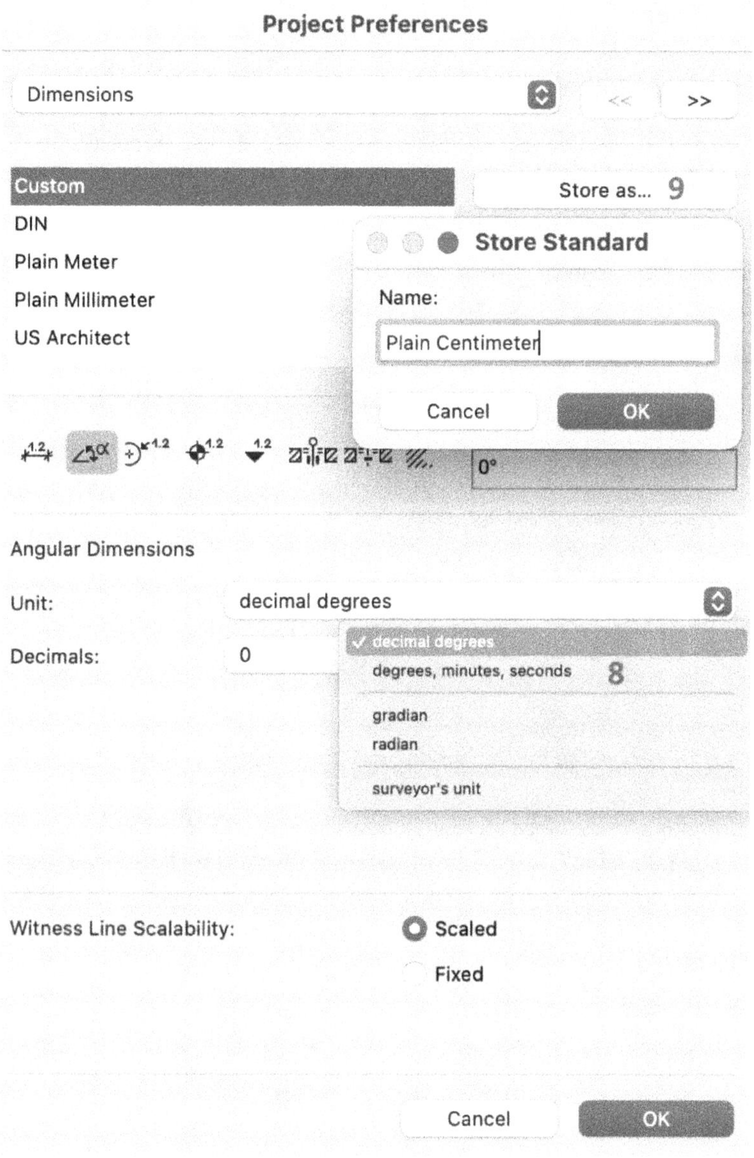

Project Preferences

Dimensions

<< >>

Custom

Store as... 9

DIN

Plain Meter

Plain Millimeter

US Architect

Store Standard

Name:

Plain Centimeter

Cancel   OK

0°

Angular Dimensions

Unit: decimal degrees

Decimals: 0

✓ decimal degrees

degrees, minutes, seconds   8

gradian

radian

surveyor's unit

Witness Line Scalability: ◉ Scaled

  Fixed

Cancel   OK

Figure 7.8 – Exploring other settings, saving the definition, and following the used naming convention

Your new dimension is now ready to be used when placing dimension lines. Let's take a look at the **Dimension** tool so we know how to do this in our example model.

Setting up the Dimension tool

Similar to the **Text** tool, the **Dimension Tool** also has many settings that we better check before we start placing dimensions, in order to work smoothly and efficiently. The **Dimension** tool can be found in the **Document** tab of the Toolbox (expand if necessary) or through **Document** > **Documenting Tools** > **Dimension/Level Dimension/Radial Dimension/Angle Dimension**. Let's start with the (linear) **Dimension** tool.

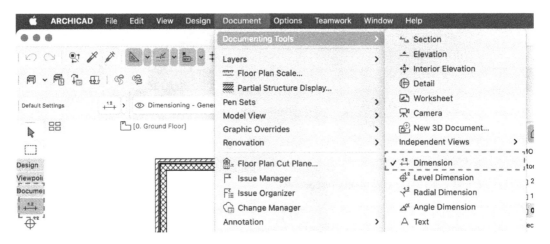

Figure 7.9 – Accessing the Dimension tool – the (linear) dimension is marked

Make sure you are working on the small residential project we have been developing, open **Dimension Default Settings**, and take a look at the following settings (you can choose your own settings for the project):

Figure 7.10 – Dimension Default Settings folded out completely: lots of settings to go through!

1. **Dimension Type**: There are several types of dimensions available within the **Dimension** tool (as a Construction Method):

   A. **Linear Method**: This is the most commonly applied dimensioning method.

   B. **Cumulative Method**: The length value is added each time, starting from the dimension line origin (= the first node).

   C. **Base-Line Method**: The measure relative to the origin is also given, similar to the **Cumulative Method**, but without marking the first node.

   D. **Elevation Dimension**: An indication of elevation heights to be used in an elevation or section view.

> **Note**
>
> Specific dimensions such as **Angular** or **Spot** dimensions (height on plan) have their own specific tools!

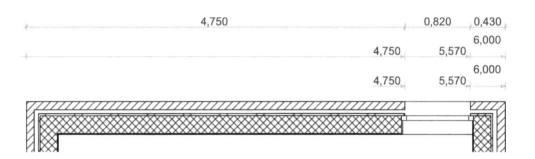

Figure 7.11 – The three construction methods for a linear dimension
show the same information in different ways

2. **Marker Type** determines the appearance of the dimension line end. Depending on the type of dimension (**1a**, **1b**, **1c** or **1d** in *Figure 7.10*) you see different options.

Here are the first set of options:

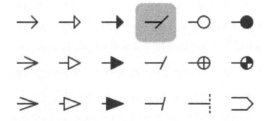

Figure 7.12 – Marker types for the first three dimension types

Here are the second set of options:

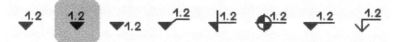

Figure 7.13 – Marker types for Elevation Dimension

3. For each type, you also have options for either the Extension of **Witness Lines** of linear types or the appearance and Direction of the **Elevation Marker**.

Figure 7.14 – Extend witness lines beyond the dimension line or not

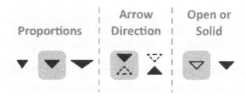

Figure 7.15 – Change Proportion, Direction, or Appearance of the Elevation marker

4. For **Witness Lines** there are four options, apart from them being extended or not:

   A. **None**: No Witness Lines are shown.

   B. **Sized Height**: You can set the marker size/Witness Line Length in the corresponding tab (bis).

   C. **Custom Height**: Define a fixed length in the tab (bis) below. The length for each witness line (or for all witness lines in one dimension line) can also be edited graphically after placement using the pet palette, as explained in the next section.

   D. **Dynamic Height**: Define a fixed gap size between the node (the point where you will click when placing the dimension line) and the dimension line (bis). Again, this can be edited afterward, but when moving the dimension line, the witness lines will change length, keeping the gap's size as defined.

5. If you check the **Static Dimension** box, a dimension line with a *fixed value* will be created. In other words, there will be no connection between the object and the dimension line! Of course, this is not the usual method within BIM, but it may be necessary to visualize a fixed design size or regulation limitation, for example.

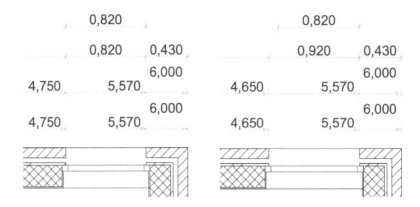

Figure 7.16 – The top dimension in the image is not changed
after editing the window (making it 10 cm wider)

6. **TEXT STYLE**: In addition to the usual settings for text, such as font, pen, layout, ...(a), you also need to indicate for dimension lines whether the text should be above, within, or below the dimension line (b). Similar options are available if you use **Elevation Dimensions** (without or with +/- notation).

7. A pointer is also known in other software as a "leader." It is only available for linear dimension types. Although you can set it in the default settings, it is only possible to activate it when one or more segments and their accompanying textboxes are selected, through the selection settings, for example, when the width of a wall is not readable due to fills. In one of the next subsections in the chapter (*Editing dimension lines using the pet palette*), we will explain how to select segments.

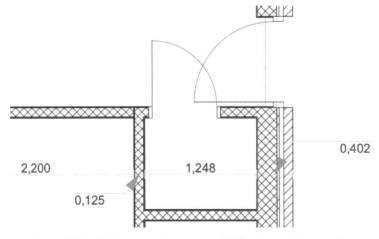

Figure 7.17 – Pointers can improve readability in certain scenarios

8. In **DIMENSION DETAILS**, you can choose whether or not you also want to display the height of joinery (a), including multiple options for this. Here, you can also choose whether you prefer dimensioning skins manually or automatically (b), the latter also allowing you to choose which parts of a composite should be dimensioned (composite parts are explained in detail in *Part 2 – Chapter 8*). Finally, this tab also has a setting for where the dimension number may appear in relation to the dimension line (c), for example, when there is not enough space between the end lines of a dimension line. A detailed explanation of this setting can be found in a two-part video on the *ARCHICAD YouTube Channel*: *Part 1* (https://www.youtube.com/watch?v =vg7H9b8iloM&list=PLnXY6vLUwlWVl8fLABOSLCE-1ONHihRxe&index=17) and *Part 2* (https://www.youtube.com/watch?v=naxX8XxStzo&list=PLnXY6 vLUwlWVl8fLABOSLCE-1ONHihRxe&index=18). Depending on your choice, additional options are also available to better (automatically) control the placement of dimension lines. This works well for most cases in the default Archicad 26 Template.

9. Don't forget to *choose the right layer* in the **Dimension** settings or through the Info Box! You may not want to see all your dimensions from your 1:200 scale floor plan also on your 1:20 scale details, and you may also use a different dimension line for details...

Many – but not all – of the above settings are also available through the Info Box. Remember that you can slide this user interface element sideways by scrolling when your mouse pointer is hovering over this area:

Figure 7.18 – If the Info Box was to show all of the settings within the available window space, the height would be too small to read everything

Placing dimension lines using the Dimension tool

Archicad allows the manual and automatic creation of dimension lines. Even after automatic dimensioning, the dimensions can still be adjusted manually. Manual dimensioning is therefore a good basis to fall back on. In this part, we will see how this works.

In your project, open a 2D Floor Plan view and activate the **Dimension** tool. You can use the settings you created in the previous part – we will be adding linear dimensions first.

In the *Info Box* you can choose the appropriate Construction Method (Linear) and Geometry Method (see *Figure 7.19*):

1. **X-Y Only**: Only orthogonal dimensions can be placed, even if the nodes you click are not on a straight or orthogonal line.

2. **Any Direction**: The dimension line can be placed in any direction. During placement, you will have the chance to choose a line along which to place your dimensions.

3. **Arc Length**: Serves to dimension curved elements and shows the length of an element along its curve.

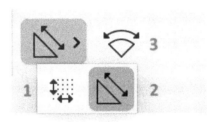

Figure 7.19 – Geometry methods for the Dimension tool in the Info Box

To start dimensioning, follow these steps:

1. Choose **Any Direction** and move your mouse pointer to the first point of the elements you want to dimension. We will start by dimensioning the wall at the north elevation (at the top of the screen) on the ground floor.

> **Pro tip**
>
> Watch the shape of your mouse cursor when adding nodes. When you don't click on a point of an element, this part of the dimension line will not be *associative*, meaning the values will not follow future design changes. Archicad provides feedback for this: an associative point is shown as a circle with a crosshair, while *fixed* points are shown as a square with a crosshair... subtle, but important!

2. Click to confirm the position of the first node. Continue clicking for each point you want to add to the chained dimension line. The order in which you add points is of no importance, but do pay attention to the first two nodes, as these will determine the available options for the direction when finishing the placement of a dimension line.

3. You can zoom in and out and pan as much as you want during this process. Do not worry if you add a wrong point or pick a fixed point by accident – we will learn how to correct this in the part about *Editing dimension lines using the pet palette*.

4. Unfortunately, there is no option to use *Backspace* to "go back one step" in the process of adding dimension nodes as is the case when modeling walls, slabs, or any other 3D element or drawing a 2D drafting element – try it!

Figure 7.20 – Adding dimension nodes. Can you spot the fixed node?

5. To finalize the placement of your dimension line, *double-click* on an empty spot in the view or press *Fn + Backspace/Delete*. The cursor changes into a black *hammer* and a preview is shown of the dimension line with witness lines connected to the clicked nodes. If not all points are on one horizontal or vertical line, a *directional selection circle* appears, because we chose **Any direction** for Geometry Method, indicating that the direction of the measuring line can still be determined.

6. Move the hammer cursor in the right direction to dimension the elements horizontally – vertically or diagonally is not the right choice for this particular wall.

7. Click again to position the dimension line with the hammer.

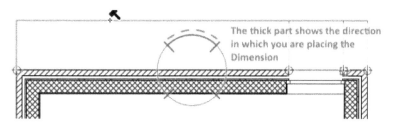

The thick part shows the direction in which you are placing the Dimension

Figure 7.21 – Positioning the dimension line horizontally

Some pro tips

Pro tip 1: When positioning the dimension line, you can use numeric input to place it at an exact distance from your wall. This way, you can easily create a neatly annotated document. For example, hover over the outside face of the wall, press *y*, enter 500- as the value, and confirm with *Enter* if you are moving outside the building as in the last image. If you originally were on the inside, the value is 500+.

Pro tip 2: You can also position a dimension line aligned to any other element by hovering over that element with the hammer icon, which then changes into the so-called *Mercedes* cursor. Click the element, and then position the dimension line as described above. Of course, this only works for the **X-Y Only** Geometry Method!

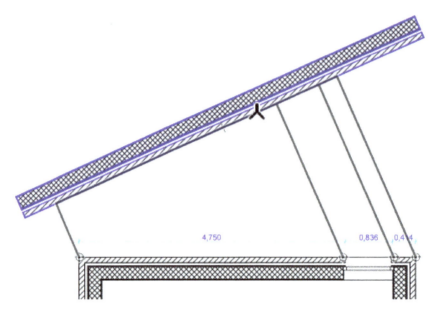

Figure 7.22 – Aligning to another element (the three-pointed Mercedes cursor appears)

In the following figure, you will see the resulting dimension.

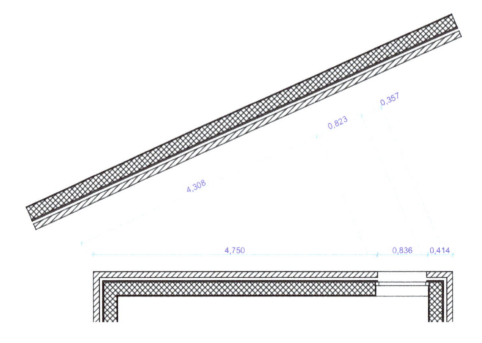

Figure 7.23 – The resulting aligned dimension (with dynamic witness lines)

Apply what you have learned to add dimensions to all the walls of the ground floor (exterior and interior). Now that we have placed some dimensions, it is time to see how we can edit some of these dimensions and maybe correct some mistakes.

Editing dimension lines using the pet palette

Dimension lines contain many parts and have many hotspots: the dimension line itself, nodes with witness lines, textboxes, and so on. As always, the options presented in the pet palette are determined by which part you click. We will go over several scenarios for what you may want to change about dimension lines.

To move a node in a single or chained dimension line, for example, the fixed node we created in our first example, follow these steps:

1. Click the node you want to move (watch for a check mark near the cursor).

2. Click the selected node again, choose **Move Node** from the pet palette, and start moving. You will see a preview of the perpendicular witness line. At the end, you will see a square (fixed node) or circle (associative node).

3. Finish by placing the node at the desired point:

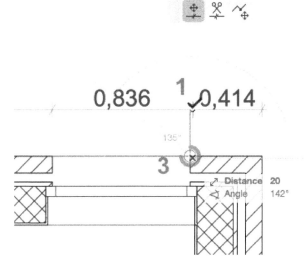

Figure 7.24 – Moving a fixed dimension node to an associative point of the model

When clicking a node, three options are available in the pet palette:

1. **Move Node**: This has been explained in the above steps.

2. **Split Dimension Line**: This cuts the dimension line at the clicked node, asking you which side to keep selected. The selected side can then, for example, be moved. You can add a point or any other option (click the dimension line and choose the appropriate option in the pet palette).

3. **Break Dimension Line**: This does more or less the same as **Split Dimension Line**, but after selecting a side, this option immediately lets you reposition this part (the selection and pet palette are skipped).

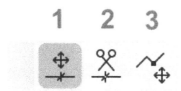

Figure 7.25 – All options for dimension nodes

To delete a node in a chained dimension line, follow these steps:

1. Click the node you want to remove.

2. Press *Fn + Backspace/Delete*.

OR

1. Select the whole dimension line (watch for the Mercedes cursor).

2. *Cmd/Ctrl + click* on the obsolete node

To add an extra node in a single or chained dimension line, follow these steps:

1. Select the entire dimension line.

2. Click it again and choose **Insert/Merge Dimension Point** from the pet palette.

3. Add a node by clicking on the reference point in the model.

OR

1. Select the entire dimension line.

2. *Cmd/Ctrl + click* directly on the reference point in the model to add a node in the dimension line without using the pet palette.

To delete one segment of a chained dimension line, follow these steps:

1. Select the obsolete segment by clicking its midpoint (watch for the check mark).

2. Press *Fn* + *Backspace/Delete*.

Clicking the dimension line itself gives you a seemingly different set of options in the pet palette, but all of the Adjust commands give a similar result as the options you had for the nodes, for example, **Move Dimension Line Segment** lets you move one segment while at the same time splitting up the original dimension line where needed. Explore these options on your own, while keeping an eye on the feedback in the Status Bar (the bottom-left corner of the screen). The Move commands of the pet palette should be familiar by now.

Automatic dimensioning

Archicad dimensions are not only mostly annotative (meaning they adjust automatically during design changes) but we also have some specialized tools at hand to quickly add dimensions for a whole plan. **Automatic Dimensioning** is available through **Document** > **Annotation** > **Automatic Dimensioning** > **Exterior Dimensioning…** or **Interior Dimensioning….** In order to use these dimensioning wizards, you first need to select the elements you want to dimension.

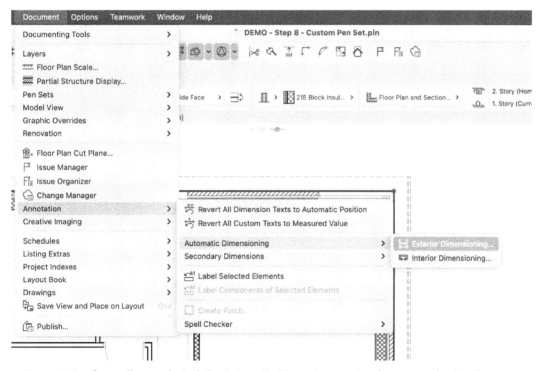

Figure 7.26 – If no walls are selected, the Automatic Dimensioning wizards are grayed out in the menu

Let's start by dimensioning all of the exterior walls on the first floor automatically:

1. Check the settings and construction method for the **Dimension** tool to make sure the preferred default settings will be used for the automatically generated dimension lines.

2. Navigate to the first floor and select all exterior walls.

3. Start the **Automatic Dimensioning** wizard.

4. The default settings are quite all right and mostly self-explanatory – we just change the settings for the doors (a) and windows (b) using the drop-down menus and check the **Place Dimensions on Four Sides** box (c). The value for the distances between the dimension lines (numbers 1 to 4 (d) as shown at the top right) was set to 500 (e). Confirm with **OK**.

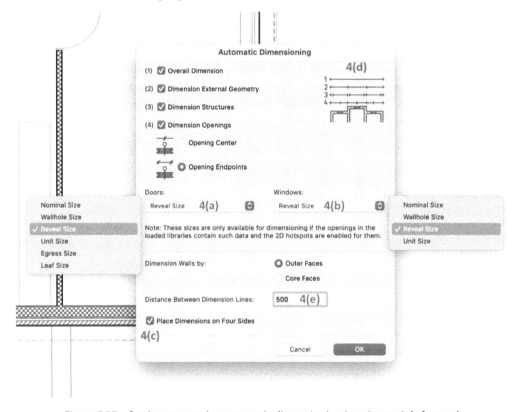

Figure 7.27 – Setting up exterior automatic dimensioning is quite straightforward

5. Click the lower horizontal wall in the floor plan view to determine the direction. This is a more crucial step in non-orthogonal projects. In that case, you would *not* check the **Place Dimensions on Four Sides** box.

6. To position the innermost dimension line, use numeric input so this line is at the same distance from the wall as the distance between the lines.

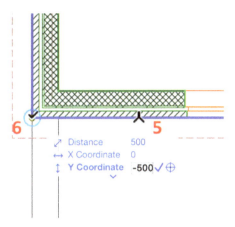

Figure 7.28 – Continue to use the skills you've learned in the first chapters!

7.   Zoom out and delete any obsolete (double) dimensions and add points for missing dimensions using the techniques described in the previous part, for example, the corner window doesn't get dimensioned automatically.

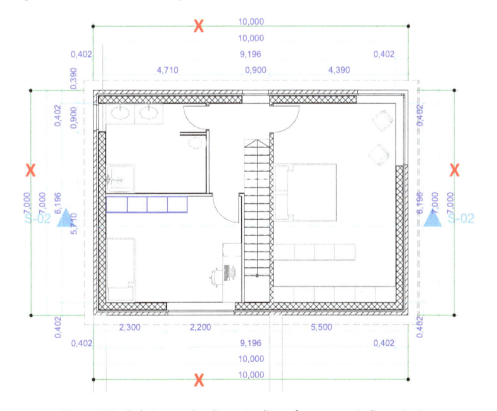

Figure 7.29 – Deleting surplus dimension lines after automatic dimensioning

As shown in *Figure 7.29*, using automatic dimensioning often results in double dimensions, mostly due to choices made under 4(d) in combination with the shape of a design, but deleting them is easy!

We can also automatically dimension interior walls, albeit one dimension line at a time. The process is quite similar, and there are fewer settings:

1. Check the settings and construction method for the **Dimension** tool.

2. Select the three "vertical" interior walls (a) and the east and west exterior walls (b) on the floor plan of the first floor. You could also add the slab if you would like to add dimensions for the stair opening as well…

3. Open the settings dialog through **Document** > **Annotation** > **Automatic Dimensioning** > **Interior Dimensioning…**.

4. You can adjust the settings to your liking. The default settings should work – dimensioning all skins (a) or making a specific choice (b) for the composite parts (explained in *Part II – Chapter 8*) will clutter the view in this preliminary phase. Confirm with **OK**.

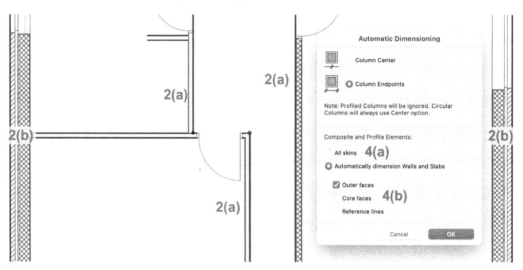

Figure 7.30 – Selecting the necessary walls and choosing settings

In *Figure 7.30*, the exterior walls are included in the selection, so we will have an interior dimension between the exterior and interior walls as well.

5. By drawing a polyline, we can tell Archicad which walls have to be dimensioned within the selection. Pay attention that this polyline is perpendicular to the selected walls for the parts that cross these walls. End the polygon input by clicking twice.

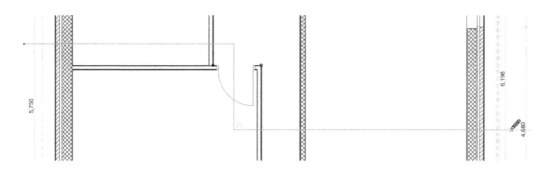

Figure 7.31 – In an orthogonal floor plan, drawing the perpendicular polyline is easy using snap guides

6.   Position the dimension line and use numeric input or reference points if wanted.

7.   Zoom out and delete any obsolete dimensions. For example, the thickness of the exterior walls is already dimensioned at the exterior of this floor.

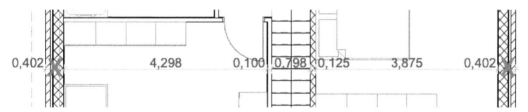

Figure 7.32 – The result for the interior automatic dimensions

You can now complete dimensioning your project with these techniques. After finishing, it is time to learn one more annotation technique at this moment: labeling – another way of extracting and visualizing some of that BIM data!

Exploring labels

Labels appear like text with some linework to add a box and an arrow. However, they are integrated with the virtual building in Archicad. Their strength lies in bringing model information into our drawings. This not only speeds up our workflow but also guarantees consistency: everything you see in a label is model information. It is synchronized with the model. When a label displays the name of a composite, it gets updated when the composite changes. You are sure that all labels display the same information that is in the model and that may have been displayed in other views or extracted documents.

> **Tip**
> Use labels as much as possible whenever you want to display a piece of information from the model. This will help you to avoid errors and contradictory information in your drawings.

Associative labeling

The Label tool can be found in the **Document** section of the Toolbox, just like the annotation tools. Let's add a label to one of the columns we modeled:

1. Select the **Label** (**1** in *Figure 7.33*)tool.

2. By default, **Geometry Method** is set to **Associative**. We prefer not to use the **Independent** method unless we want to add a static text comment.

3. The default for **Label Type** is **Text / Autotext**.

4. Hover the cursor over the column until the column is highlighted in blue and click on it.

Archicad adds the label, which displays the ID of the column (in this example, CRE-001). We will learn how to change IDs and add other information later on. The Archicad 26 INT template includes the semi-automatic generation of IDs. This explains why we see a name we did not add ourselves…

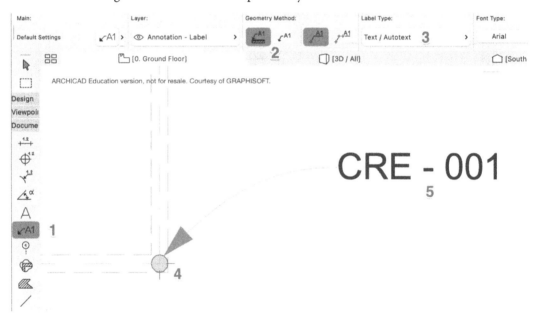

Figure 7.33 – Adding a Text / Autotext label to a column

When you open **Label Selection Settings**, you see a slightly different dialog layout than what we are used to:

1. **TYPE AND PREVIEW** gives access to either the **Text** / **Autotext** type or one of the *object-based label types*. These are actual GDL scripts, which can contain quite elaborate combinations of text and linework and extract info from the objects with which they are associated.

2. **TEXT LABEL** gives the configuration specific to the **Text / Autotext** label type, which includes **Label Orientation** and **Label Text**. This text is set to **#Element ID**, which is an Autotext entry.

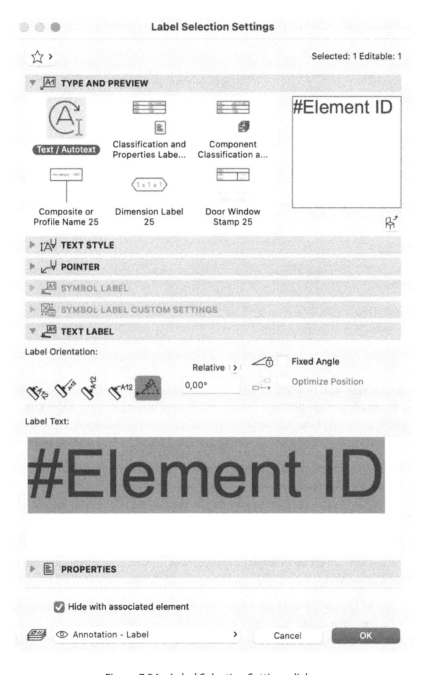

Figure 7.34 – Label Selection Settings dialog

If you want to include another string, you can start editing the label by double-clicking on its text. You then enter a regular Text Edit field, where you can type what you want, like a regular rich text editor. Use the button at the top left to include Autotext entries, which are values that Archicad calculates for you. The strength of such a field is that it gets updated when its source changes.

Figure 7.35 – Opening Autotext categories while editing the label text

There are countless entries you may select, grouped in categories. There are entries for the project, site, or building. Many of them are oriented toward the setup of layouts and drawings. Others relate to the associated object, including **General Parameters**, **Column** (in this example), **Classifications**, and **Properties**. The default entry #Element ID can be retrieved from the **General Parameters** category.

Autotext can be used in a few places in Archicad and we will make more use of it when setting up title blocks for layouts, in *Chapter 14*.

Object-based label types

You can switch the label type to one of the many object-based alternatives. These are in fact parametric GDL scripted labels, which are set up to display summaries of information, extracted from the objects with which they are associated.

Here is an example of **Generic Label 26**, which displays not only the **ID** field but also **Name**, **Dimensions**, and a few other fields, including **Surface Area** and **Volume**. Beware that we dragged the label to the side after placement, to not overlap the model drawing. Dragging does not disconnect the association with the object.

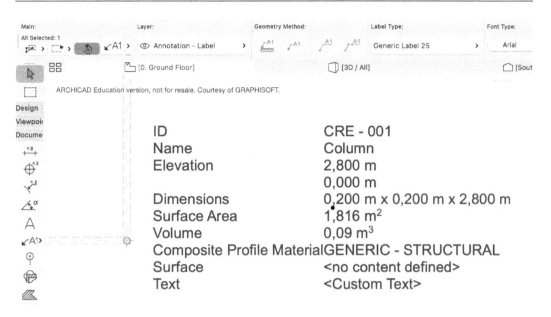

Figure 7.36 – Generic label associated with a column

> **Power tip**
>
> The strength of the BIM approach is that all these text values are automatically updated when the associated object is modified. As a user, you only have to worry about what to display and where to position it in the view, while Archicad ensures that the values are up to date.

We have now annotated our project, resulting in views that can be used as actual construction drawings. Sometimes, we also need to refer to the documents of others to be able to finalize a design. Before we learn how to create our own 2D output, we'll take a look at how we can use external references in Archicad in the next section.

Linking reference drawings

Just like CAD software, Archicad allows a designer to use existing drawings as a reference when modeling or to compare with. Especially for renovation projects, this is a good approach, as it allows you to place the existing context directly inside your modeling environment, at full scale.

However, with experience, we also learn that the unforgiving nature of an integrated 3D/2D environment tends to bring problems upfront, for example, it is common that discrepancies between a plan and section remain unnoticed in a traditional drawing-based approach, yet they need to be dealt with in a full 3D modeling approach. And to be totally clear, this is a strong added value of the BIM approach: avoid errors and contradictions between the different representations.

There are a few commands you can use to add an external drawing to your project:

1. Go to **File** > **Open**. While this would appear to be straightforward, it actually starts a new Archicad project (and closes the current one – so save your work first!) using the drawing as a template, which means that you'll get the layers and line types and hatches from the drawing, rather than from your Archicad template. This is really not recommended, as you'd have to start almost everything from scratch: pens, line types, fills, layers, and so on. The navigator is also empty, apart from a single story. Not something you would do for fun and a lot of extra work.

2. Go to **File** > **External Content** > **Place External Drawing**. This can only be called when you are already working on a project that started from a template. This way it does not disturb the Archicad project structure. The whole drawing becomes a single object, which you can scale, rotate, and move around. This is recommended for keeping the import "as is," so we prefer to use this for PDF and images, rather than for CAD drawings. You can still use snapping to the lines if the original file was vectorial (for example, a DWG saved as PDF). You can also use the context menu and **Explode into Current View** if you want to turn it into editable lines as if they were drawn in Archicad.

3. Go to **File** > **External Content** > **Attach Xref…**. This is a concept derived from the widely used *AutoCAD* software. **Xref** is shorthand for *eXternal REFerence*, which helps us to include a drawing without messing up our current Archicad project structure. This is the approach we prefer, as it gives us more control over the drawing structure, such as its layers, line types, and hatches. This is also recommended to align with other teams using other CAD software, where 3D model exchange is not required or possible.

Let's use the last technique to add an external reference DWG-file (a *DraWinG* file format conceived by Autodesk®, but widely used within the design industry for exchanging 2D and 3D drawings) to our project.

Example of using Attach Xref

We'll show the use of an *AutoCAD* external reference drawing for the famous *Villa Savoye*, by *Le Corbusier*, but you can use (almost) any DWG you want, due to the excellent DWG compatibility of Archicad.

Download the DWG called `Autocad drawing Villa Savoye - Le corbusier - ground floor plan dwg` `(https://ceco.net/autocad-blocks/architecture/top-plan-view/autocad-drawing-villa-savoye-le-corbusier-ground-floor-plan-dwg-dxf-270)` or use any other DWG of an architectural project if you have access to one.

We will work on an empty project, starting from the default template. By now, you should know how to launch a new project. Open any floor plan view and follow these steps to add the DWG of your choice:

1. Open the **Attach Xref** dialog (**File** > **External Content** > **Attach Xref...**).

2. Use the **Browse...** > **From File** pop-up button to get a regular **Open File** dialog, filtered on either *DWG* or *DXF* files. In our example, we use a DWG file of the ground plan of the Villa Savoye.

3. We need to set the drawing transformation using three sets of input values. Using the option checkboxes, they can all be configured to **Specify On-Screen**, meaning that we can graphically control every one of those settings on the drawing screen (instead of using numbers and percentages):

    A. **Insertion Point**: We prefer to set **X** and **Y** to 0,00, so have the **Specify On-Screen** option checkbox unchecked. This way the drawing is using the same origin as Archicad does. We can always move it later if needed.

    B. **Scale**: Here, we start with **X** and **Y** set to 1,00. This value depends on the units used in the DWG. Archicad uses meters internally, so if your DWG also uses meters as units, you are good to go. Otherwise, this is where you can set a scaling value.

    C. **Rotation**: Finally, this helps when the orientation of the drawing does not match the orientation of your Archicad project. We prefer to keep it at 0,00°.

4. We keep everything else as it is. If needed, you can refine **Anchor Point**, **Place on story**, and **Translator**, which gives you (very) deep control over the conversion. Press the **Attach** button to perform the attachment.

Figure 7.37 – Attach Xref dialog, to configure drawing transformation and translation

5. There are a few more dialogs that open, requiring more information from the user:

    •   **DWG/DXF Partial Open** dialog: To confirm the layers to include in the attachment. With large drawings, this may help to limit what you reference.

    •   Depending on the drawing, a few more dialogs may pop up to point to **SHX** files, containing vectorial CAD fonts. Let's skip them all unless you have them laying around from older CAD drawings.

    As a result, you get the DWG attached to the ground floor as a group of lines, arcs, fills, and possibly text.

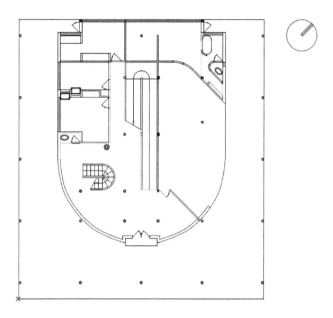

Figure 7.38 – Attached Xref, with the Archicad origin at the bottom left

6. To confirm that the scale is correct, use the **Measure** tool (shortcut *M*) to measure one of the door leaves. In this example, it reads as 800 mm. If the drawing used another scale, this is typically an order of magnitude off, such as 8 mm, 80 mm, or 8000 mm. In that case, undo and use another scale in the **Attach Xref** dialog.

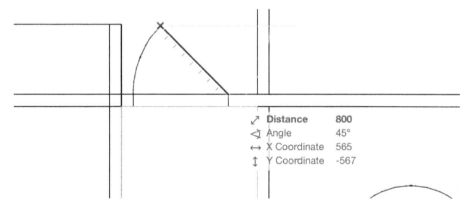

Figure 7.39 – Measure the length of a line representing a door leaf

Now the drawing is here, you can start modeling, using the drawing linework for snapping and guides, speeding up the modeling considerably: measure the thickness of a wall, set up a wall composite, and start modeling.

When you have multiple drawings, you can attach one for each story. If you have sections, you could do the same. As an alternative, we sometimes attach a drawing to an Independent Worksheet and use **Trace and Reference** to display it in a section or on a ground floor plan, which gives more flexibility.

> Tip/note
>
> To protect the drawing, the linework is a group and the lines are locked. You can use **Xref Manager** (**File** > **External Content** > **Xref Manager...**) to add more Xrefs or to remove them when you no longer need them.

Let's now return to our own annotated project and learn how we can create our own 2D output.

Basic export and printing (output)

Finally, it is necessary to know how to get our design into a drawing we can print or send to others as a PDF or DWG file. We distinguish between printing directly or using a layout, which gives you more control, but also requires more preparation.

Direct printing from a drawing window

At any point in time, you can print the content that you currently see. Archicad supports any installed printer but also has additional support for large-scale printers ("plotters"). If you want, you can print the result directly, which is mainly a matter of picking the right printer driver.

We'll show the macOS dialogs, but rest assured that the Windows version shows familiar printing dialogs. To make a print, follow these steps:

1. **File** > **Print...** (*Cmd+P/Ctrl+P*) is as straightforward as it gets. You get a **Print** dialog, set to the default printer, and use the configuration from your operating system.

2. When you select **Show Details**, apart from the regular options such as paper size, two-sided printing, and so on, there is a series of Archicad-specific options:

   A. For **Print Area**, we select **Entire Drawing**, to get everything, even the parts that are currently not visible due to **Current Zoom**, unless that is what you want.

   B. **Scale** is an important one. The default is the **Original** architectural scale that is active on this view, but you can set a **Custom** scale or even have Archicad rescale it to **Fit To Page** if all you need is a quick print.

   C. **Arrangement** is how the printout at the set scale is spread over multiple pages, which is often required if you print an architectural drawing on a regular office printer. Archicad adds cut marks to help you reassemble the pages into a large sheet.

3.   If you are pleased with the setup, click the **Print** button.

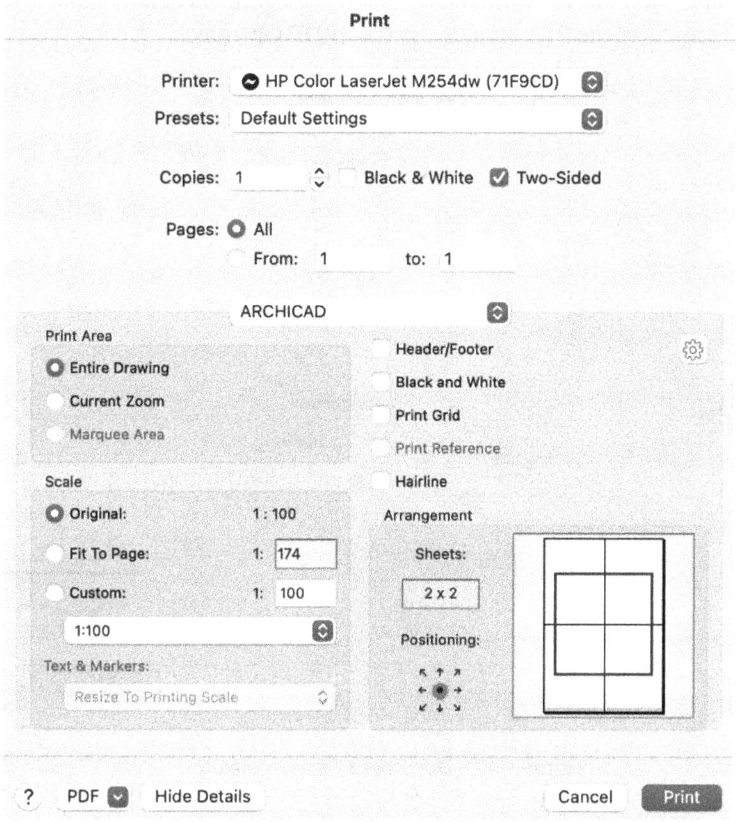

Figure 7.40 – Print dialog (on macOS)

For more advanced control, we advise using a layout, which is discussed further in the subsection on *Using a Layout*.

We'll keep it simple for now and assume you want to turn your model into a 2D drawing that you can share digitally.

Printing or exporting to PDF

The *Portable Document Format* or *PDF* is the standard format developed by *Adobe* to share documents and illustrations faithfully, including the right color, font, and scaling.

To create PDF files, follow one of these options:

- The first option is to *print* the view to a PDF file, by selecting one of the available PDF printer drivers from your operating system, like we did when printing.

- An alternative is to save or export your file to PDF (**File** > **Save As...** > **PDF File**). The **Document Options** dialog is just like the **Print** dialog but skips the printer configuration. If you need more in-depth control over the PDF generation, click **PDF Options...**.

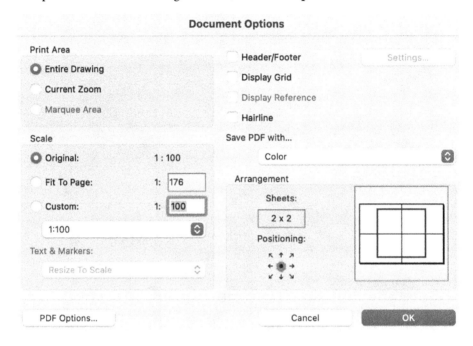

Figure 7.41 – Document Options dialog

Saving as DWG

When you want to share a drawing that needs to be edited, you may want to share it as a *DWG* file, which is the native format for *AutoCAD*. This format is widely supported by most CAD and BIM software and is often requested in projects.

Since there are countless settings that control how Archicad will translate its internal drawing entities into an external DWG file, you can store and recall the configuration using a so-called **Translator**. This helps you guarantee that you can repeat the process without having to remember every setting each time you export a drawing:

1. We need to save the file as DWG (**File** > **Save As...** > **DWG File**).

2. Apart from setting the folder and filename, you need to decide on the translator to use. The list of available translators depends on the template you started from.

3. The INT template has a few configurations, but you can create your own. Don't do this now, unless you are very familiar with the DWG format. We simply select the **01 For further editing** option, which works well with the recommended setup of an AutoCAD drawing.

> **Pro tip**
>
> To be honest, the configuration of a drawing, including the correct line weights, line styles, and scaling of patterns and fills, involves (many) more steps and decisions.

Using a layout

Printing directly seems easy but is in fact not the recommended approach. Archicad has a very advanced **Publisher** system, which allows you to create large sets of printouts and exports, allowing you to prepare a whole package of documents to be generated in a single operation. This will be the subject of *Chapter 14* .

For now, we'll go through the minimal steps:

1. From the **Navigator** palette, select the third tab at the top, called **Layout Book**. Now the hierarchy no longer shows the project viewpoints, but the structure of the **Layout Book**, containing **Layouts** and **Master Layouts**.

2. Click the **New Layout...** button, just underneath the hierarchy.

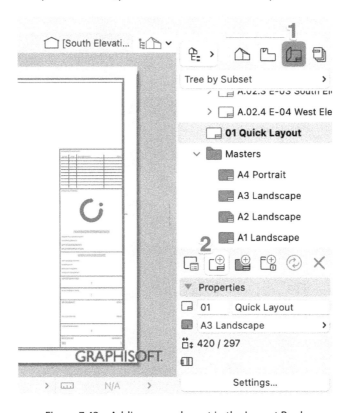

Figure 7.42 – Adding a new layout in the Layout Book

3.   The **Create New Layout** dialog opens.

4.   We'll keep it simple and set our layout number in **Custom ID** to 01 and **Layout Name** to Quick Layout.

5.   You need to indicate the page setup from the **Master Layout** popup. We will use **A3 Landscape**, which has been preconfigured in the template.

6.   Click the **Create** button.

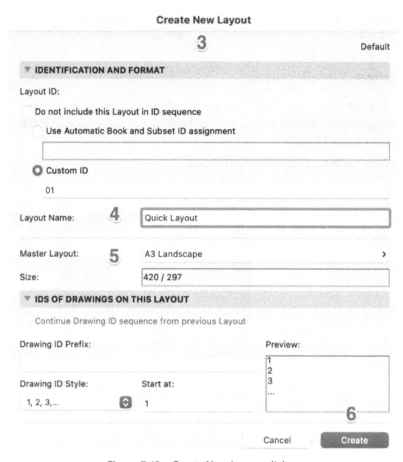

Figure 7.43 – Create New Layout dialog

Our layout is again a 2D window, with an empty layout, which already contains a title block. We are not going to configure the title block here, but understand that the title block was created on the Master Layout we selected. So, in this case, the **A3 Landscape** Master Layout.

The final step is adding our view to the layout. This can be done from **Navigator** as well:

1. Ensure that **01 Quick Layout**, which is empty, is still active in the **Layout Book** tab of **Navigator**.

2. Go to the **Project Map** tab of **Navigator**.

3. Drag the **0. Ground Floor** story onto the layout. Be careful not to double-click this story, since this would hide the layout again.

   Archicad places the ground floor onto the layout.

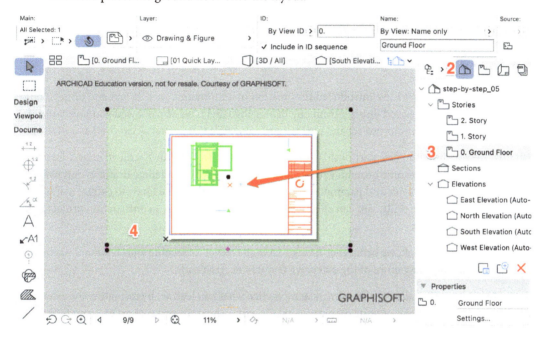

Figure 7.44 – Drag a viewpoint onto the open layout

What just happened?

What we just did in the steps in *Figure 7.44* gives us the desired result, but can be confusing for new users, so let's have a closer look at what happened when dragging a Viewpoint on the Layout:

- When you dragged the story onto the layout, Archicad prepared a **View** for you, which captured the current display settings, including scale, active layers, and much more. Basically, everything that's required to reconstruct what you see in the 2D window.

- This view got dragged onto the layout and became a **Drawing** object.

- A drawing references the view and can be placed and cropped onto a layout.

Why is this so involved? Archicad helps you to keep the configuration of different drawings. You can continue to work on the model and you can toggle different display settings, but every time you return to your view, it recalls its stored settings. When you return to your layout, it recalls the drawing and the view it references, which gets refreshed automatically.

Archicad guarantees that, at the time of printing or exporting, your layouts will retain the set scale, colors, and position, but the content of the views will get updated to reflect the latest state of the model.

This may seem quite a lot, but we will return to the published workflow in *Chapter 14*.

Let's summarize the concepts we just introduced.

Summary

In this final chapter of *Part 1*, we introduced the creation of annotations such as text and dimensions and added information derived directly from our BIM using labels. We also showed you how you can include a CAD drawing in your project and how to send a drawing to a printer or export it as a PDF file.

With this, we conclude the first part of this book. You may not feel like an Archicad expert just yet, but we have covered quite some ground: you are familiar by now with modeling, drafting, and using a variety of Archicad commands and parametric objects. You have also learned how to extract a drawing from a model. With these skills, you can effectively start with Archicad in a construction project and produce usable drawings.

In the next part of the book, we bring everything to a higher level to become a professional Archicad user and learn about many more of the extensive features of Archicad.

This would be the perfect time to take a break and practice what you learned by recreating the example model or applying it to one of your own projects. Or, you can try to recreate an existing building design as an Archicad model.

Part 2:
Becoming an Archicad Professional – Learn About Archicad Tools and Settings to Create and Publish Any Type of Project in Full Detail

In this part, we will dive deeper into the specialized tools and underlying mechanisms in Archicad. The model that was started in *Part 1* will be further expanded using a variety of tools and workflows, allowing you to complete a developed design with documentation, data extraction, and 2D and 3D visualization. By the end of this part, you will feel confident in modeling, documenting, and visualizing your designs using Archicad in a professional way.

This part comprises the following chapters:

- *Chapter 8, Using Advanced Modeling Tools for Developed Design*
- *Chapter 9, Using Advanced System Tools for Designing Stairs and Curtain Walls*
- *Chapter 10, Using the Mesh Tool and Wizards to Finalize a Design*
- *Chapter 11, Using Advanced Attributes and the Renovation Tool for a Wider Design Range*
- *Chapter 12, 2D Construction Drawings and 3D Views with Linked Annotations*
- *Chapter 13, Data Extraction and Visualization*
- *Chapter 14, Automating the Publication of BIM Extracts*
- *Chapter 15, The Various Visualization Techniques in Archicad*
- *Appendix: Some Final Tips and Tricks*

Using Advanced Modeling Tools for Developed Design

In this chapter, we'll expand on the tools we already know about and learn how to apply them in developed design, requiring more advanced modeling methods. This includes the creation and manipulation of Composites for Walls, Slabs, and Roofs. We will get a grip on the way building materials define priorities to create automatic element connections.

In this chapter, the following topics are covered:

- Creating and using advanced Composites for Walls, Slabs, and Roofs
- The different Archicad priorities and how they influence the way elements connect
- Creating more elaborate Roofs and adding skylights

Advanced Composite Walls, Wlabs, and Roofs

In *Part 1*, *Chapter 6*, we already got to know the importance of Archicad attributes and how tools, tool settings, and view settings depend on them, but also how they depend on each other. We learned how complex *and* powerful attributes can be. In this section, we will further explore one of the most advanced types of attributes: **Composites**.

Knowing the different parts of a Composite definition and the attributes involved

Composites can be used as a definition for four tools: the **Wall**, **Slab**, **Roof**, and **Shell** tools. Where a *basic structure* for these tools is just one material, a Composite is in essence a layered structure consisting of multiple Building Materials, functioning as Skins, separated by Lines, which are called **Skin Separators**.

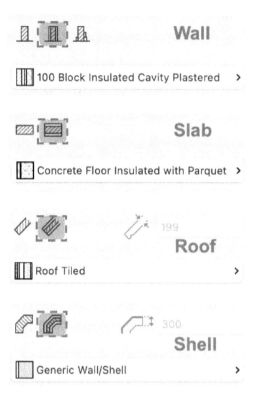

Figure 8.1 – Elements able to use composite structures

As shown in *Figure 6.16* in *Part 1, Chapter 6,* the **Building Material** attribute itself is defined by other attributes: **Section Fill** and **Surface** – which in itself is further defined by a **Vectorial Fill** and some **Pens**. This means that we are not able to directly change the appearance of the individual materials from within a Composite definition; we can only choose the materials to be combined.

We can access **Composites…** in several ways:

- Through the menu: **Options** > **Element Attributes** > **Composites…**. This option is used when you have nothing selected and you want to define a new Composite or edit an existing definition.

- Using the contextual menu: *RMB* on a selected wall, slab, roof, or shell that has a Composite structure and selecting **Edit Selected Composite/Profile…**. This method is intended for editing the definition of the selected element but it can also be used to access these attribute settings more quickly. It brings you to the same dialog as using the menu but shows the definition of the selected element and is often quicker than browsing through menus.

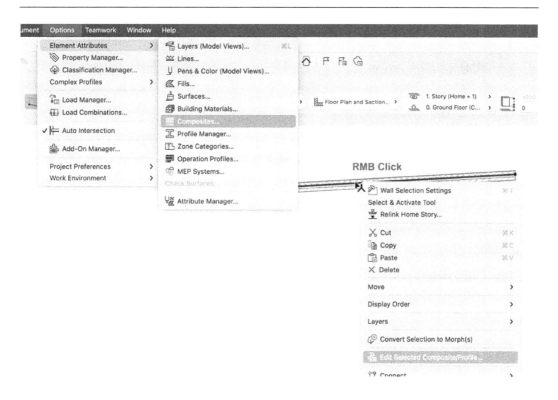

Figure 8.2 – Both ways of accessing Composite definitions, combined in one image

In the **Composites** dialog, we clearly see the two main parts of its definition (see *Figure 8.3*):

- **Skins**: These are the Building Materials that are added as layers in a Composite. They take on the physical, formal, and graphical properties of the building materials making up the Composite. Each layer has its own thickness (**1a**) (see *Figure 8.3*). Added together, this gives the total thickness of the Composite (**1b**). You can add (**1c**) an unlimited number of Skins to a Composite. Of course, you can also delete superfluous Skins (**1d**). Skins can be moved up or down using the small arrows (**1e**). Changes in the thickness, order, or Building Material of Skins are shown in the preview (**1f**) to the right, showing the cut fill of the Composite and the surface color (explained later).

- **Skin Separators**: These separation lines are optional. You can disable them with the check mark (**2a**). They form the boundaries between the fills defined by the chosen Building Material for each Skin. They are individually defined by a line type (**2b**) and a pen (**2c**). The end lines of each Skin can also be switched off using a checkbox (**2d**) – the **Air Space** end line is turned off in *Figure 8.3*. The contour lines (**Outside/Top** and **Inside/Bottom**) within a Composite structure are of bigger importance, as it is possible to show only the contours of an object.

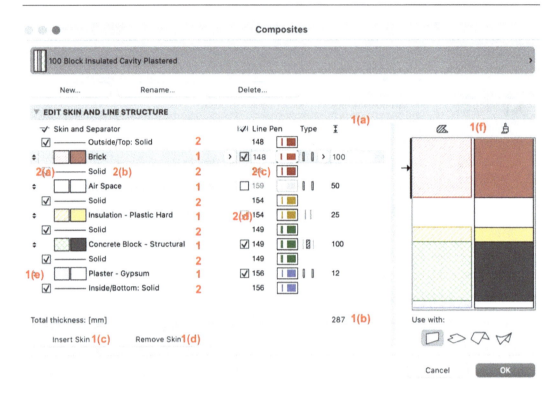

Figure 8.3 – The Composites dialog with Skins and Skin Separators

Now that we know the basic ingredients, let's create our own Composite definition.

Creating a new Composite

To create a new Composite, open the **Composites** dialog through **Options** > **Element Attributes** > **Composites...** and follow these steps:

1. Click **New...** (**a**) and choose **New** (**b**) in the popup. Enter My own composite as the name (**c**) and confirm with **OK** (**d**).

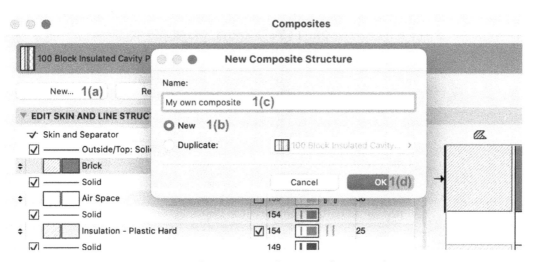

Figure 8.4 – Creating a new Composite from scratch

2. The newly created definition contains one Skin (**Brick** by default). We are going to create an external cavity wall, so change **Brick** to **Brick - Finish** (**a**), as this is the **Outside/Top** (**b**) layer. Keep this default order in mind when creating external Composites! It is of less concern in Composites used for interior elements.

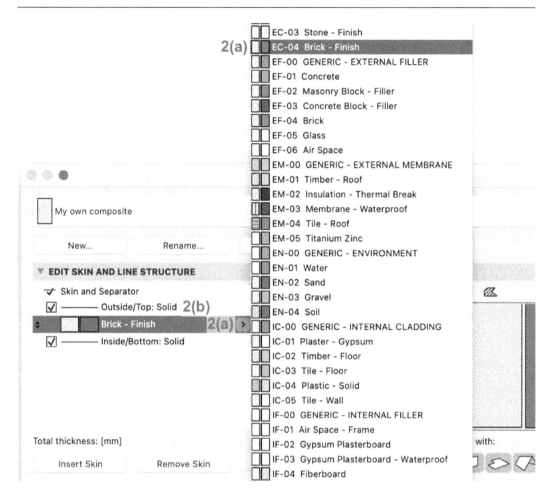

Figure 8.5 – Changing the first Skin

3. Add three more Skins by clicking **Insert Skin** three times (**a**). Change every material (**b**) in accordance with *Figure 8.6* and adjust the thickness (**c**) of each Skin accordingly.

4. Set the Skin Separators (**a**) and End Lines (**b**) as shown in *Figure 8.6* and set the Line Pen (**c**) for both.

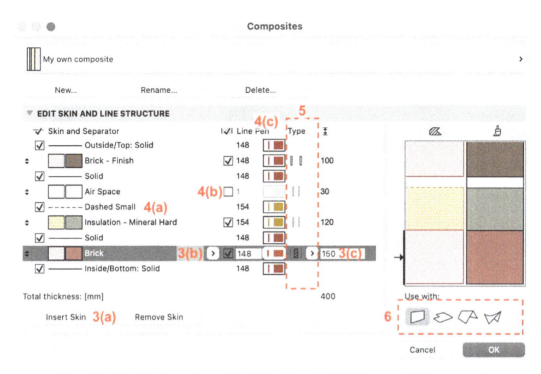

Figure 8.6 – Set Skin Separators and End Lines to create the 2D representation you want

5. For each Skin, we should also set **Type** as shown in *Figure 8.6*. There are three options available: **Core**, **Finish**, and **Other**:

-   **Core**: This is typically used for the structural core of a Composite. Multiple layers can be set as the core, but multiple cores always connect to each other and thus *cannot* be separated by the **Other** or **Finish** Skin Types. We can also create Composite walls without a core. This is useful, for example, for lightweight partition walls (plasterboard walls), when we do not want to show them in the **Core Only** option for **Partial Structure Display** (see *Chapter 14*).

-   **Finish**: This is typically used for the finish layers of a composite. Multiple layers can be set as finishes, but multiple finishes are always "extended" uninterruptedly to the inside or outside of a Composite – they are always adjacent to at least one contour line. The inside and outside can be separated from each other by **Core**(s) and/or **Other**(s). *Figure 8.7* shows some possible combinations.

-   **Other**: This is used for anything that does not fit the definition of **Core** or **Finish**, and is mostly used for air layers, membranes, foils, insulation, and so on. Multiple layers can be set as **Other** – they are not automatically connected to each other.

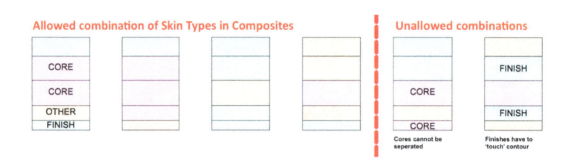

Figure 8.7 – Archicad will not allow just any combination of Skin Types

6. To finish our Composite definition, we have to set which elements they will be available for. As already mentioned, Composites are available as an attribute in four of Archicad's modeling tools: Wall, Slab, Roof, and Shell. Set your definition to be used with the **Wall** tool only.

That's it, you just created your first Composite! Close the dialog by clicking **OK**, and use it to draw some Walls. Open the definition again and change a few settings and see how the modeled walls change accordingly. You could even change the Pen Set and see how this affects the drawing.

> **Pro tip**
>
> It is good practice to use **Duplicate** instead of **New** in *step 1* (*Figure 8.4*), especially when you are just starting your Archicad journey. This applies to basically any attribute you want to create: using what is already present in the provided template helps you tackle the complexity of the interaction between attributes. Once you have some more experience, you will easily create Composites and other attributes from scratch.

By now you understand that attributes are very powerful, but you need to know how they interact to completely master Archicad. Practice is key, but going back to the introduction we gave in *Chapter 6*, and especially studying *Figure 6.16*, might help you to better understand the relationships.

Now that you know how a Composite is defined and how you can create one, it is time to see how you can combine multiple Composites in elements in your model and how they interact and make connections using Archicad priorities.

Archicad priorities

Using the Composites we created in the previous section, *Advanced Composite Walls, Slabs, and Roofs*, opens up a lot of new possibilities for more detailed construction documents. How different elements connect, however, also becomes increasingly complex. To learn how this works, we have to dive into one of Archicad's settings that provokes great enthusiasm when starting out, followed by possibly some disappointment and frustration when you evolve your skills, and hopefully ending in a very efficient workflow once mastered: priorities. There are in fact four different types of priorities in Archicad, within a priority hierarchy. Arranged from highest to lowest priority within this hierarchy, these are as follows:

- **Layer Intersection Priority**: Prevents or allows connections between elements on different layers

- **Building Material Priority**: Regulates the connections between different elements based on the Building Materials used in these elements and their embedded priorities

- **Wall and Beam Junction Order**: Allows you to prioritize a connection where more than two Wall or Beam elements meet using the same material and located on layers with the same Layer Intersection Priority

- **Element Priority**: A sort of "natural hierarchy" from one tool to another, this only comes into play when all other priority types are the same for the different elements involved in the connection, as their priorities do not have an influence anymore

Let's look in detail at each of these priorities and their uses.

Layer Intersection Priority

Simply put, elements on two layers with a *different* Layer Intersection Priority are not able to have a connection of any kind. In other words, if you want a connection between a wall and a slab (based on their materials), they *can* be on different layers, but these layers *need* to have the *same* Layer Intersection Priority!

As we explained in *Chapter 3*, in the *Modeling tools in general* section, layers are used to group elements in a logical way, for example, external walls versus internal walls, or annotations versus dimensions. Using layers, we can control the visibility of the elements on these layers by switching certain layers on or off. **Layers (Model Views)** is an Archicad attribute, and can be accessed through **Options > Element Attributes > Layers (Model Views)...**, through **Document > Layers > Layers (Model Views)...**, by using the button in the **Quick Options** bar, or by using the shortcut key *Cmd/Ctrl + L*. We will take a detailed look at this dialog in *Chapter 14*. For now, it is enough that we understand the influence of the Layer Intersection Priority. Let's take a quick look at how this is set in the **Layers (Model Views)...** dialog:

1. Open the **Layers (Model Views)** dialog.

2. Scroll down to the **Shell - General** layer and click the number **1** in the fourth column. This is the current layer intersection priority of this layer in the active Layer Combination. Change it to 2.

3. Check whether the **Structural - Bearing** layer has a different layer intersection priority (by default, it should be set to **1**).

4. Confirm with **OK**.

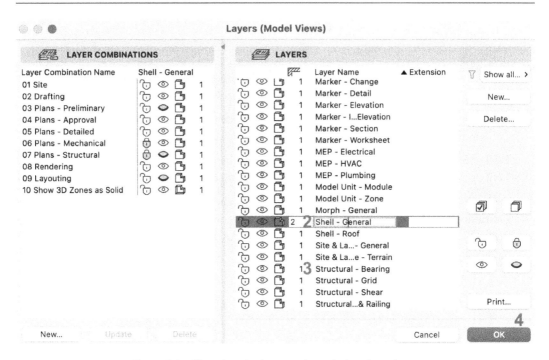

Figure 8.8 – Changing the intersection priority of one layer

5. With this custom Layer Combination, activate the **Wall** tool and draw some Walls using the same Composite. One or more of these Walls should be on the **Shell - General** layer, with others on the **Structural - Bearing** layer. Walls on the former layer will *not* connect with walls on the latter anymore.

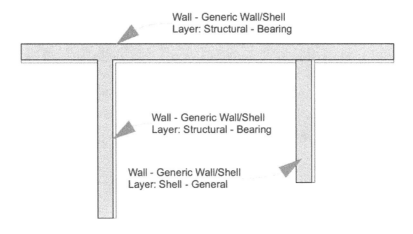

Figure 8.9 – Walls with equal materials on different layers do not always connect

The wall to the right in *Figure 8.9* is not connected to the one that is horizontal in this view. It has the same definition but is placed on a layer with a different intersection priority.

> **Note**
>
> When changing the layer intersection priority of some layers, we did not "save" this in a (named) Layer Combination in the **Layers** dialog. Archicad then created a temporary custom combination, which we used to test the principle of layer intersection priority. Layers will be explained in full detail in *Chapter 14*. To revert to a default Layer Combination from the template, just pick one from the dropdown in the **Quick Options** bar where it says **Custom**.

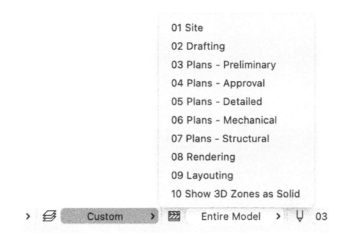

Figure 8.10 – The Quick Options Bar showing the temporary Custom Layer
Combination. Choose another one to revert the changes made

6. If you change the active Layer Combination to **05 Plans - Detailed**, you will notice that all Walls connect again. In this combination, the layer intersection priority is the same for both layers used in the example in *Figure 8.9*.

Building Materials and priority-based connections

In the *Advanced composite walls, slabs, and roofs* section, we learned how we can combine multiple Building Materials into a multilayer Composite. In the previous paragraphs, we have seen how a different layer intersection priority can avoid the connection between elements using the *same* Composite definition. This also implies that elements *will* connect when they are on layers with equal layer intersection priorities. How they connect depends on their Building Materials and the **Building Material Priority** (**BMP**) of those individual materials.

Every Building Material in Archicad has a priority value ranging from 1 to 999. A higher number gets priority over a lower one. This applies to **Basic Structures** for elements such as Walls, Slabs, Columns, or Beams as well as to **Composite Structures** in Walls, Slabs, Roofs, and Shells. To see and edit the BMP, take the following steps:

1. Open the dialog through **Options** > **Element Attributes** > **Building Materials…**.

2. Select any material and try changing the **Intersection Priority** value back and forth using the slider (left = **Weak** priority, right = **Strong** priority) or using numeric input (a higher number = stronger priority).

3. Notice the list on the left in *Figure 8.11* graphically shows the priority for all materials in the list. You can sort by priority by clicking the column header.

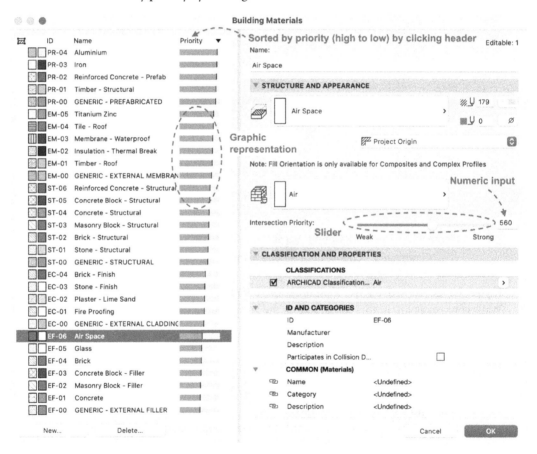

Figure 8.11 – Building Materials dialog with the Intersection Priority slider

This can get quite complex, especially in places where Composites with multiple Skins meet. The examples in *Figure 8.12* will probably clarify this for you:

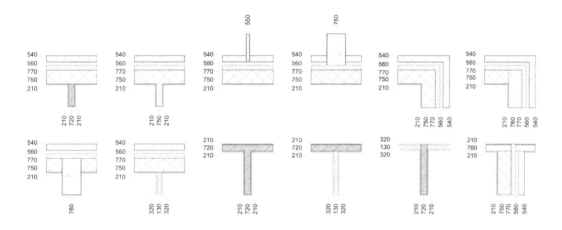

Figure 8.12 – Some examples of how Composite Walls can connect using Building Material priorities

The general principles we can learn from *Figure 8.12* are as follows:

- A material with a higher BMP will go "through" a material with a lower BMP.

- A material with a lower BMP will be "stopped" by a material with a higher BMP.

- Materials having the same BMP will connect seamlessly.

- When a material with a BMP of 130 is encapsulated in two (or more) Skins with a BMP of 320 (or higher), the whole Composite will go "through" any Skins with a BMP lower than 320. All Skins will be "stopped" by a material with a BMP higher than 320.

- When a material with a BMP of 770 is encapsulated within multiple layers having a BMP between 210 and 750, this one layer (BMP of 770) will go "through" any layers with a BMP lower than 770, while all other layers will be "stopped" by a material with a BMP of 760.

How this connection is actually "created" and whether a connection is even possible is determined by the location of the **Reference Lines** in the connecting elements, as we already briefly explained in *Chapter 3*, in the *Modeling tools in general* section. When we model an element that "touches" another element, Archicad automatically creates a connection, but we should add some nuance. The connection is only created when the following prerequisites are met:

- Elements are on layers with the same **Layer Intersection Number**. As we have seen in the section *Layer intersection priority*, a different Layer Intersection Number prevents elements from connecting (no matter what).

- **Auto Connection** is active when connecting: through **Options** > **Auto Intersection**, we can switch this off – by default, this is on and is best left on. **Auto Intersection** is only turned off in exceptional cases when we explicitly do not want an automatic connection and instead need more control over how a connection is made.

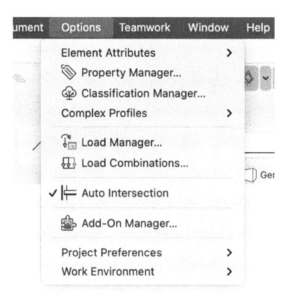

Figure 8.13 – Leave Auto Intersection on!

- **Reference Lines** can connect. When we model two Walls and their geometry "touches," Archicad will try to connect the Reference Lines of both Walls, prolonging them (automatically) if necessary. Let's see what this means in a simple example:

I.   Model a Wall, using **215 Block Insulated Cavity Plastered** as a Composite. The Reference Line should be on **Outside Face**. (Shown in **1** in *Figure 8.15*).

II.   Model a second wall (with the same Composite and Reference Line on **Outside Face** as well) perpendicular to the first one and finish this wall with a node on **Inside Face**. (Shown in **2** in *Figure 8.15*).

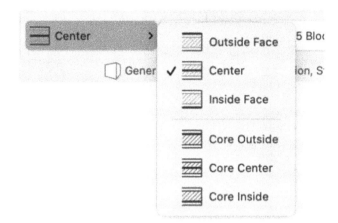

Figure 8.14 – Choose the position of the Reference Line using the dropdown in the Info Box

III. Select both Walls. You will notice how the Reference Line of the second wall was automatically extended until it touched the Reference Line of the first wall. How it connects is determined by the BMP (see *Figure 8.12*). (Shown in **3** in *Figure 8.15*).

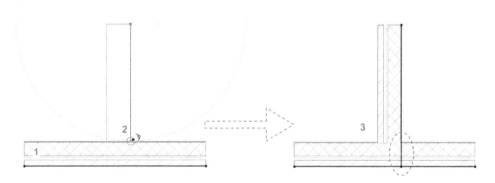

Figure 8.15 – Default automatic connection of Reference Lines in the case of two Walls

There are multiple scenarios where the Reference Lines of two Walls cannot connect, the most basic example being when the Reference Lines of the Walls are parallel. In *Figure 8.16*, we see several steps of the modeling process of such a situation:

1. Model two Walls with the **215 Block Insulated Cavity Plastered** Composite, forming a corner. Pay attention to the position of the Reference Lines: one is on **Inside Face**, the other on **Outside Face**.

2. Model a wall with the **Brick Double Plastered** Composite, perpendicular to the horizontally placed wall you modeled in *step 1*, to the left of the red dotted line in *Figure 8.16*. This Wall will *not* connect to the Wall modeled in *step 1*.

3. Model another Wall, but now to the right of the red line – this Wall will automatically create a connection based on material priorities.

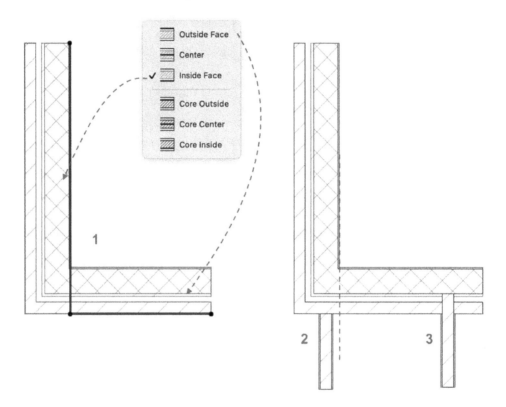

Figure 8.16 – Modeling a scenario where not all Walls are connecting the way we might expect them to

Solving this issue can be a challenge and it might mean that you have to split up Walls into several segments. For our example though, we can use a technique that will prove useful in many cases:

1. Select the vertically placed wall and navigate to **Edit** > **Reference Line and Plane** > **Modify Wall Reference Line…**.

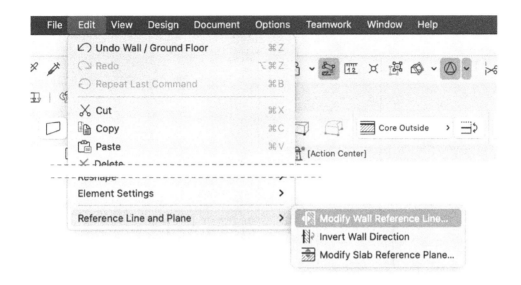

Figure 8.17 – Modifying the location of the Reference Line

2.   Choose **Outside Face** instead of **Inside Face** and confirm with **OK**.

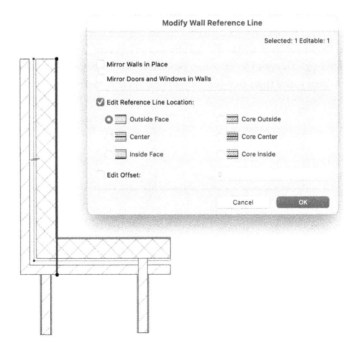

Figure 8.18 – Make sure you know where the Reference Line is located before you
start the command – the interface does not show the current state!

3.   The Walls now connect properly – select them to see how the Reference Line influences connections.

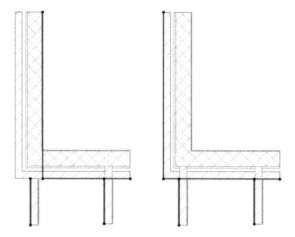

Figure 8.19 – Before and after modifying the Reference Line
location – with the Reference Lines highlighted

Junction order for Walls and Beams

Walls and Beams have an extra option to control (complex) connections. **Junction Order** can be found in the Wall and Beam settings and is defined by a number ranging from 0 up to and including 16. Higher numbers have priority over lower ones. Numbers do *not* have to be equal to connect. This only comes into play where more than two (equally defined) elements meet.

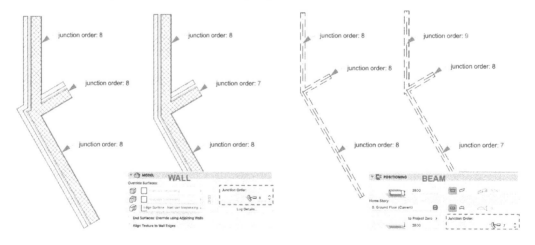

Figure 8.20 – Wall connection to the left and beam connection
to the right, showing the use of Junction Order

> **Note**
> In *Figure 8.20*, the Reference Line for the Walls is on the inside face!

Element Priority

Finally, when all previously explained priorities do not come into play, so when elements are on layers with the same layer intersection priority, have materials with the same BMP and Wall and/or Beam junction order are set the same, there is one more type of priority: **Element Priority**. There is a hierarchy for the different Archicad tools, arranged from high-priority elements to low:

- Columns
- Beams
- Slabs
- Walls

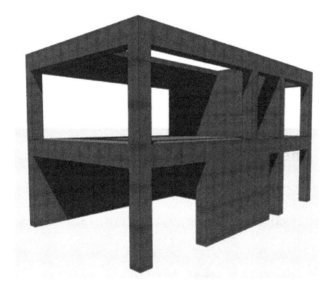

Figure 8.21 – The default 3D representation shows various elements
in the same material, seamlessly connecting, but...

In a default representation (*Figure 8.21*), there is not always enough distinction between elements to see the Element Priority. In *Figure 8.22*, this is explicitly shown.

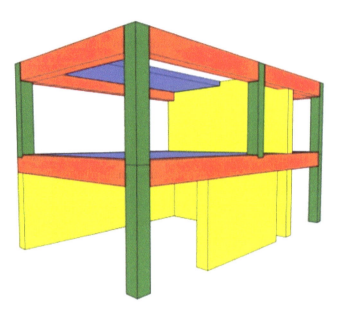

Figure 8.22 – …in fact, Element Priority creates a hierarchy when
other priorities do not regulate connections

Now that we have learned how to create and use more complex definitions for Walls, Slabs, and Roofs and how this influences the way these elements connect, it is time to learn how we can create more complex geometries with the **Roof** tool.

Complex roof geometry and skylights

Although we already saw how to model Single- and Multi-plane Roofs in *Part 1 – Chapter 4*, there are many more shapes and types that can be modeled. This section goes a bit deeper into the options and possibilities of several tools for modeling more complex roof geometries and we will also show how to add a skylight.

Sloped flat roofs using the Mesh tool

In Archicad, it is possible to create roofs using multiple tools. We already discussed the **Roof** tool, but you can also use **Slab**, **Shell**, **Mesh**, or even **Object**. Just be aware that the tool defines the geometric possibilities and constraints.

For plain flat roofs, **Slab** may be sufficient, as we showed in *Part 1 – Chapter 3*, since it simply defines a contour and a Composite. However, Slabs cannot have a slope in Archicad. If you need a slope, you should use the **Roof** tool, which adds a pivot line and allows you to set the slope direction and angle.

An alternative that is sometimes applied is using the **Mesh** tool. This is commonly used to model terrains, but its main geometric approach works for flat roofs with a top slope angle too. This can help in modeling sloped Slabs or Roofs or include the drainage slopes on flat roofs in more detail if you want to have it geometrically correct. Modeling a mesh with a slope in one direction works as follows:

1. Using the **Mesh** tool, you go on as usual when modeling a slab: you draw its contour.

2. However, specific to the mesh, you can adjust the vertical position or Z-height of individual nodes of the geometry. Click on the contour edge to open the pet palette.

3. The last option is **Elevate Mesh Point...**.

4. In the **Mesh Point Height** dialog, you can set the vertical height, which is expressed by default relative **to Mesh Reference Plane**. Set it to 500.

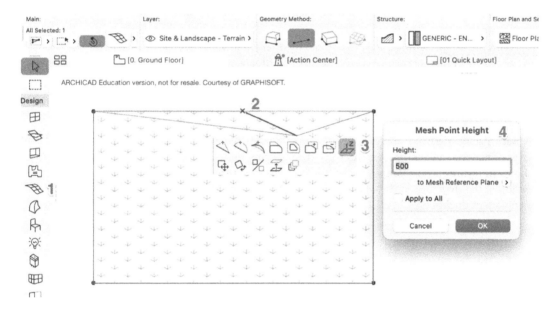

Figure 8.23 – Elevate nodes or edges of a mesh from the pet palette

5. As a result, the edge you clicked on will be moved vertically. This isn't obvious in the 2D window, so switch to a 3D window or any section or elevation to better see the results. When you select the mesh, its reference contour gets highlighted and it becomes easier to see the elevation of nodes.

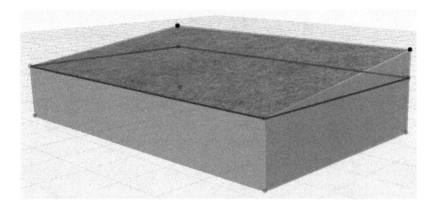

Figure 8.24 – Mesh with adjusted nodes to create a sloped volume

6. With the default settings, the mesh is clearly oriented toward the modeling of terrains, but you are free to assign other materials and surfaces if needed. The tool is deliberately not named the **Terrain** tool.

Pro tip

Beware that the **Mesh** tool does not support Composites, so it only makes sense for a homogeneous layer of material. Off course, we can solve this by using multiple tools for one roof buildup as shown in *Figure 8.25*: a Composite slab for the (flat) structure, a mesh for the (sloping) screed on top of the structure, and roofs for the insulation and waterproof membrane (sloping with a consistent thickness).

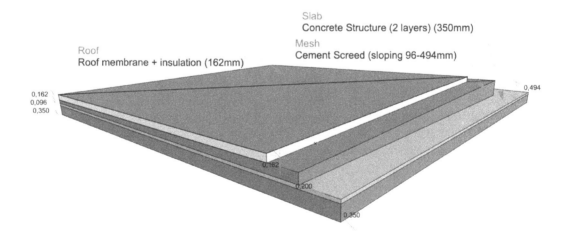

Figure 8.25 – Using three different tools for a realistic model of a flat roof

Complex Multi-plane Roofs (gables, overhang...)

We learned about the various roof types you can create with a Multi-plane roof in *Chapter 4, Building a Basic Residential Model: Adding Roofs, Zones, Beams, and Columns*. However, as with most Archicad geometry, you can further refine roof geometry using the pet palette. There are countless minor and major changes, so we will highlight a few of the most important ones here:

1. Clicking on the *outer roof contours* allows you to manipulate the overhang, with Archicad ensuring that the adjacent overhangs stay connected.

2. Clicking on an *edge of the pivot line* allows you to move it horizontally, with Archicad adjusting the connection to the adjacent roof planes.

3. Clicking on the *ridge*, allows you to move it horizontally or vertically, again with Archicad adjusting the adjacent roof planes.

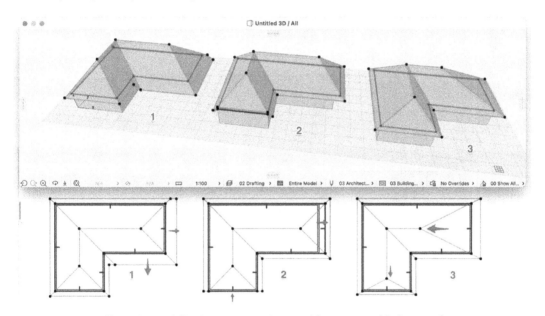

Figure 8.26 – Adjusting contour, pivot, or ridges on a multi-plane roof

4. Moving the ridge edge just above the pivot line turns that plane into a *gable*. As an alternative, you can directly click on one pivot edge and use the last option in the pet palette. This opens **Custom Plane Settings**, which allows you to alter a few settings for the plane linked to the pivot edge you just clicked:

    •   A plane at 90° becomes a gable and thus there will be no roof plane

    •   You can also adjust **Top Surface**, **Bottom Surface**, and **Cover fill** for this plane, but not when you make it a gable, as there is no plane geometry anymore

- You can apply changes to **Clicked Plane**, **Clicked Level** (if there are more levels), or **All Planes** at once

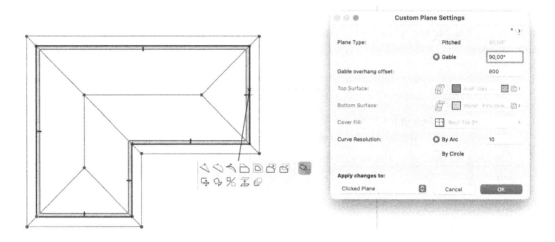

Figure 8.27 – Customising the clicked plane of a Multi-plane roof

5. When you turn the pivot edge or the contour edge into an *arc*, you can even get a rounded plane edge.

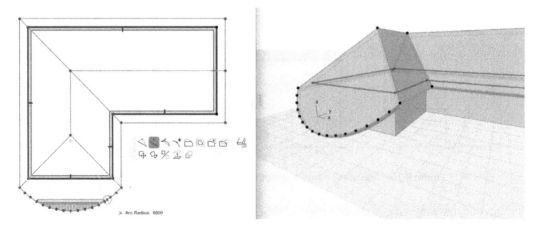

Figure 8.28 – Turning an edge into an arc

Now that we have seen some options for editing Multi-plane roofs, let's see what options the seemingly simple Single-plane Roof has to offer.

Creating complex Single-plane roofs

While there is deep control over a Multi-plane roof, you can fall back on individual, Single-plane roofs in many cases, for ultimate control over each individual plane.

Adjusting contours and pivot lines in a 2D window works very well, but having the ridge lines meet properly is quite hard. And here, Archicad has a nice manipulation method available for you:

1. We start with two Single-plane roofs that need to meet with a clean ridge line. You select one of the roofs first.

2. With the roof selected, you can press *Cmd/Ctrl + LMB* on the roof you wish to adjust by clicking on the ridge line of the second roof. When the cursor becomes an arrow with bold scissors, you are in the right spot. The first roof plane is extended and its ridge line is oriented to fit the second roof plane.

3. To complete this operation, you repeat it for the second roof plane: select it and *Cmd/Ctrl + LMB* on the first roof ridge line to connect to it. The two ridges are now perfectly aligned and you can further refine the other edges.

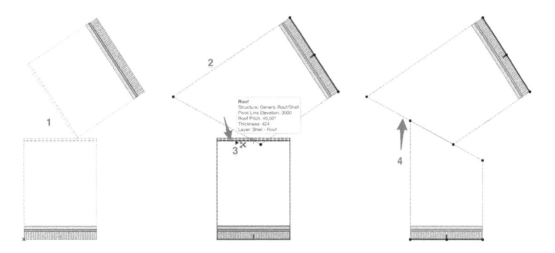

Figure 8.29 – Adjusting the top roof edge for two Single-plane roofs

---

Pro tip

This technique can also be applied in a 3D view but it only works when the two edges that need to have a common, clean ridge line are *not* yet touching before this operation. So sometimes you have to model the roof planes too short or use the pet palette to offset both edges a bit away from their future common ridge location.

Inserting a skylight

Inserting a skylight into a roof is very similar to inserting a window or door into a wall:

1. Activate the **Skylight** tool.

2. Check **Nominal Dimensions** in the **Info Box**. Here, they are set at a width of 600 cm and a height of 1600 cm.

3. Click on the roof where you want to place the skylight. This works best in a floor plan, as you can position it more easily.

    The skylight appears as an opening in the roof and is accessible in a 3D view as well.

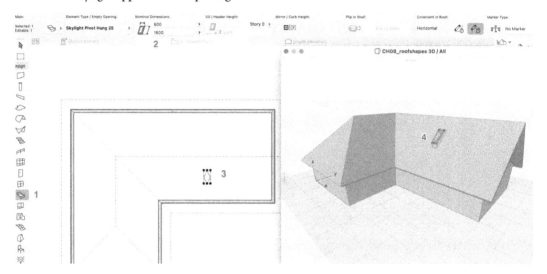

Figure 8.30 – Placing a skylight in a roof

When you open the **Skylight Selection Settings** dialog, you will immediately recognize the configuration of a Library Object:

1. There are a few variants in the library, but we'll stick to the default **Skylight Pivot Hung**.

2. By default, the skylight has **Constraint in Roof** set to **Keep Horizontal Position**. This helps to maintain its position on the floor plan, even when the roof angle is adjusted. The skylight pivots from its lowest edge.

3. In the **SKYLIGHT SETTINGS** panel, you can further adjust **Lining, Edge Angle...**.

4. **Header Angle** defines the direction of the opening at the top side of the skylight and is set as **Horizontal**. **Sill Angle** defines the direction of the opening plane at the bottom side and is set as **Vertical**. You can select **Perpendicular**, **Horizontal**, **Vertical**, or **Other** when you want to define a custom angle in relation to the roof plane angle.

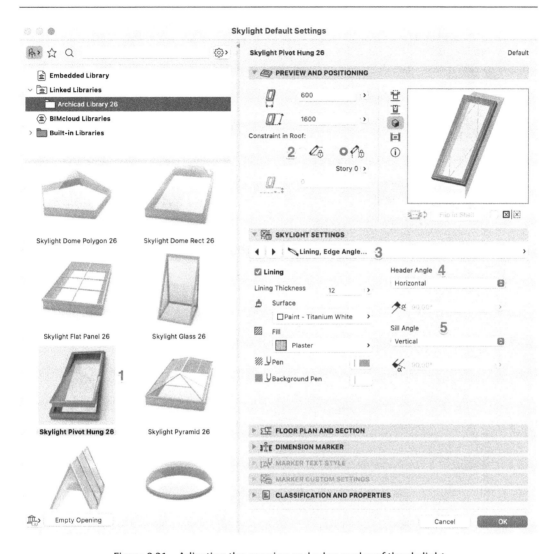

Figure 8.31 – Adjusting the opening and edge angles of the skylight

Notice that, just like with the roof it is placed into, the skylight contour is shown as a projection from the 3D model in the 2D plan view. In *Figure 8.30*, we see a symbolic projection, as the roof is positioned above the currently active Story.

A closer look at trimming walls to roofs

In *Part 1 – Chapter 4,* we already explained how to trim a wall to a roof. We go a little further in the next example.

When you trim a wall against a roof (or a shell), it takes into account the **Trimming Body** of the roof. This is a volume extruding upwards or downwards from the roof, which is used to remove or trim parts of other objects that pass through. For a more controlled way of trimming to a roof, follow these steps:

1. Select the elements you want to trim, such as Walls.

2. Launch the **Trimming** command from the Toolbox or the **Design** > **Connect** > **Trim Elements to Roof/Shell** menu.

3. With the *Trim cursor*, click on the roof plane you want to use to perform the trimming.

4. Click on the side of the roof where you want your elements to remain. In this case, we click below the roof. Archicad shows a preview of the operation in a *bold blue wireframe*.

   The final result is that the Walls are trimmed against the roof.

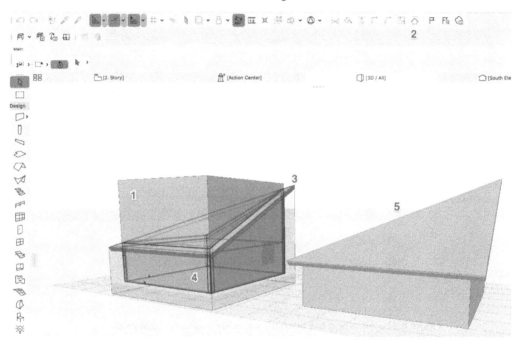

Figure 8.32 – Trimming walls to a roof, with trimming body in red

> **Dynamic operations**
>
> Trimming is a *dynamic* operation. When you modify the Walls or the roof, the operation is recalculated automatically. Try changing the roof slope angle or moving a wall.

You can still clear the trimming operations. When you select the roof, a small **Connect** button appears. If you click on it, you get a context menu giving you an overview of all the elements that have been trimmed by this roof. You can clear operations one by one or even all at once from this menu, with Archicad highlighting the element that will be reset.

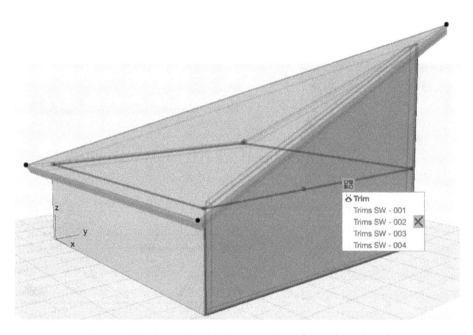

Figure 8.33 – Accessing the trim operations for a selected roof

From the **Roof Selection Settings** dialog, you can also access the **Trimming Body** configuration from the **MODEL** panel. The trimming body can be defined as **Pivot Lines down** or **Contours down**.

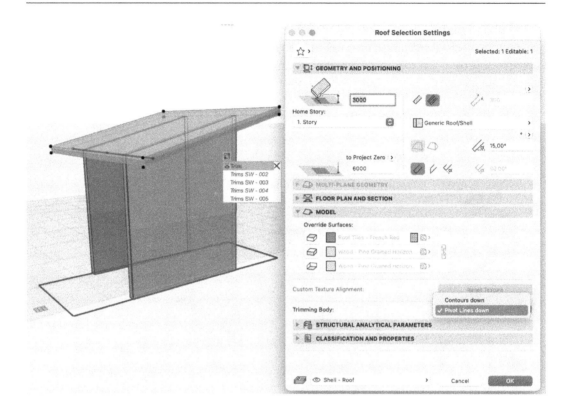

Figure 8.34 – Trimming Body settings

> **Elements (partially) disappearing?**
> Beware that every part of the wall that passes outside of the trimming body is also removed!

Now that you have a deeper understanding of the trimming function, it is time to see some other ways of connecting elements in Archicad, starting with Merge.

Merging elements with roofs, shells, or morphs

We have already discussed the priorities of Building Materials and how they automatically define the connection between Walls and Slabs. However, this does not apply to all possible elements. In case elements overlap with a roof, shell, or morph, priorities can be used to remove these overlaps when you apply a **Merge** connection (**Design** > **Connect** > **Merge Elements**).

In the following example, we created a Morph (**1** in *Figure 8.35*), a Column (**2**), and a Wall (**3**) that overlap with a Roof. After you select all elements and apply the **Merge Elements** connection, you will notice how the Building Material priorities are applied to clean up any overlap. This is one of the methods in Archicad to prevent geometric clashes.

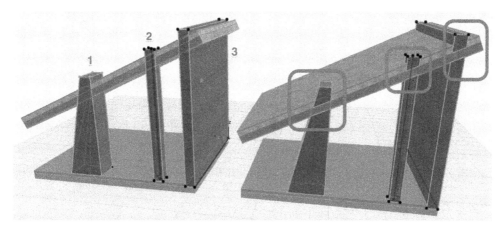

Figure 8.35 – Removing overlaps with the Merge Elements connection

Using Solid Element Operations (SEOs)

In a **Solid Element Operation** (or SEO), we use one object's geometry to influence another one. This is another geometric operation provided by Archicad to ensure elements can be adjusted in various ways.

In *Chapter 5*, we applied *Boolean* operations between morphs to create new morph shapes. SEOs provide similar operations between regular Archicad objects but have a few additional tricks up their sleeves.

When you select an object in Archicad, you can call the different SEOs from the context menu or from the regular menu (**Design** > **Solid Element Operations…**). In an SEO, you assign two sets of objects:

- **Target** is the object that will be impacted and have its shape changed
- **Operator** is the element that impacts the target

And finally, you have to select an appropriate operation:

- **Subtraction** is quite self-explanatory: the volume of one or multiple objects is removed from the volume of the other selected object(s)
- **Subtraction with upwards extrusion** is quite unique: it assumes that you use a subtraction volume consisting of (an) object(s), but expanded or "extruded" vertically so it involves everything above the object(s)

- **Subtraction with downwards extrusion** is the exact opposite: it considers the volume of the object(s) and everything below
- **Intersection** is the Boolean overlap between two selections of objects
- **Addition** is the union of two selections of objects

The combination of a subtraction with an extrusion is a unique modeling feature as far as we know, and this Archicad command comes in handy in many situations.

Example of an SEO: connecting walls to stairs

In the following example, we want to connect Walls against the bottom side of a staircase. We model these elements in overlap, which means that the Walls pass through the staircase. To correct this using SEO, do the following:

1. Select the Walls and open the **Solid Element Operations** palette (**Design** > **Solid Element Operations...**). With the Walls selected, click **Add as Target**: these objects will be modified by the operation.

2. Select the staircase and click **Add as Operator**: this object will be used as the tool to modify the Walls.

3. Select the appropriate operation: **Subtraction with upward extrusion**.

4. Click **Execute** to perform the operation.

   The result: the Walls are cut by the volume of the staircase itself and an invisible volume above the staircase.

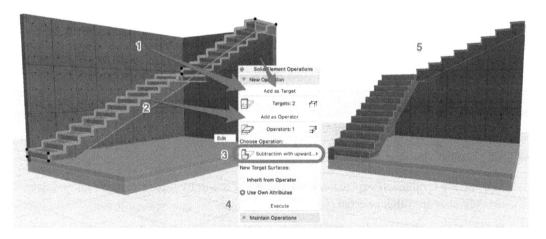

Figure 8.36 – SEO operation between Walls (target) and stairs (operator)

Choosing between an SEO and trimming

Using an SEO has a different impact than using a trimming operation on how elements are combined. We'll redo the example we saw earlier but now we'll use the **Subtraction with upward extrusion** operation, with the roof as **Operator** and the Walls as **Target**. Now the part of the Walls that extends outside of the roof will remain unchanged. Compare *Figure 8.37* to the previous example (*Figure 8.34*) where we used a trimming operation and elements outside the trimming body were cut off (disappeared).

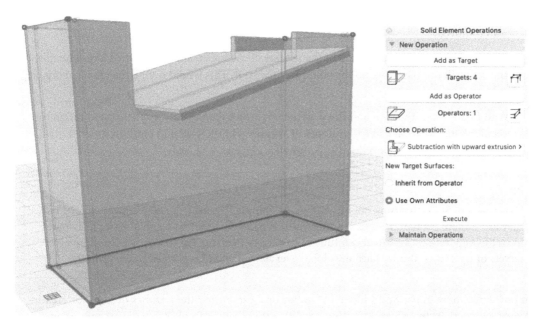

Figure 8.37 – SEO instead of trimming between overlapping roof and Walls

Choosing between an SEO or Morph Boolean

You may remember when we discussed the **Morph** tool (*Custom shapes with the Morph tool* in *Chapter 5)*, that we introduced you to **Boolean Union**, **Subtraction**, and **Intersection**. That looked very similar to Solid Element Operations.

Why do we need similar methods in Archicad? When would you use a Boolean operation on a Morph and when would you fall back on regular Solid Element Operations? Let's see the answers:

- When the shape you act upon needs to remain fully parametric, an SEO is preferable, as it keeps the original objects editable. This is the case when we connect Walls to a sloped roof or when an element needs to be cut by a slope.

- When the final shape is all you need and it does not need further intelligence, a Morph is a more lightweight operation, as Archicad does not have to remember and re-evaluate the generation of the SEO each time. You can do this if the object is completely standalone and the information content is not relevant to the project.

So, to return to our example, when connecting a wall against a roof, the subtraction with upwards extrusion is a very good fit:

- The SEO is *dynamic*, so it is evaluated again when, for example, the roof slope changes or the wall is moved.

- The *upwards extrusion* ensures that the volume of the roof plane is expanded large enough so the wall will not pass beyond the volume of the roof. This would not be the case in a regular subtraction operation.

- And finally, both the roof and the wall *remain fully editable and parametric* and retain their regular behavior: you can insert doors or windows in the wall and you can insert a skylight into the roof and maintain them as parametric objects.

> **Pro tip**
>
> One aspect of SEOs that may not be obvious, initially, is that they can appear everywhere, but they are sometimes hard to retrieve afterward. When you select an element that participates in an SEO operation, Archicad helpfully displays a pop-up button to access a list of involved targets or operators. But we lack any global overview of all SEOs applied in the project.

You now know how to use Solid Element Operations, expanding the number of tools for creating the right connections in your model. This concludes this chapter and has taken your basic modeling skills to a more expert level.

Summary

In this chapter, we learned about a few additional, more advanced methods to create the elements in our building model.

First, we deepened our understanding of composite structures, giving us full control over the different element Skins and their building materials.

After that, we learned how Archicad has a nifty system of priorities linked to materials, which allows us to have elements that join and connect as construction elements should. This brings a lot of power to control building element connections at a fairly detailed level.

Next up, we looked at a few more options to model complex roofs, including how roofs interact with other elements.

In the next chapter, we will look into three "system-based" modeling elements: stairs, railings, and curtain walls.

Using Advanced System Tools for Designing Stairs and Curtain Walls

In this chapter, we will learn how to model stair systems in more detail (including using railings as a second system tool). We will also explore the versatile **Curtain Wall** tool, which is also a hierarchical system tool.

In this chapter, the following topics are covered:

- Modeling complete Stair systems

- Adding railings to our Stairs

- Modeling basic Curtain Walls

Complete stair systems with railings

While we already know from *Chapter 5*, how a basic stair is set up, there still is a lot to learn about this complex architectural element. In this section, we will cover editing the shape, changing the 2D representation, and creating specific stairs as well as adding railings – safety first!

Editing the stair settings

Archicad's **Stair Tool** is quite flexible – you can use it for a plain, straight stair, but it can contain multiple flights and landings and be controlled in depth.

In *Chapter 5*, we already explained the importance of the stair baseline and the contour. However, if you select a stair, you have two ways to control it – via its settings and graphically.

Using the **Stair Settings** dialog, you get an elaborate dialog, displaying the stair as a hierarchical system, which you can control top-down:

- At the highest level, you control the main **Stair** setup, including **Stair Height**, **Width**, **Risers**, **Goings**, and **Rules and Standards**. We already used these when we created the basic straight stair in *Chapter 5* .

- The **Structure** branch gives you access to the **Flight** structure and the **Landing** structure. You can use the dialog to pick the main structures and further refine them in the branch underneath, which is updated depending on the structure you pick.

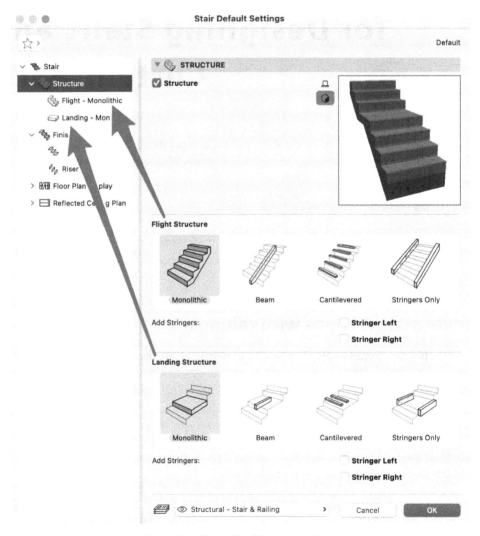

Figure 9.1 – Stair – the Structure settings

- Similarly, you can control **Finish** in the next branch, where you can toggle the finishes for **Tread**, **Riser**, or both. You can also disable the **Finish** sub-element altogether. Let's see what **Tread** and **Riser** do:

    - The **Tread** branch controls a parametric **Geometric Description Language** (**GDL**) object, which has its own parameters for thickness and nosing, including dimensions and used building materials. It even has its own **CLASSIFICATION AND PROPERTIES** section for when you need to fully specify the treads in the model information.

    - The **Riser** branch gives you similar control.

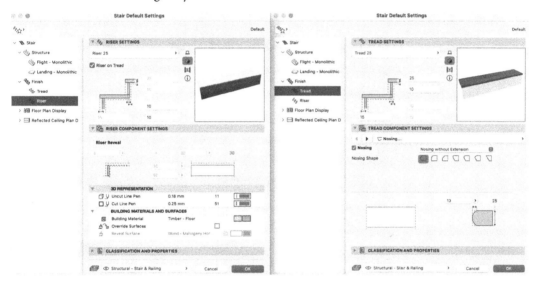

Figure 9.2 – Tread and Riser settings

- **Floor Plan Display** gives you detailed control over the 2D representation – **Grid and Structure**, **Break Mark**, **Walking Line**, **Numbering**, **Up / Down Text**, **Description**, and **Tread Accessories**.

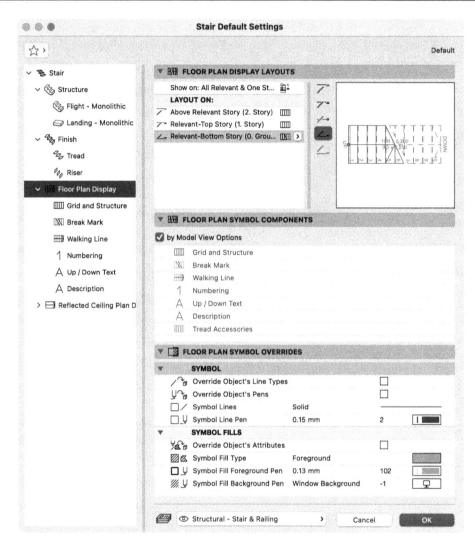

Figure 9.3 – Stair – Floor Plan Display

However, since there are too many things to control, Archicad suggests you rely on **by Model View Options** (**MVO**). This implies that you can control the display of multiple stairs globally and ensure they all look coherent, which improves the professional display of your drawings.

To control these settings, close the **Stair Default Settings** dialog by clicking **OK**, and open the **Model View Options** dialog (**Document** > **Model View** > **Model View Options...**). On the left side (**1** in *Figure 9.4*), you can select one of the option combinations, while on the right side, you control different sets of options. Here, we will only look at **STAIR OPTIONS** (**2**). You can toggle the various items to display or hide in the floor plan (**3**). We will return to **Model View Options** in *Chapter 14*, in the *Understanding quick options/view filters* section.

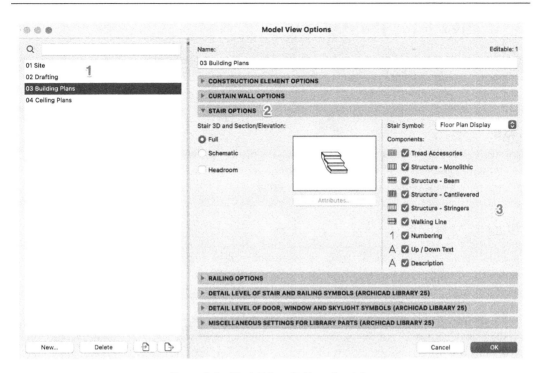

Figure 9.4 – Model View Options for stairs

The next example illustrates how different stairs receive the same representation style, controlled by the MVO stair options.

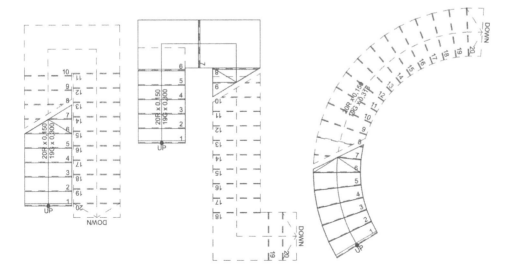

Figure 9.5 – A coherent floor plan display of various stair shapes

Graphically editing the stair

As a stair is a complex geometrical shape, you should not be surprised by the fact that the graphical representation of this element has lots of options too. This section will give you insight into this editing approach while trying not to let get lost in the various options and combinations.

> **Pro tip**
>
> We strongly advise you to manage stairs with a top-down approach – start from the baseline to define the global layout, adjust the settings to define the structure, finish, and floor plan display, and finally, refine the layout using the graphical editing methods.

Editing the baseline

For the first set of editing methods, we can rely on our well-known Pet Palette editing techniques. For stairs, this contains options to adjust the baseline, as with any linear element (inserting points or making an arc).

In the following example, we click on the baseline (**1**) and select the last option in the top row (**2**) to access the **Select Segment Type** dialog, where we can turn a **Flight** segment into a **Landing** segment, or even make it a winder segment, with equal angles (**Winder with Equal Angles**) or equal goings (**Winder with Equal Goings**).

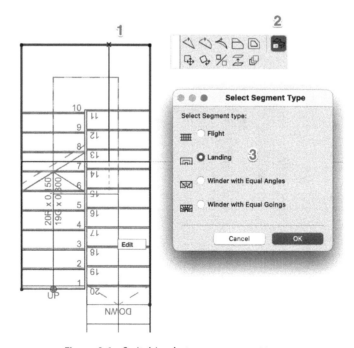

Figure 9.6 – Switching between segment types

Editing stair symbol and treads

A selected stair also shows an **Edit...** button, which gives you access to all the graphically defined items – the baseline, control of the contour (which defines flights and landings), the distribution and shape of treads, and the position of railings.

In the top corner of the view, you can switch between the **Schematic** and **Symbol** display of the stair, giving access to different parts of the stair:

- In the **Symbol** display, you can access the break mark, the walking line, and the up/down text.

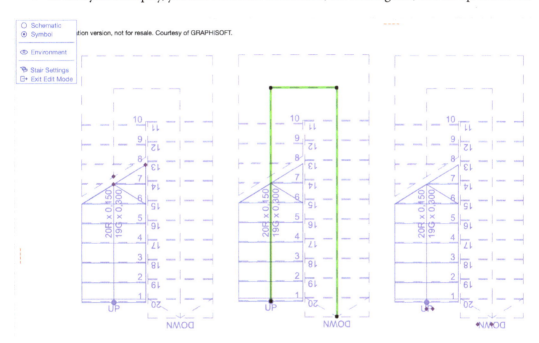

Figure 9.7 – Editing the symbol (Break Mark, Walking Line, and Up/Down Text)

- In the **Schematic** display, you can access the treads. Each tread is individually editable, and when you select a tread in this editing mode, you are greeted (again) by a familiar Pet Palette. It looks a lot like the Slab or Fill Pet Palette, so you will already know how to edit the contour of a tread.

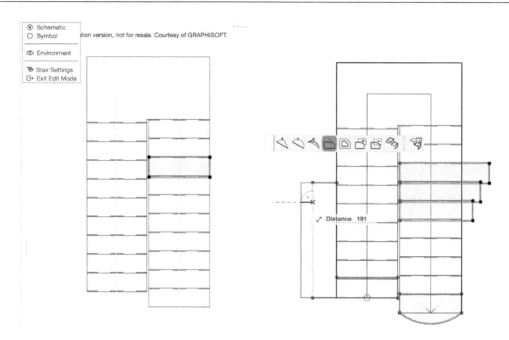

Figure 9.8 – Editing the Schematic of the Stair (adjusting the contour of Treads)

Beware that when we view the stair in 3D, the preceding edits only affect the **Tread** finishes. To adjust the main stair structure, we rely on editing the baseline.

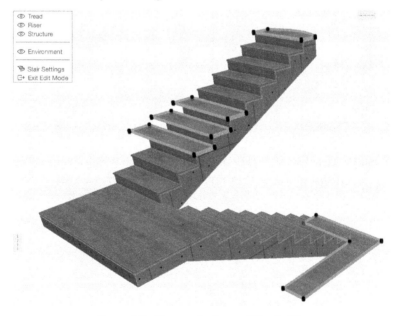

Figure 9.9 – The result of our editing in 3D

You can also select individual treads (in 2D or 3D) and access their settings (*Cmd/Ctrl + T*), which allows you to change the style and materials of individual treads.

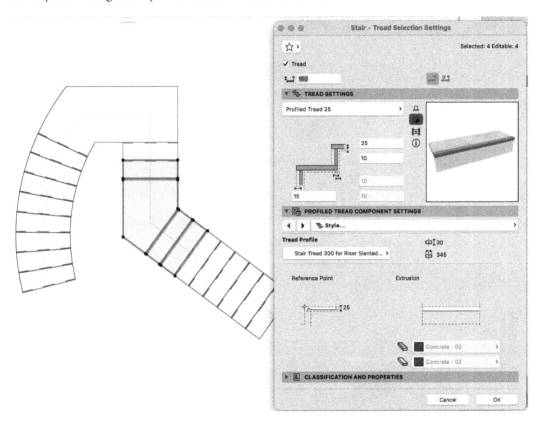

Figure 9.10 – The editing settings for individual treads

After you make such changes, the main **Stair Selection Settings** dialog warns you that some treads are customized and gives you the **Reset Custom Treads** option.

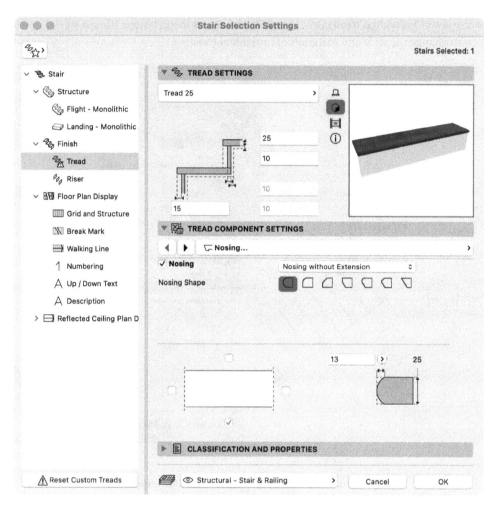

Figure 9.11 – The Reset Custom Treads button

This kind of warning is common in Archicad for elements and objects in which you have created custom settings for a part or a sub-element. Examples include the Edge of a Roof, parts of a Curtain Wall, or, as shown in *Figure 9.11*, sub-elements of Stairs.

A multi-story display of stairs

While most buildings are fairly straightforward when you think in terms of building stories, stairs are used to move between those stories. In a traditional drawing, there is a convention that a plan view is a vertical projection from a horizontal cut through the building at eye level. For stairs, this leads to a series of representation conventions, which can be replicated in Archicad with the **multi-story display**.

You place the stair on the story of its lowest tread, where the stair starts to rise. In the story above, this stair becomes visible as well. You can switch between different **Floor Plan Display** configurations in the **Stair Settings** dialog. Normally, objects are only visible on their own story, but this can be controlled from the stair settings, where **Show on: All Relevant & One Story Up** is chosen.

Modeling multiple stacked stairs

In a stairwell, you often have multiple repeated stairs stacked one above the other. As a single stair is not intended to move across stories, you are advised to work with copied and adjusted stairs, each passing from their Home story to the story above. Via the multi-story display, you can control how they appear in a floor plan.

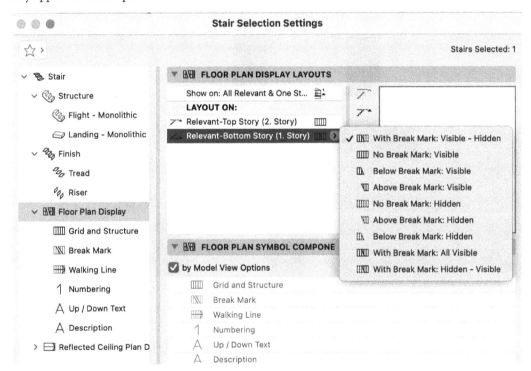

Figure 9.12 – Controlling the display of stair flights above and below the break mark

There are three common situations:

- When a stair has no stair underneath, the part above the break mark is shown with a hidden line type (dashed) – **With Break Mark: Visible - Hidden**

- When a stair is placed above another stair (on the story below), you hide the part above the break mark for the first stair and set the stair on the story below to only display the part above the break mark – **Above Break Mark: Visible**

- When there is only the stair from the story below, it is displayed with no break mark at all – **No Break Mark: Visible**

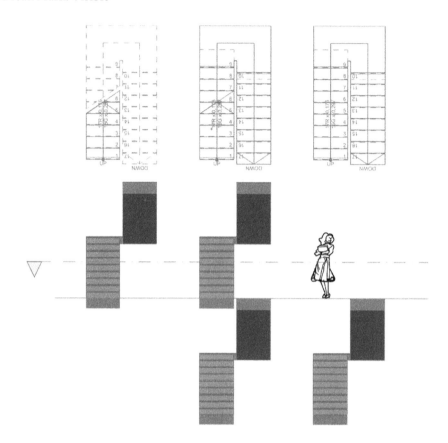

Figure 9.13 – Different representations of stairs in the floor plan (top) and in a section (bottom)

Figure 9.13 shows three examples – break mark and hidden (left), break mark and visible (middle), and no break mark (right). With these three situations, you are quite flexible and don't have to rely on 2D drawing techniques to mask parts of stairs or drawing auxiliary 2D lines to fill in parts that are not displayed.

Creating round and spiral stairs

Two special types of stairs are the round stair and the spiral staircase. It may not be immediately obvious how you would model those, but there are a few reliable methods available in the Pet Palette.

For a *round* stair, you can switch to the **Arc by centerpoint** option in the Pet Palette when defining the stair segment. The arc is still editable afterward if you need to refine it or add more segments. You can also start from a straight segment and turn it into an arc, just as you did for walls, beams, or slab editing.

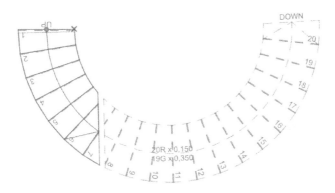

Figure 9.14 – A stair using an arc segment

To make a *spiral* stair, the **Arc by centerpoint** option (last in the bottom row of the Pet Palette) can be used to set the center point at the same distance as the stair width (e.g., 1000 mm) and then complete the arc. Do this by first hovering above the starting line so that the stair is going clockwise at the inside of the baseline, and then hovering further in a clockwise direction to indicate the length of the segment arc. Use the preview to get it right so that the stair becomes the smallest possible spiral stair.

Figure 9.15 – The spiral stair at the inside of the baseline arc segment

As an alternative, you may want to draw a circle first so that you have a guideline when drawing.

The resulting stair may still need further adjustments, either in its structure or by using graphical editing.

Adding railings

While railings initially were only available inside the library or within the older stair system, we now have a dedicated **railing** tool, which is a system tool, just like the stairs and the curtain walls inside Archicad.

Railings can be placed and linked to stairs, but they can also be used as separate horizontal items, which can be linked to the edge of a slab. As independent objects, they can even represent other objects, such as fences.

Adding railings to stairs (and slabs)

Activate the **Railing** command (**1**), and check that it is set as **Associative** (**2**) so that it maintains a link to the stair to which it gets attached. You can now use familiar placement methods to select the edges of stairs (**3**). You are free to select the edges you need, and you don't even have to select the full stair edge (**4** and **5**). And, as expected, the railings also have a 3D representation.

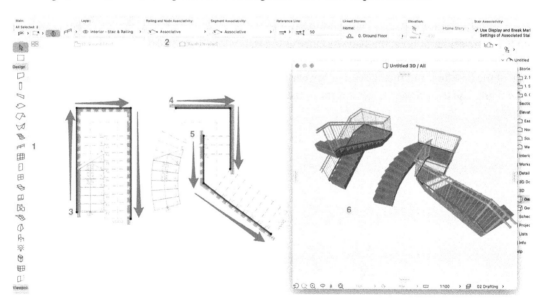

Figure 9.16 – Adding associative railings to stairs

Since you associated the railing with the stair, it will follow changes in the stair baseline. You can even adjust the railing baseline separately, but don't overdo it...

The exact same method is used to associate railings with slabs, so you can ensure that the railings for stairs and slabs use the same structure. This gives you more coherent control than having the railing controlled somewhere in the settings of a stair and requiring a separate tool for horizontal railings, which was the case in older Archicad releases.

Railing settings

When you select the Railing Tool and open its **Railing Selection Settings**, you get a familiar system tool dialog. Railings, just like stairs, are defined hierarchically:

- At the top level, you have the global settings of the railing, including its **Home Story** and the *associativity* to nodes and segments. We prefer to keep everything associative, so the railings follow the design changes for stairs and slabs.

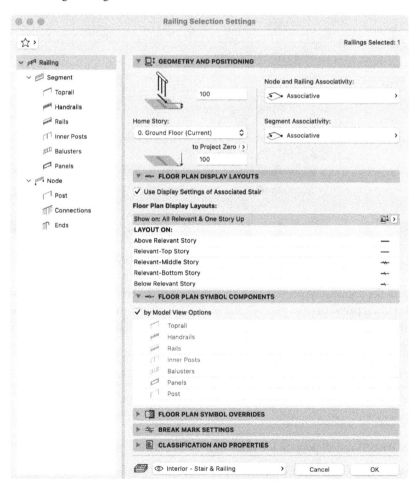

Figure 9.17 – The Railing Selection Settings dialog

- The level below gives access to the **Segment** and **Node** branches. They allow you to control in great depth the different parts of a railing. You can also navigate to the correct branch by clicking the part in the preview at the top right of the **Railing Selection Settings** dialog.

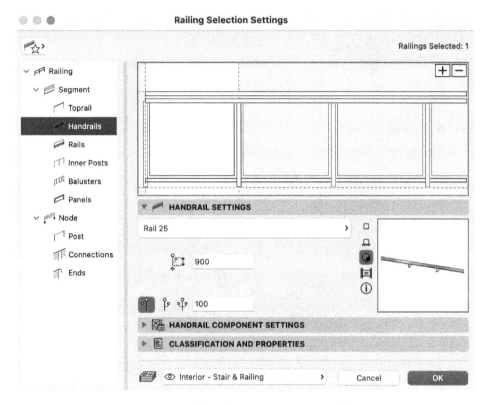

Figure 9.18 – The railing segment pattern editor

- For each of the segment types, you get a pattern preview where you can add one or sometimes multiple railing parts. Use the plus (+) and minus (-) icons at the top right to add segments to one of these types. Each of the segments can be positioned and then configured in its shape and size:

  - **Toprail**: Zero or one rail at the top side of the railing

  - **Handrails**: Zero, one, or two rails at one or both sides of the railing

  - **Rails**: Zero or multiple rails

  - **Inner Posts**: Vertical posts at the nodes

  - **Balusters**: Pattern of vertical balusters per segment

  - **Panels**: Plain panels per segment

Here are a few examples of railings and how they can be configured to become very different objects, shown in *Figure 9.19*:

- Railing number **1** has a **Toprail**, no **Handrail** or **Rail**, **Inner Posts** (at 1000 mm, but without a **Profile**), no **Balusters**, and **Panels** (with a **Top Offset** of 25 mm and a **Bottom Offset** of 0 mm, with a Fixing).

- The second Railing has a **Toprail** (with a Vertical Fixing), **Handrail** (on the left side, at 900 mm), **Rails** (at 700 mm), **Inner Posts** (at 1000 mm, with a 100 mm **Top Offset**), and **Panels** (with a **Top Offset** of 100 mm and a **Bottom Offset** of 45 mm, with a Fixing).

- At number **3,** we see a **Toprail**, **Handrail** (at 900 mm), **Rails** (at 70 mm), **Posts**, and **Balusters** every 100 mm.

- Number **4** is a Railing with two **Rails** (at 150 mm and 850 mm), **Inner Posts** every 1500 mm, and **Balusters** (with a 50 mm **Vertical Offset** above and below).

- The last example at number **5** has a **Toprail**, **Handrail**, **Rails** (on the bottom side), a single **Post** every 1000 mm, and **Balusters** every 100 mm.

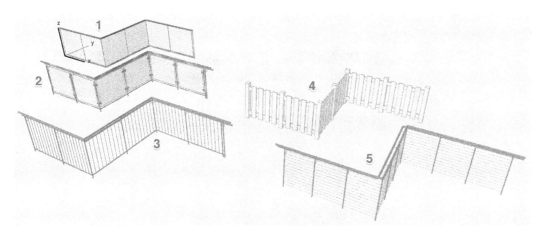

Figure 9.19 – Five differently configured railing styles

Just like with the stairs, you can work from the settings and the geometry, with an **Edit...** button to dive into fine-tuning the geometric components.

The Railing Tool is a very deep tool. It also introduces a large series of new object categories, which indicates how significant this is in the Archicad virtual building model structure.

The basics of modeling curtain walls

Curtain walls are assemblies of panels and members, which are often used for the façade of office buildings. Functionally, they behave like walls, as they define an enclosure between the interior and the exterior environment, or between spaces. They are not really windows or doors, but they can contain opening doors for some of their panels.

In Archicad, curtain wall is (again) a systems tool, with a hierarchical structure. It has a lot of settings, on many levels. With this section, we aim to give you a good insight into how the tool works and how you can create basic curtain walls with it.

Creating a standalone curtain wall

When you select the **curtain wall** tool, you can model it starting from its reference line, which can contain multiple segments or even be curved (depending on the geometry method you choose). You do not need a wall to place this joinery element into. Let's go through the steps to create a single curtain wall:

1. Activate the **curtain wall** tool from the Toolbox or through **Design** > **Architectural Tools** > **Curtain Wall**.

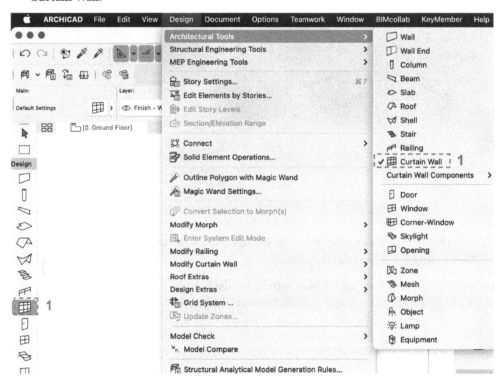

Figure 9.20 – Activating the Curtain Wall tool

2. With **Geometry Method** set to **Single** (**a**), click two nodes to create a curtain wall by drawing its reference line (**b**). As always, you can use the Tracker for numerical input (**c**) or Snap Guides if you prefer working graphically.

Figure 9.21 – Modeling a curtain wall in Archicad works largely the same way as modeling a Wall

3. The placement method resembles what we do when modeling a wall – for example, to position the curtain wall members on the other side of the reference line, you can use **Flip**.

4. Select your curtain wall and admire it in 3D (*RMB* > **Show Selection in 3D** or select the element + *F4/F5*). The different components of this hierarchical element (panels, members, and so on) are clearly distinguishable. Let's see how this is set up:

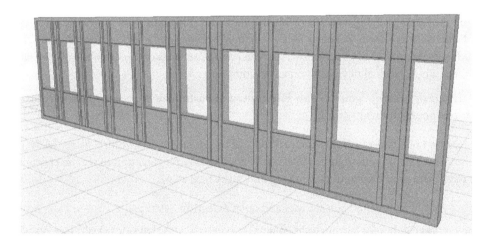

Figure 9.22 – The result in 3D, showing the different curtain wall parts

The created element consists of panels (glass or another material), placed within a Scheme, with frames as separators.

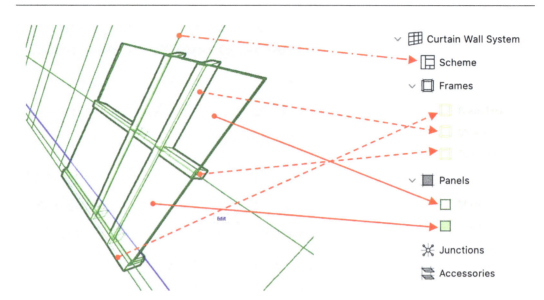

Figure 9.23 – The main hierarchy of curtain walls as it is shown in the
Settings dialog, amended with sub-components (green)

When you select a curtain wall, it can be further refined in two ways:

- From the **Settings** dialog, you can define the hierarchical structure and any of the parts shown in *Figure 9.23* top-down. All parts follow a defined Scheme, and every part within the Scheme follows its definition. In other words, you can edit the settings of the curtain wall as a whole, without any individual deviations on a sub-component level.

- In the 2D or 3D view, you have an **Edit** button to customize the geometry and adjust any sub-component individually.

Let's explore how both ways of editing can work. To edit the curtain wall definition using the **Settings** dialog, follow these steps (you might want to confirm with **OK** after each step to see the effect of the edits you made, and then reopen the settings again for the next step):

1. Select your newly created curtain wall and open **Selection Settings**.

2. The settings window for curtain walls contains two parts. On the left side (**a**), all elements within the hierarchy of a curtain wall can be selected. On the right side (**b**), the chosen sub-component can be configured. The menus on the right side differ for each system component. In the first section, **Curtain Wall System** (**c**), the global settings of the curtain wall are set up. As with any basic modeling tool in Archicad, the **GEOMETRY AND POSITIONING, FLOOR PLAN AND SECTION, CLASSIFICATION AND PROPERTIES**, and layer settings are available here. In the **MEMBER PLACEMENT** tab (**d**), it is specified how and where the components of the system will be placed.

3. In the **GEOMETRY AND POSITIONING** tab, the **Nominal Thickness** (**a**) and **Home Story** (**b**) settings of the curtain wall are indicated. The nominal thickness is the space occupied by the curtain wall (the boundary where other elements must stop or connect). You can position the **Reference Line** (**c**) setting relative to some characteristic surfaces of (parts of) the curtain wall. Since frame parts are also defined relative to the panel center, it is best to keep this setting as it is. Change the total curtain wall height (**d**) setting to 4500 mm and the slant angle (**e**) setting to 60,00° (the Sun symbol indicates the outside; thus, our curtain wall will be leaning inward).

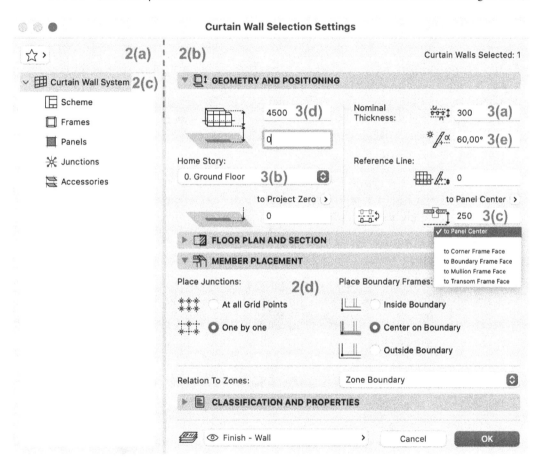

Figure 9.24 – The general settings for the curtain wall system

Curtain Walls and Stories

Note that, although a curtain wall has a **Home Story**, you do not have the option to link this element to another story (e.g., at the top with a certain offset). The way we want or don't want a curtain wall to stretch in an architectural design is somewhat different from a "normal" wall. Maybe you want every element to stretch when the story height is larger, or maybe you want to add extra elements within the pattern. As there are so many options, it is better to leave this kind of design choice to the user, instead of letting your software offer you a proposal that has more chance to be wrong than right. In general, a curtain wall is also one of the few exceptions to the general convention of modeling elements *per story*.

4. The Scheme in *Figure 9.26* can be seen as a pattern grid, drawn at the position of the glass line and intended to set up a curtain wall (as a base); it remains invisible on the drawings. With the + and - buttons (**4a**), columns (at the top) and rows (at the bottom) can be added or removed. These can also be moved with the double arrows (**4b**). The pattern can be cleared with the **X** button (**4c**) in the top-right corner of the preview. With both columns and rows, you can choose how they should be divided (**4d**). There are three options:

    - **Fixed Sizes**: The exact sizes of the grid cells are defined. Panels at the ends of the curtain wall will be cut off and filled in with custom panels, depending on various options sub-settings, such as **Pattern Origin** and **Infill with**, which only appear when you choose this type of division.

    - **Best Division**: You can set sizes for each grid cell and mark (**4e**) which of those are fixed sizes. For both rows and columns, at least one item has to remain *unlocked*.

    - **Number of Divisions**: With this setting, you set the number of repetitions of the defined pattern. You need at least one size of the columns or rows to be flexible (*unlocked*).

    Whatever the choice of division type, you can always determine what the last column/row in the repetition should be (**4f**).

    To fully understand all of these options and possible combinations, refer to the *Archicad Help Center*: https://help.graphisoft.com/AC/26/int/index.htm#t=_AC26_ Help%2F001_ACHelpIntro%2F001_ACHelpIntro-1.htm

5. The preview in *Figure 9.26* to the right shows the resulting pattern proportionally, as well as the types of panels and frames. By clicking on a panel/frame, the panel/frame class can be changed (**5a**). Note that clicking a panel in the preview also highlights the corresponding Column and Row (and vice versa). Panel/frame classes are defined in the panel/frame components (**5b**) and accessible through the navigation on the left or by using the shortcut button.

Select the **A1** and **A3** panels in the preview and set both to **Main Panel**, as shown in *Figure 9.25*.

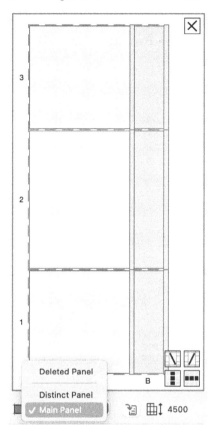

Figure 9.25 – You can select multiple panels with Shift + click

6. In the **PATTERN ORIGIN** tab, it is indicated how the pattern and, thus, the frames will be positioned relative to the reference line. The starting point can be set both horizontally and vertically at different points. It is also possible to indicate numerically where the pattern starts in relation to the point of origin. Set up the pattern as shown in *Figure 9.26*.

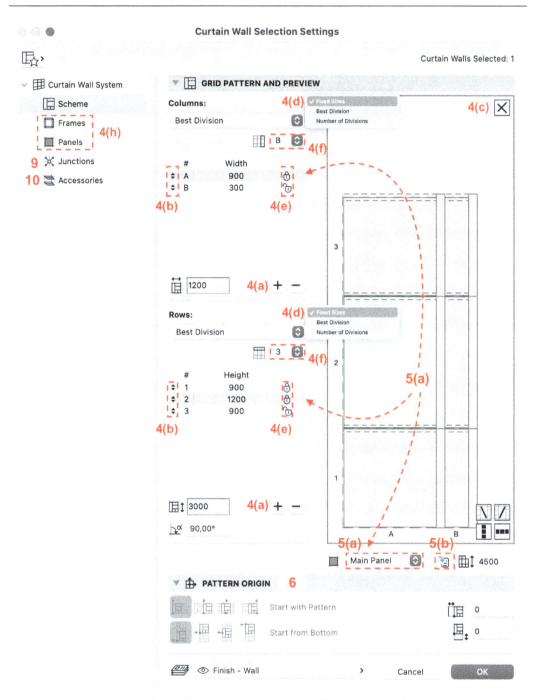

Figure 9.26 – The grid pattern settings and interactive preview

7. In the **Frames** settings, the type and dimensions of the frame classes are set. A class is a type of frame – for example, a Mullion, Transom, or Boundary frame. Here is a detailed overview of the frame settings:

*   Sizes (**7a**), Materials (**7b**), and **Frame Intersection Priority** (**7c**) are each set separately; however, the settings windows are similar for every class (depending on what frame is chosen from the library).

*   If all profiles happen to be identical, you can copy settings by clicking the frame class name with the RMB and choosing **Copy...** on the source and **Paste ....** on the target (**7d**). The profiles are not linked and can, therefore, be changed individually.

*   The profile type can be selected from the library at the top in the **Frame Class Name** and **Type** window (**7e**), or on the **Mullion/Transom/Boundary/Corner Frame Type and Geometry** tab (**7f**). Depending on the selection, the parameter list and the preview window change.

*   You have the option to select **Profiled Frames**, for which a parametric profiles attribute is used. This attribute is covered in *Chapter 11*, in the *Creating parametric complex profiles and using geometry modifiers* section.

*   In the parameter window, the profile height (**7g**), profile width (**7h**), and position of the glass line (**7i**) are indicated. Then, the depth (**7j**) and width (**7k**) of the panel inset are determined. **Frame Intersection Priority** (**7c**) gives a priority to styles. This setting causes styles with a higher priority to pass through those with a lower priority. Note that there is again a (smaller) preview in the top-right corner, showing what part of the Scheme you are editing. Although it is a complicated tool, Archicad gives you all the information you need to keep track of what you are doing.

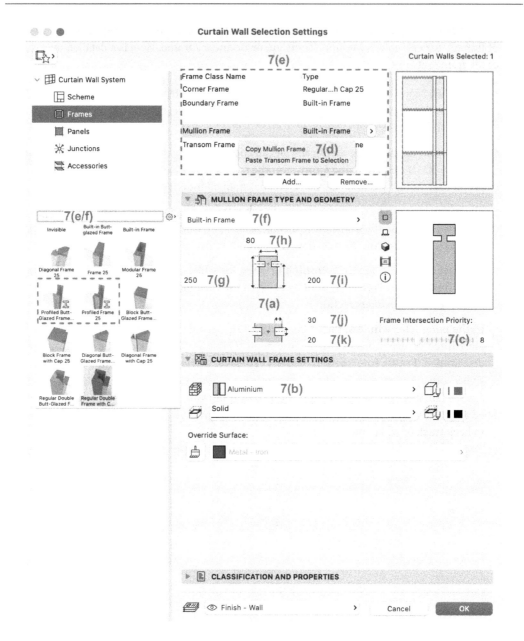

Figure 9.27 – Typical frame settings

- Change the Mullion settings according to *Figure 9.27* (changing a few dimensions) and copy and paste this definition to the Boundary class (*Figure 9.28*).

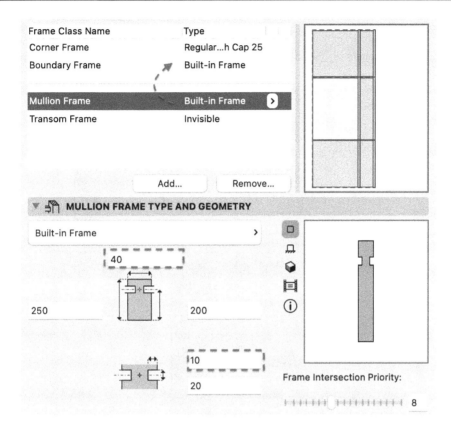

Figure 9.28 – Creating slender Mullion and Boundary frames

- Choose **Invisible** as the type for the Transom class – this mimicks panels glued or siliconed together. Set the depth of panel insert to 0 so that no gaps appear in the result, as shown in *Figure 9.29*.

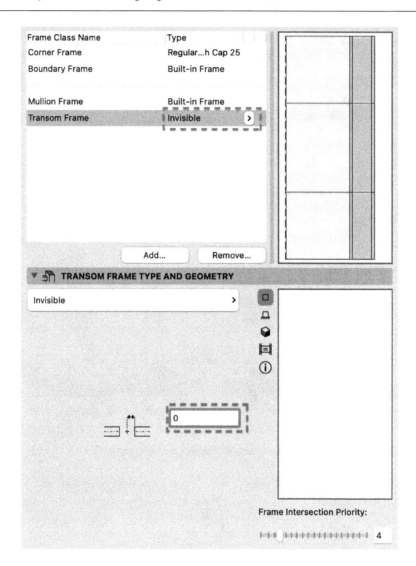

Figure 9.29 – Creating an invisible joint between panels

8. The panel settings are very similar to the frame settings. By default, you have two panels – **Main Panel**, set as **Glass**, and **Distinct Panel**, with the same default material as the frames – **Aluminium**. You can create more panels with **Add...**, or remove panels that you don't need. Any of the defined panel classes can then be applied in the Scheme.

Figure 9.30 – Panel and frame settings are very similar

9. Curtain walls without Profiles use Junctions, connecting glazed elements to a standalone structure. If in the Scheme on the **MEMBER PLACEMENT** tab, **At all Grid Points** is selected for placing Junctions, they are placed immediately and confirmed by clicking **OK**. If **One by one** is selected, the junctions must be placed manually afterward. The connections are set via the parameters of the Junction Object in that part of the **Curtain Wall Settings**. Junctions are normally combined with all frame classes set as invisible.

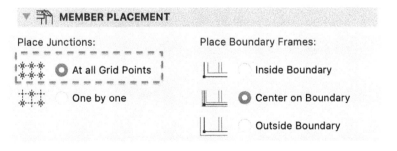

Figure 9.31 – Placing junctions at all grid points to model them all at once

10. Accessories such as sun blinds are not positioned automatically. These can be added to the Curtain Wall in Edit mode, discussed next.

11. Confirm your changes with **OK** and compare your result to *Figure 9.32*.

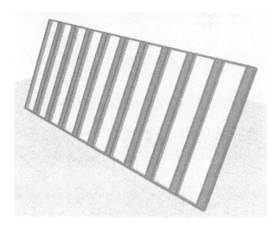

Figure 9.32 – Editing through the settings changes the curtain wall as a whole

As explained before, we can also edit the individual elements of a curtain wall. To do this, we will use Edit mode (which is similar to what we learned how to do for stairs and railings):

1. Select the result from *Figure 9.33* and click the small **Edit** icon that appears. We are using the 3D viewport for convenience, but options are also available in 2D views.

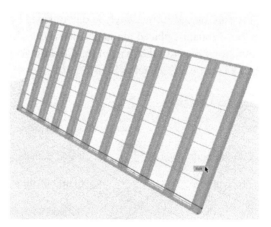

Figure 9.33 – Sometimes you have to look a bit to find the Edit icon...

2. Select one panel in the row highlighted in *Figure 9.34*, and click the small icon to the right so that the whole row is selected.

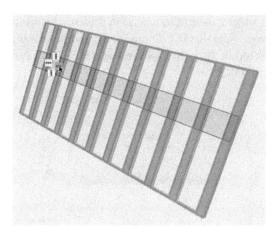

Figure 9.34 – More small icons let us easily select certain groups of components

3. Change the panel class of the whole row to **Distinct Panel** in the info box:

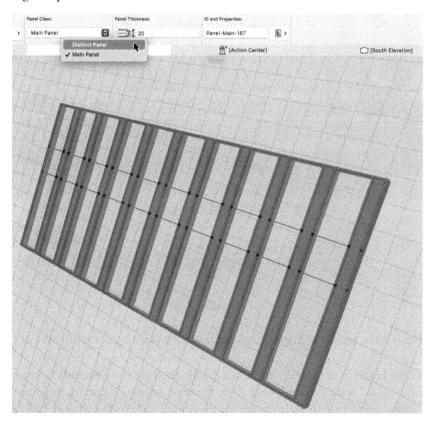

Figure 9.35 – The info box now works on a component level

4. Select only half of the wider panels (alternately) in this row, by using *Shift* + click, and open **Panel Selection Settings**. Note that in Edit mode, any component is a tool on its own. In the settings, choose **CW Window 25** as the **Custom Panel** type (**a**) and set **Opening Type** to **Top Hung** (**b**). Confirm with **OK**.

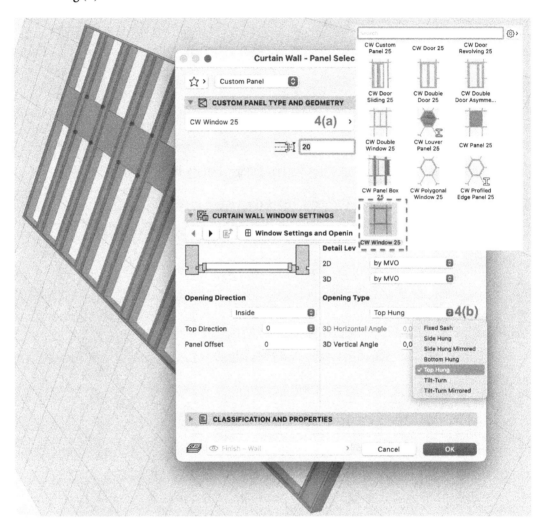

Figure 9.36 – You can add curtain wall Windows (and doors) to your curtain wall

5. Activate the Frame Tool (**a**) from the (curtain wall) Toolbox. Show **Scheme Grid** by clicking the eye cursor (**b**) in the top-left corner. Enter the first node by clicking the grid line endpoint (**c**), as shown in *Figure 9.37*. Finish by adding the second node on the other end of the curtain wall. Note the settings in the Info Box – we used a **Mullion Frame** class.

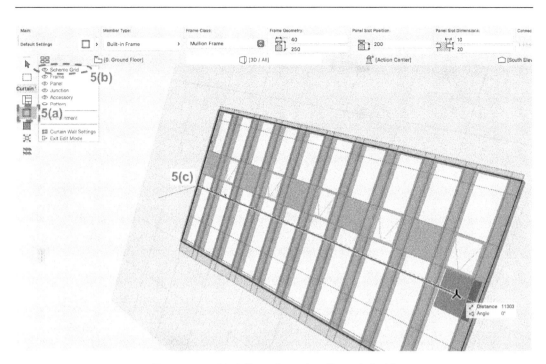

Figure 9.37 – Modeling a frame in edit mode in 3D

6. Activate the Accessory Tool (**a**) and model a sun-shading overhang (the only accessory available in the template we use) over the whole length of the frame we created in *step 5*. Enter the first node (**b**) on the left side and continue to the right. Note that you can only place complete pieces with a length equal to a full column width in our Scheme (in contrast to modeling frames).

7. After entering the second and final node, you are asked to choose a side (inside or outside) with the eye cursor, the direction shown by two light lines pointing one way or the other – click when pointing outward! Select all the newly modeled accessories (*Cmd + A/Ctrl + A*) and open **Accessory Selection Settings** to change the vertical rotation to the slant angle (**a**) we set in *step 3* of the first editing option (through **Settings**). Confirm with **OK** (**b**) and click **Exit Edit Mode** (**c**) to check the final result.

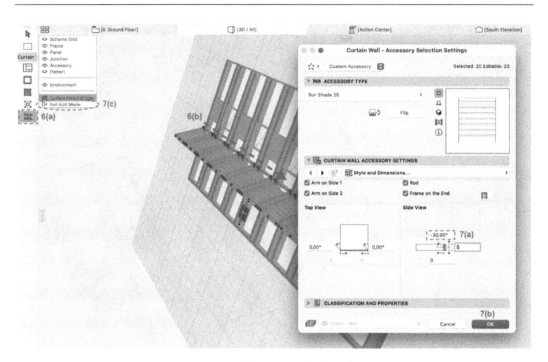

Figure 9.38 – Adding a sun-shading device

Although we have already explored a wide range of editing options, there still are more! You can create a completely new pattern through Edit mode, for instance. This and all other remaining options can be explored on your own, once you know the basics well enough. As always, the Graphisoft Help Center is a good place to start further exploration.

Integrating a Curtain Wall into a façade

While you may define the complete façade of a building using curtain walls, in many cases, it needs to be connected and integrated with adjacent walls.

A good approach to creating a curtain wall that fits into an already modeled façade is by using the Boundary Geometry Method. This allows you to graphically define the contour in which the facade pattern will be replicated, and works as follows:

1. Activate the **Curtain Wall** tool.
2. Select **Boundary Geometry Method**.
3. Ensure that **Plane Input Method** is set to **Manual**, as this will give you the freedom to define your working plan as you see fit.

4. When asked to define the **Working Plane**, pick three points, such as the corners of the adjacent walls (**4a**, **4b**, and **4c**). This relocates the blue working plane vertically, in the plane of the façade.

5. Now, you can draw the boundary edge of the curtain wall using the walls as reference points.

6. The resulting curtain wall can still be edited as usual.

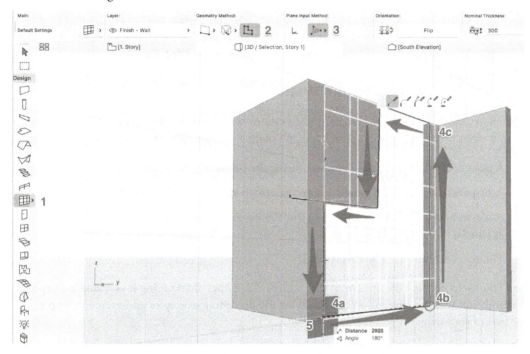

Figure 9.39 – Using the Boundary method in 3D for tailor-made Curtain Walls

A second method to integrate walls with curtain walls is by making a **join connection** between a Curtain Wall and a regular Wall (**Design** > **Connect** > **Join Wall(s) to Curtain Wall…**). This creates a dynamic link between a series of selected walls and one curtain wall of your choice, based on their reference lines.

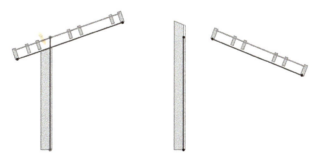

Figure 9.40 – A join between a curtain wall and a wall

> **Note**
> Beware that the total curtain wall thickness defines the cutting of the wall. Also, note that the wall may be extended beyond its endpoint to meet the cutting plane. This join method is especially useful when joining slanted elements.

To finish off such a join, you may need to graphically edit the curtain wall to ensure that profiles are positioned at a suitable intersection point. As with other connections, you can still remove the connection afterward from the pop-up button displayed when you select the Curtain Wall.

Creative uses of the Curtain Wall

As with most tools in Archicad, you can use and get creative with them however you see fit. When looking around at other users and projects, a variety of alternative uses for curtain wall can be encountered:

- A glass roof with panels and members (using the Boundary method)
- A suspended ceiling with panels and profiles (with specific profiles and custom panels)
- A plasterboard room separation wall (for the internal structure, with or without insulation as panels)
- Timber construction (wood skeletons)

When you focus on how the system of this tool works, rather than seeing it as a single-purpose command, you will, for sure, be able to use it for the preceding examples and even come up with some uses of your own!

Summary

In this chapter, we further developed our skills for modeling stairs, adding railings as a second system tool. We also learned the basics of the versatile **curtain wall** tool.

In the next chapter, we will return to the **Mesh** tool for the modeling of a terrain and we will be introduced to two semi-automatic wizards: one for truss structures and one for roof structures.

10
Using the Mesh tool and Wizards to Finalize a Design

In this chapter, we will take a closer look at the **Mesh tool**, which we introduced in *Chapter 8*, in the *Sloped float roofs using the Mesh tool* section. Besides that, we will learn how to use wizards such as **TrussMaker** and **RoofMaker**, allowing us to semi-automatically generate geometry based on the relatively simple geometry we already modeled.

The topics covered in this chapter are as follows:

- Creating a detailed sloping site model with the **Mesh** tool and surveyor data
- Using the TrussMaker and RoofMaker wizards to generate geometry semi-automatically

Learning these Archicad skills will improve your overall productivity and broaden the range of specific elements you will be able to model.

Creating a site model using the Mesh tool

Although *mesh* is a generic term, often used in many 3D modeling applications for a collection of triangles connecting points, in Archicad, it is a specific tool for creating site models in the first place (and can also consist of one or more rectangles, instead of only triangles). The tool was already introduced in *Chapter 8*, for developing complex roof geometries, but now we will focus on its main application.

Basic mesh

A **mesh** is a contour-based element, just like a slab or a roof, but it has an important geometrical distinction: its top nodes can vary in height, allowing you to define multiple slopes in a single element. They are a natural fit for model terrains, but there are a few creative uses for them, as we already showed in *Chapter 8*, in the *Complex roof geometry and skylights – sloped flat roofs using the Mesh tool* section.

Drawing a mesh contour

Since you already know how to model a slab and a roof, modeling a mesh is almost exactly the same: you draw a contour, using the Pet Palette and the Tracker. The main difference is that you define an overall height and then adjust each individual node, with Archicad taking care of the connecting faces.

Editing a mesh

If you click on a node of the mesh, you get the regular editing commands you know from slabs and roofs, but also an additional **Elevate Mesh Point…** command, which opens the **Mesh Point Height** dialog, where you can set the height of the selected point relative to a reference plane. You can even toggle **Apply to All** to set all points at this height.

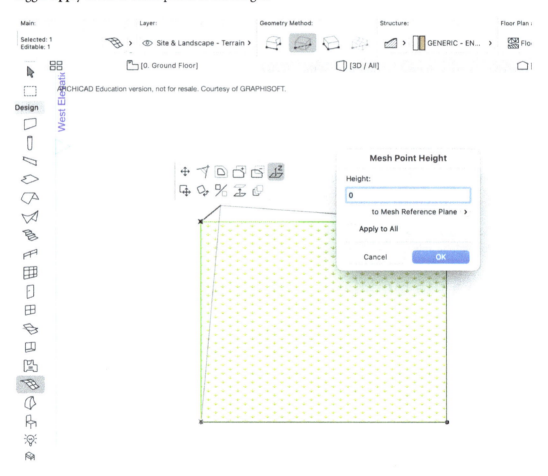

Figure 10.1 – Setting Mesh Point Height

If you open **Mesh Selection Settings**, you can see that the reference plane of the mesh is the initial top level of the mesh (*Figure 10.2* (**a**)), against which all other points are measured vertically. It is easier to see in 3D as the reference plane gets highlighted (*Figure 10.2* (**b**)). Points can be displaced vertically (*Figure 10.2* (**c**)) using the method shown previously if you work in 2D view. In 3D view, it is more interactive, with a Tracker to allow you to enter exact values.

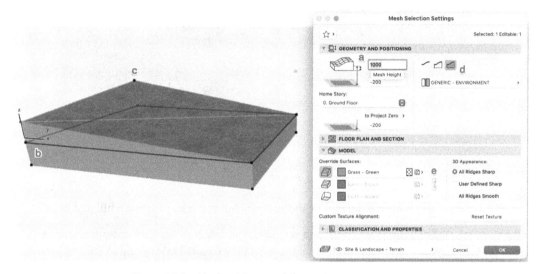

Figure 10.2 – Mesh settings and the various components

Be aware that you do not have this control for the bottom of the mesh as it is always horizontal. If you click on one of the bottom nodes, you can only stretch the height, moving all of the bottom points together.

The mesh structure can contain either the top surface only, contain the top with skirts (vertical side planes), or be presented as a solid body (*Figure 10.2* (**d**)). It is also assigned a building material, which in this example is **GENERIC - ENVIRONMENT**. Moreover, you can override the different surfaces if needed. **Default Mesh Settings** override the top surface using the **Grass - Green** surface (*Figure 10.2* (**e**)).

Editing the nodes of a mesh is not difficult. We already showed moving nodes vertically using the Pet Palette, but you can move them horizontally as well. Be aware that the top and bottom nodes are linked together so the side skirts are always vertical.

Clicking on an edge of the mesh gives you the same Pet Palette operations to offset an edge, turn straight edges into arcs, or insert new points, in addition to changing the height of the edge (which is the same as changing the two adjacent points). Notice that the reference plane stays as it is and is always horizontal.

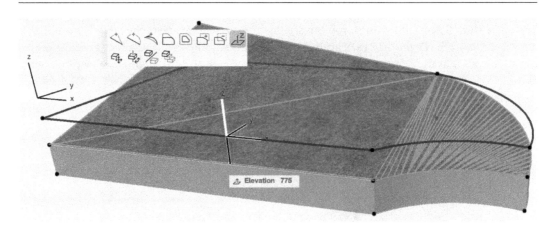

Figure 10.3 – Editing a mesh edge with the Pet Palette

Adding new points or openings to a mesh

What if you need to add nodes in the middle of the mesh, for example, to create a hill or a hole?

With a mesh selected and the **Mesh** command activated, you can draw another contour on top of the mesh, which displays the **New Mesh Points** dialog. Select the **Create Hole** option to use the contour as an opening or **Add New Points** to insert new nodes and edges. Archicad will take care of making the faces fit to the other edges.

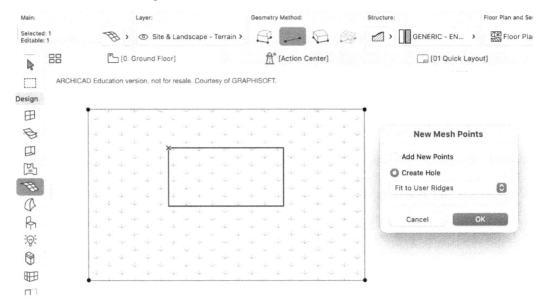

Figure 10.4 – Adding new mesh points on an already selected mesh

After clicking **OK** in the dialog for **New Mesh Points**, select your mesh and view the result in 3D.

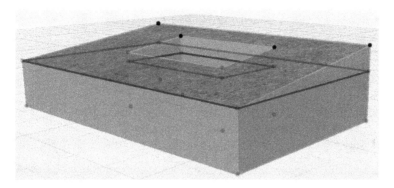

Figure 10.5 – Mesh with a hole created with new Mesh Points

Up until now, we have created quite simple-looking terrain models (flat, singular slope, etc.), but in reality, a terrain is often much more complex. Luckily, Archicad provides some handy methods for developing more realistic terrain models, which we will explore next.

Creating a full terrain model

The same technique (adding new points) can also be used to easily create a terrain model, using an underlay containing the contour lines (or lines of equal height):

1. For this method, a contour drawing is needed. You can download one from your preferred topographic service website, or maybe you already have one from a surveyor. In this example, we simply used a Line (**Rectangular** Geometry Method) and some Natural Splines, drawn in an Independent Worksheet (combining techniques learned in *Part 1* of *Chapter 6*). Your drawing should be similar to the one shown in *Figure 10.6* – ours fits within the limits defined by the elevation lines in the default Archicad 25 INT template.

Figure 10.6 – Contours to be used to create a terrain model – drawn with splines

2.   Open your **Ground Floor** Plan View and show the contours you created as a Trace Reference (on the Worksheet in your **Navigator**, *RMB* > **Show as Trace Reference**). Make sure to open the **Trace & Reference** palette as we learned in *Part 1* in *Chapter 3*, and adjust the transparency and color as needed.

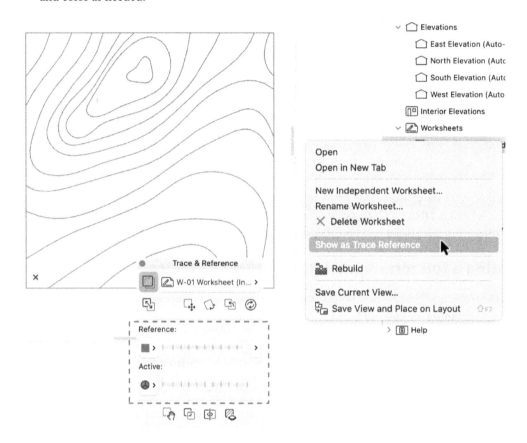

Figure 10.7 – Setting up Trace & Reference

3. Model a mesh on top of your contours, using the Rectangular Geometry Method.

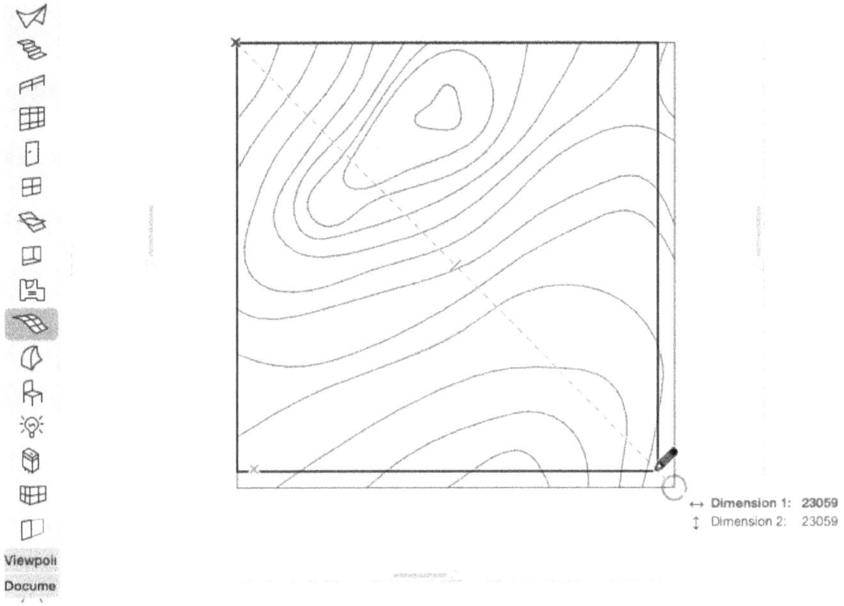

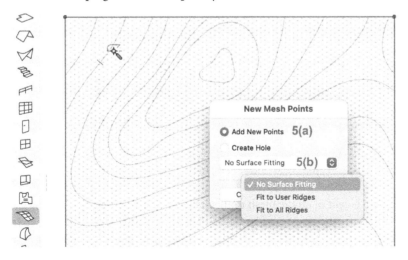

Figure 10.8 – Modeling a simple mesh

4. With the **Mesh** tool still active, select the mesh you just modeled.

5. Using the Magic Wand (*spacebar + LMB*), click on one of the lines in the Trace Reference. A dialog pops up. Choose **Add New Points** (**a**) and select **No Surface Fitting** (**b**). As we have modeled a non-sloping mesh, what option you choose has little to no effect.

Figure 10.9 – Adding new points

6. Do this for all the underlying lines. You might have to fiddle a bit with the surface fitting, depending on how dense and how curved your contour lines are. Sometimes the Magic Wand works better if you copy the splines from the Worksheet to the Floor Plan View.

7. Select the mesh (**7a**) as shown in *Figure 10.10* – be sure to click between the contour lines and corresponding points, not directly on them, when selecting (otherwise, you will select sub-elements). Then, click a point on a contour line (**7b**) – tip: we started with what is the lowest line of height in our example. Choose **Elevate Mesh Point...** (**7c**) from the Pet Palette that pops up. In the dialog box, fill in the height (**7d**) for this first line – 500 in our example. Ensure the working units you are using are as described in *Chapter 7*, in the *Setting working units and dimensions* section. Be sure to check the box (**7e**) next to **Apply to All** – after all, these are lines of equal height, so all the points on such a line should be moved to the same elevation. Confirm with **OK**.

Figure 10.10 – Changing Mesh Point Height

8. By changing the height of all the mesh points on a contour line, you are creating a slope. The resulting mesh starts showing triangles in the Plan View, which gets confusing, so try to keep track of where you are clicking. The heights used in our example are added in *Figure 10.11*, showing the final result in 2D.

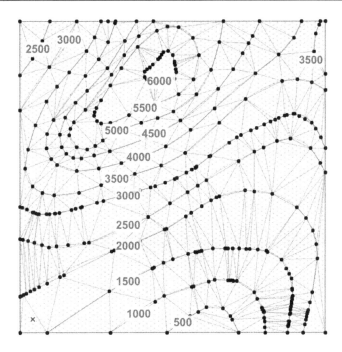

Figure 10.11 – Final mesh – at least in 2D

9. Select the mesh and show it in 3D. The corner points of the initial mesh are not necessarily located on a contour line and probably need to be moved as well. You can easily do this in 3D, by clicking the point and using the Pet Palette. In contrast with the 2D view, there is no **Apply to All** option – points are moved individually. Make a good guesstimate graphically for the height or use the Tracker if you know the height.

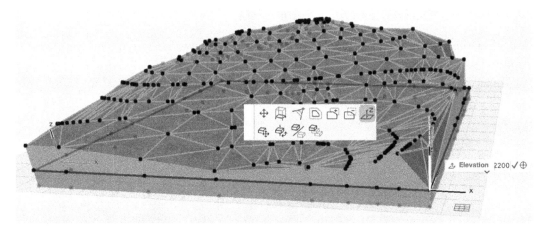

Figure 10.12 – Moving the corner points of the original mesh is easier in 3D

10. The finished terrain model shows quite a fluid shape in both 3D and 2D views, such as a section.

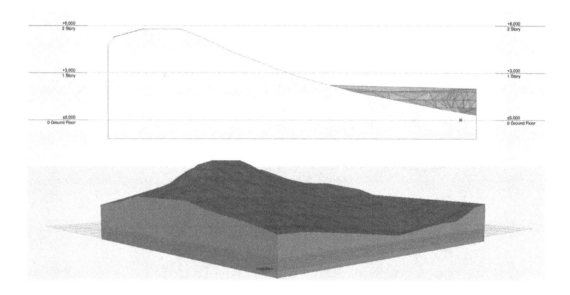

Figure 10.13 – Final result in a 2D section and in 3D

> **Note**
> Using the method with contour lines described here works well but can cause issues for large terrain models if you use splines. In such a case, polylines will work better. Although these are less organic in shape, they generate much fewer points when using the Magic Wand.

Integrating with foundations and groundwork

Now that you can model any terrain, it is probably time to see how you can fit your building project *into* this terrain...

The following example has a sloped mesh that overlaps with the building. Using the method to create a hole in the mesh will not give adequate results, since the hole will have strictly vertical sides. Here, we prefer to use **Solid Element Operations** (**SEO**) to ensure all pieces of the foundation are used to properly cut out the mesh:

1. Ideally, we would create a foundation with a Beam as a **Complex Profile** with **Modifiers**, as we will learn in the next chapter. You could use two beams, one on top of the other as well (good for now). For the SEO, the Beam is used as the Operator and the Mesh as the target, and the operation is **Subtraction**. This cleanly cuts out the foundation volume from the terrain mesh.

2. The slab is used as the Operator with the mesh again as the target, but this time, we use **Subtraction with upward extrusion** to ensure that every part of the mesh that extends above the slab is removed as well. Otherwise, we would have a part of the terrain inside our house.

3. Finally, we use the walls as the Operator and the mesh as the target, with a regular **Subtract** operation.

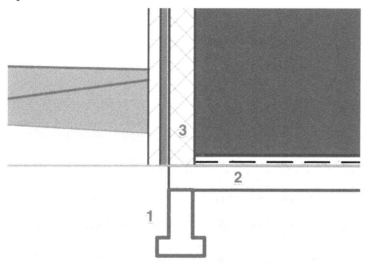

Figure 10.14 – Integrating a mesh with the building using SEO

> **Pro tip**
>
> If you want to take it a step further, you may want to work in multiple steps, with a first operation to make the overall excavation hole from the mesh and adding another mesh to the hole to represent the ground fill. From this fill, you'd then use SEO to cut out the foundation soles, slab, and walls, where needed. This requires more setup and takes more effort to prepare, but it will allow you to have an accurate volume estimation of the excavated and filled ground.

Importing surveyor data

If you have access to surveyor data, in the form of a list of the x, y, and z coordinates of the measured terrain, you have the option in Archicad to generate a terrain mesh based on these coordinates. For this to work, you need a list of coordinates that is formatted correctly:

*   The file should be in .xyz or .txt format. The first one is a typical file format for surveyor data and is also used for point clouds, for example. In our example, we will use the more generic .txt format.

- The `.txt` file should only contain X, Y, and Z coordinates, with the three coordinates of one point on a line and a space between each coordinate value.

- The units in which your file is written should be known. Otherwise, the resulting mesh will be the wrong size.

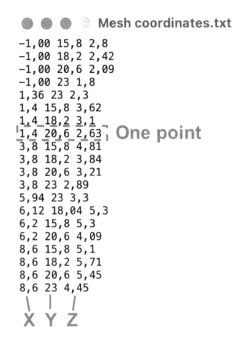

Figure 10.15 – Example of a .txt file with X, Y, and Z coordinates defined in meters

To generate a mesh using this kind of list, you only need to apply the following simple steps:

1. Starting from a Floor Plan View, through **File > Interoperability > Place Mesh from Surveyors Data…**, you can access a file browsing menu where you can navigate to your `.txt` file.

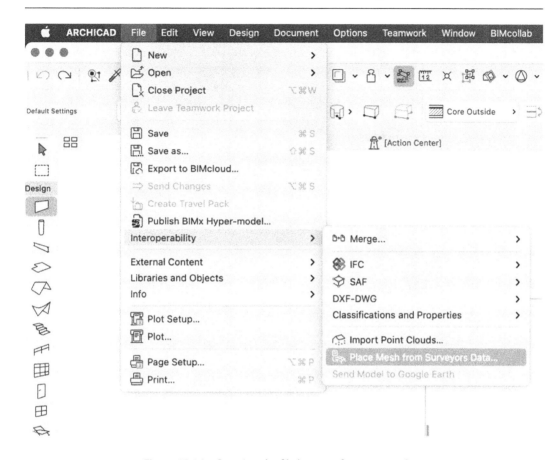

Figure 10.16 – Opening the file browser for surveyor data

2. Choose the `.txt` file you want to use and choose **Open**. If you do not have a file, you can copy the coordinates from the list in *Figure 10.15*.

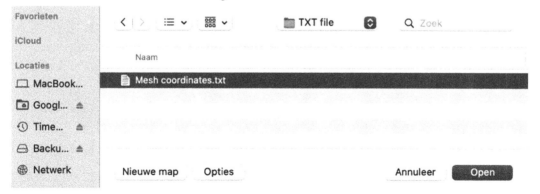

Figure 10.17 – Navigate to your file with X, Y, and Z coordinates and click Open

3.   A dialog pops up where you should set **Surveyors Unit** (**a**) to **meter**. For **Placement** (**b**), we chose **Original location** in the example. This choice depends on the source of the data. If the coordinates are in real-world coordinates (or in other words, geo-referenced), this option could result in a placement several kilometers away from your origin. In that case, **Define graphically** could be a better option, allowing you to position the mesh manually. Using **Zoom to the new mesh** ensures we will see the placed mesh, should it be generated outside our current view and zoom area. Confirm with **OK**.

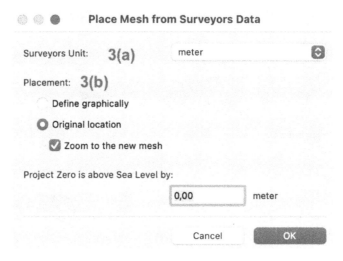

Figure 10.18 – Choose the right settings in the dialog

4.   Look at the resulting mesh in 3D. The mesh has quite a large solid body underneath the surface, but you can correct this if needed through its settings.

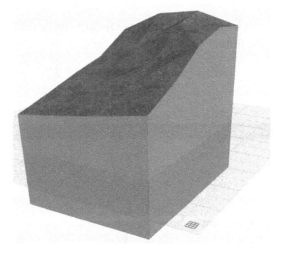

Figure 10.19 – The 3D result of the generated mesh

By now, you know how useful the **Mesh** tool is for creating a terrain model, ranging from simple flat or sloping terrains to fully developed models based on real surveyors' data. But there also are some other uses of this tool that make it more versatile.

Refining the display of the mesh

You have a lot of control over the appearance of a mesh:

- With **Override Surfaces**, you can set the surface to use for the top, skirt, and bottom surfaces. Without the overrides, all planes inherit the surface from the applied building material. Use it to distinguish between different zones in a garden or playground.

- For the structure, you can switch between top surface only, with skirt, and solid body. You may use the top surface only to model other objects, such as freeform roofs or canopies, or even as a complex shape used in an SEO.

- Model ridge smoothing gives you the option to either accentuate the different ridges by making them sharper or smooth them out for a cleaner view. This is recommended when you import a terrain model. When you model from contours, you may want to have the contours be sharp (**User Defined Sharp**), while all other ridges Archicad creates to complete the mesh remain smooth.

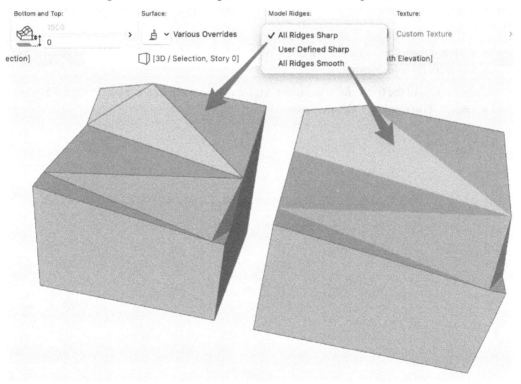

Figure 10.20 – Sharp versus smooth display of ridges

This concludes the in-depth exploration of the **Mesh** tool. You are now able to create quite complex terrain models to fit your building designs into and/or onto. Next is a section on some specific tools that will help you create rather complex structural geometries, such as trusses and roof structures, in a semi-automatic way.

TrussMaker and RoofMaker

There are a few commands in Archicad that are not really new tools but provide a series of automatic modeling steps to create a group of objects at once, based on reference geometry.

We like to call them *wizard* tools, as these commands guide you through a series of steps.

The end result of the wizards is typically a group of objects or a new Library Object, which is a lot more efficient when compared to modeling it component per component. In many cases, though, it will result in a (very) good first "sketch," and some manual adjustments can still be needed to finalize the design element(s).

TrussMaker

We will use the **TrussMaker** wizard to create geometrically complex 3D trusses using rather simple 2D elements:

1. Draw an arbitrary truss shape using lines. We do this in the floor plan, which is a suitable 2D environment for line drawing.

2. Select the lines of the truss and activate the wizard to turn them into a truss (**Design** > **Design Extras…** > **TrussMaker** > **Create Truss…**).

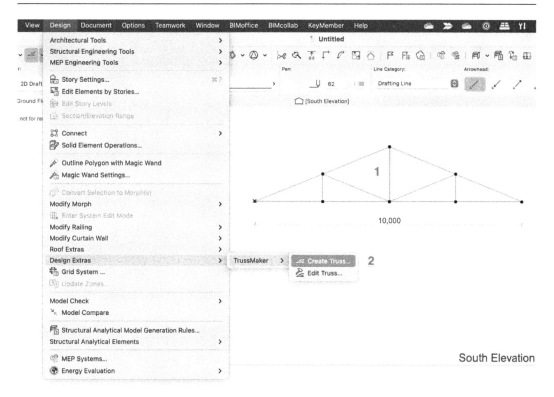

Figure 10.21 – Launch TrussMaker with selected lines

3. From the **TrussMaker Settings** dialog, you can select the main truss structure using the options on the left (**Timber construction**, **Hollow section**, or **Rolled steel profile**) and further refine **Attributes**, **Truss Profiles**, and (when you select a rolled steel profile) **Steel Junctions**.

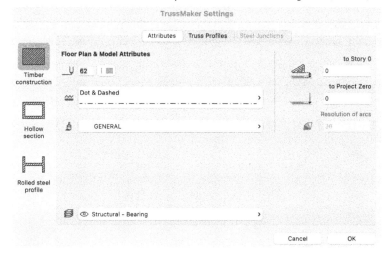

Figure 10.22 – TrussMaker Attributes

4. On the **Truss Profiles** tab, the dialog displays a preview, based on the selected linework. Here, you can edit the width and height of the cross-section, in the case of timber construction. Be aware that you can still switch to one of the other constructions.

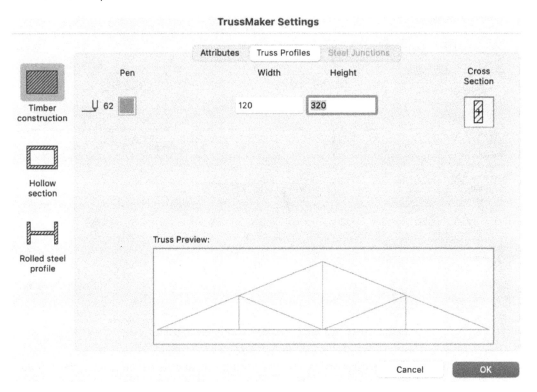

Figure 10.23 – TrussMaker truss profiles

5. If you click **OK**, you are asked to give the truss a name. It will be stored as a Library Object in the Embedded Library, which ensures that it is always available inside your project.

6. After finishing the wizard, the truss is positioned vertically, aligned with the baseline you drew. You are free to reposition the truss in any 2D or 3D window. You would typically want to place it in the floor plan and then use a 3D or Section/Elevation viewpoint to refine its position.

7. Since it is a Library Object, you can place copies of the truss wherever you need them to be. They all share the same Library Object definition.

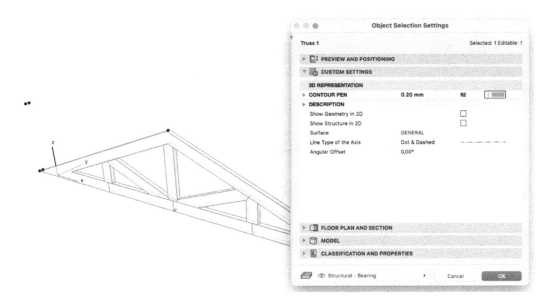

Figure 10.24 – Truss as a Library Object

In contrast with wizards in other CAD or BIM software, the generated Library Object maintains its original definition, allowing you to edit the truss from its linework, even if the lines have been deleted.

8. To edit the Truss, select it in a section of the elevation plan and go to **Design** > **Design Extras…** > **TrussMaker** > **Edit Truss…**. The lines that defined your truss are extracted and can be edited again. If you start from a floor plan, the wizard is clever enough to first ask you to navigate to a suitable section or elevation view and continue editing from the menu (**Design** > **Design Extras…** > **TrussMaker** > **Continue Editing**).

9. Make some adjustments, modify the lines, add a few new ones, and even use different pens to distinguish between profiles. When you are finished, go (again) to the **Create Truss…** menu, which reopens the dialog, and it helpfully remembers the settings used when creating the truss. You can press **Save** to redefine the truss and all its placed instances, but the dialog also provides a **Save As…** option to store the edited truss under a new name, so the other trusses are not affected.

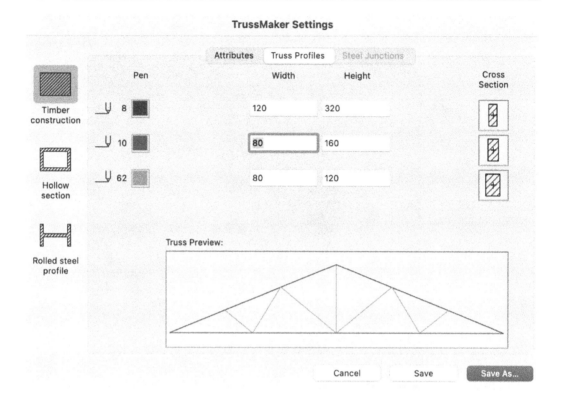

Figure 10.25 – Reopening TrussMaker Settings after some modifications

> **Pro tip**
>
> When you need multiple profiles in the same truss, you must use different pens when drawing the truss. This way, each pen number can be assigned a different profile in the wizard. This was applied in the preceding example.

When to use TrussMaker

In contrast to using regular elements and library objects, **TrussMaker** is a separate environment that assists the user with generating a series of elements that are embedded into a single Library Object. The use of TrussMaker must be understood as a convenient function, to speed up modeling. There is nothing preventing the user from creating a truss using individual profile beams and columns and grouping them. They can also be turned into a single GDL object, but in that case, you have no way of returning to the original objects. It is all a matter of preference. It works for many common situations and keeps the truss as a single object while providing some means to re-edit the truss after it was created.

RoofMaker

RoofMaker is another plugin that adds a series of utility modeling commands to help create a roof structure. You have to create a regular sloped roof first, as that forms the basis to extract theRoof Structure Elements.

> **Pro tip**
>
> You have to understand that the regular Archicad roof only contains the homogeneous layers or Skins of a roof – tiles, membranes, insulation, and finishes – but not its structure, such as rafters or purlins. Those can be created separately, using objects or beams. You'd have to set their position, slope, and configuration for the ends of the profiles yourself. With RoofMaker, these settings are picked up automatically from a roof that was already created, saving you a lot of effort.

There are two approaches to using RoofMaker to create a roof structure:

- You can add individual roof structure members to an existing roof one by one, using one of the RoofMaker creation commands in its Toolbox
- You can generate the whole structure of the roof in one go, using **Roof Wizard**

Afterward, you end up with a series of roof structure objects, grouped together, for convenient selection. You can still change all individual objects if needed, so either approach will give you comparable results, but the wizard does most of the work for you.

We will show you the different steps, starting with a fairly simple roof and then showing how to use the wizard on a quite elaborate multi-plane roof.

Adding individual members to a single-plane roof

First, create a simple rectangular roof, either by drawing individual single-plane roofs or by turning a multi-plane roof into single planes:

> **Pro tip**
>
> RoofMaker works best when the reference line of the roof is aligned with the exterior top edge of the walls underneath. The reference line is then typically located at the "bottom" of the roof. A roof with the reference line near the top of its slope can give strange results when using RoofMaker.

1. Open the **RoofMaker toolbox** (**Design** > **Roof Extras** > **RoofMaker** > **Show RoofMaker toolbox**). Here are some of the commands you have at your disposal: **Rafter**, **Hip** or **Valley Rafter**, **Trimmer**, **Purlin**, **Eaves Purlin**, **Collar Beams**, and **Tie Beams**.

2. We'll start with the **Rafter** tool. You can start by selecting a roof and run the command afterward or run the command first and pick the roof when the **RoofMaker** tool requests it. The result is exactly the same.

3. In the **Rafter Settings** dialog, you can refine the parameters of the rafter, which will be inserted as a parametric library object. The dialog uses diagrams to make the process clear. Press **OK** to start placing the rafter.

4. Pick the location by clicking on the roof plane. RoofMaker will place the rafter aligned with the slope and extending from the eave to the top of the roof plane. It even leaves a gap for the hip or valley edge of the roof. You can repeat the previous steps to place additional rafters.

5. You can also **Place Multiple Rafters** (the second column). This tool opens the same dialog, but now the **Placement of multiple rafters** section is active. The default is a fixed distance of 700 mm. Press **OK** to start placing the rafter.

6. Now you pick the zone on the roof where the rafters will be placed. We pick one corner and the other end. Four rafters are placed between these points.

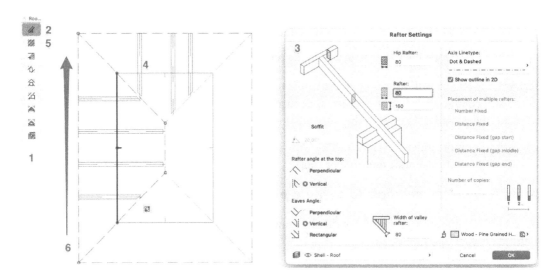

Figure 10.26 – Adding rafters

The process for other Roof Structure Elements is very similar:

- For hips or valleys, you follow the same process but have to click on the hip or valley edge of the roof. The status bar displays **Click an edge of the roof**.

- For purlins, again, follow the same process, but now you click on the top edge.

- For a trimmer, you need to pick two parallel rafters. This is a clever object as you can still move it alongside the roof plane using the interactive hotspots, which are placed to align with the roof reference line.

- For a collar beam or tie beam, you need two opposite rafters, which will be connected.

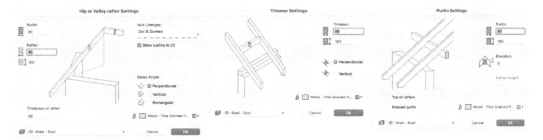

Figure 10.27 – Different roof structure dialogs

> **Pro tip**
>
> The Archicad manual recommends starting with the purlins, then the rafters or valleys, before placing the rafters. The trimmers come at the end, as you need to have rafters in place.

Removing the overlaps between the roof and structures

When the Roof Structure Elements are generated, they are positioned to overlap with the roof and are aligned to the roof planes so the bottoms of the structures coincide with the bottom plane of the roof.

We need to solve this alignment (resulting in colliding elements) by adjusting the Building Materials (and their Building Material Priorities) assigned to the Roof Composite Skins and the Roof Structures, but we had better success using an SEO to subtract the Roof Structure Elements from the Roof, ensuring no overlaps remain and all component quantities of the Roof will be accurate.

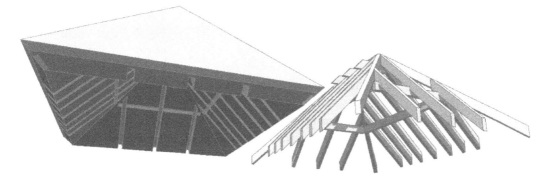

Figure 10.28 – Roof alongside RoofMaker-generated objects to show SEO subtraction

Using Roof Wizard on a multi-plane roof

Since adding all the members of a roof structure one by one is a lot of work, there is also a **Roof Wizard** feature, which automates these steps, all starting from a single- or multi-plane roof:

1. Select the roof and launch **Roof Wizard** (**Design** > **Roof Extras** > **RoofMaker** > **Roof Wizard…**). You get an extensive dialog, where you have multiple tabs for **Rafters**, **Beams**, **Purlins**, **Trimmers**, and **Ridges**.

2. Take some time to go through the different tabs. Notice how there are options to have double rafters alongside windows, and in the other tabs, you can further choose between collar and tie beams and top and eaves purlins.

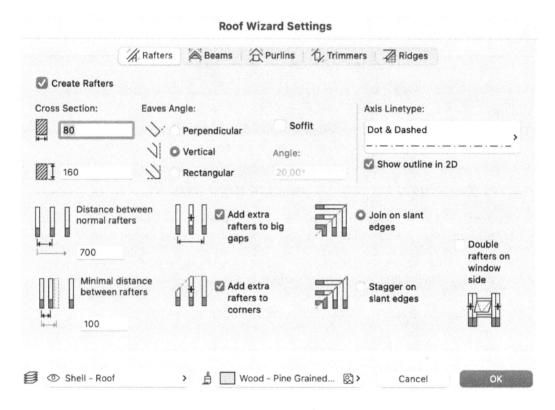

Figure 10.29 – Extensive Roof Wizard dialog

3. Pressing **OK** launches the generation of all the Roof Structure Elements, which speeds up the modeling quite a bit. They are grouped for convenient selection, but if needed, they can all be edited individually.

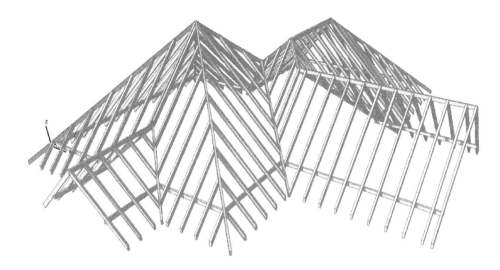

Figure 10.30 – RoofMaker result applied on a more complex multi-plane roof

In this section, you learned how to use two of Archicad's wizards, resulting in the rapid generation of 3D geometry. Let's summarize what was covered in this chapter.

Summary

In this chapter, we started by exploring the **Mesh** tool, focusing on developing a terrain with slopes, but also learned how to import surveyor data and how to properly integrate the terrain with the foundations of the building.

In the next section, we learned about two plugins to automate some complex structures: TrussMaker turned 2D Lines into a 3D Truss, with complete control over profiles, while RoofMaker added the Roof Structure Elements to already created Roofs, with the Roof Wizard turning this into a one-click operation of all rafters, hips and valleys, purlins, and trimmers.

Next up, we will learn how to use Parametric Profiles to further expand our modeling skills, and we will see how we can use the **Renovation** tool, introducing the use of data in Archicad BIM.

11

Using Advanced Attributes and the Renovation Tool for a Wider Design Range

In this chapter, we will look into the more advanced attribute called **Parametric Profiles**, which will allow us to create highly parametric and detailed linearly shaped (or profiled) elements. We will also introduce the **Renovation** tool and explain how Properties can help us in creating the necessary output for renovation projects.

Topics covered in this chapter are as follows:

- Using **Parametric Complex Profiles** for creating linear elements with a continuous cross-section
- Using **Element Properties** in combination with the **Renovation** tool to create a renovation project

Learning these Archicad skills will broaden the application range of different tools, while also adding more geometrical detail and basic data to your models.

Creating Parametric Complex Profiles and using Geometry Modifiers

Many of the objects in Archicad are set up as either line-based (walls, beams, and columns) or contour-based (slabs, roofs, and mesh). Contour-based geometry is refined with a handy set of Pet Palette commands.

For line-based elements, we have already looked at the homogeneous and the composite structures, using layers of materials. However, their core shape is always rectangular. When you need to refine the section of the object, you can switch to a so-called **Complex Profile**. This allows you to define the shape of the object from a 2D section, consisting of fills, just like you would in 2D detail drawings.

With a **Parametric Complex Profile**, you effectively get a 2D editing environment where you can add one or more fills. Each fill will then be extruded along the Reference Line, giving you a 3D shape from the profile. This method works very well for any object with a complex cross-section, such as steel profiles for columns and beams, or articulated walls, which don't have a rectangular profile. If the definition of the Parametric Profile is a Complex Profile containing **Geometry Modifiers**, the user has even more control and can set the dimensions of the used profile without having to create a new definition (e.g., change the width and height of a column profile) and thus making the Complex Profile a Parametric Profile.

In this section, we will learn how to use predefined profiles and how changing the values of the Geometry Modifiers affects the resulting shape. We will then see how you can create your own definitions and how you can define your own modifiers to make a profile fully parametric for the user.

Selecting a predefined profile

At first, you can switch line-based elements from their regular basic structure (one single material) or composite structure to a Complex Profile, from the element settings. You have two main options: you either pick a profile from a list that was already created or you create a new profile from the existing element.

It should not surprise you at this time that the predefined profiles are actually just part of the template you started from and that they can be fully adjusted if needed.

Moreover, since profiles are attributes, they can be shared among elements, so they are an effective way to control detailed geometry in one central place.

Editing a profile

Each profile is adjustable. When opening the profile for editing (using the context menu), the Profile Editor presents itself as a 2D editing environment. From that point, follow these steps (see *Figure 11.1*):

1. You can use all the drawing methods you are already familiar with, including the Tracker and the Pet Palette. However, you only have a limited set of tools at your disposal: annotation (dimensions and text), fills, lines, and arcs, and only fills will impact the generation of the 3D extrusion. The other tools help you with documentation.

2. The fill is chosen from the list of building materials, which ensures that the extruded profile can be used for accurate material takeoffs. You can draw any contour you want using the **Fill** command: rectangular, polygonal, or rounded.

3. The origin in the 2D view represents the Reference Line, and you have to imagine looking in the direction of the Reference Line to better understand where to place fills. Remember how the Reference Line was used for various tools such as the Wall Tool and Beam Tool and how it determines the positioning and connection of elements (see, for example, *Chapter 8*, in the *Building materials and priority-based connections* section).

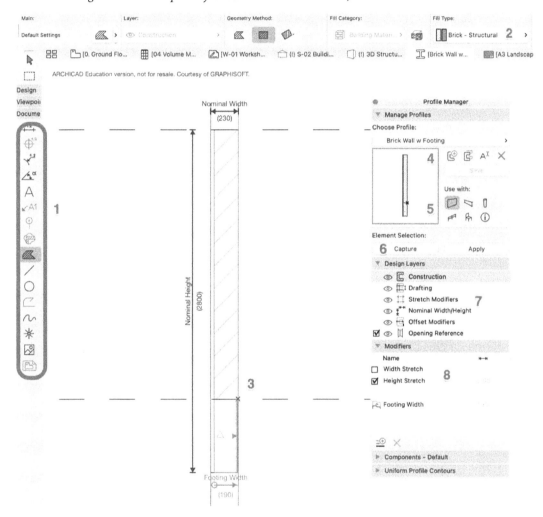

Figure 11.1 – Profile editing environment with available tools

4. In the **Profile Manager** palette, you can create new profiles from scratch, duplicate an existing profile, rename it, or even delete it from the list.

5. You can choose for which Archicad tools the profile becomes available: walls, beams, columns, railings, or certain library parts, provided they are created to accept custom profiles.

6. Alternatively, you can also **capture** the profile of an existing element, giving you a head start when refining the profile. Use this method if you have an element already modeled, but its section needs to be adjusted beyond what is possible with the Composite Structure and Priority Based Connections. A good example is when one of the Skins of a wall needs to extend or be shortened against the other Skins. After you finish editing your profile or when you picked a profile from the list, you can apply it to the currently selected element, if it is of the right category.

7. **Design Layers** present a structure for setting up the profile:

    *   Use **Construction** for the fills that define the profile.

    *   Use **Drafting** for any annotation that may help you explain the profile while editing.

    *   Use **Stretch Modifiers** to display the zone that gets stretched when you modify the main size of the element. The parts outside of the stretch modifiers keep a fixed size (e.g., you could use this to define a wall with a fixed profile on its top and a fixed profile, like a foundation sole, at its bottom).

    *   We'll discuss **Nominal Width/Height** and **Offset Modifiers** later.

    *   **Opening Reference** is used to control the position of an opening into the wall or beam.

8. **Modifiers** allow you to add parameter-driven dimensions to the profile.

The bottom side of the Profile Manager gives you control over the currently selected fill. From this menu, you can define the **Structural Function** of the fill (see the section on *Structural Function (load-bearing)* further in this chapter) and, optionally, **Override Surfaces**, if you want to have a different surface finish than what is defined in the assigned building material. There are also a few pens you can set. You can even define how the projected area of the component is interpreted in quantity takeoffs.

Figure 11.2 – Setting the display and Structural Function of the selected component (fill)

Using Modifiers

While the previous methods already allow you to define any possible profile, you may need to create many of them in a project. It is often more efficient to think of certain profiles as a series of variants, where the shape retains its overall setup but some of the dimensions can be adjusted. One Parametric Profile can thus replace a large series of fixed profiles.

Archicad already presents you with two **Stretch Modifiers** for **Nominal Width** and **Nominal Height**, which are usable in many situations. They are displayed as dimensions and you can drag their end point to define the extent of the stretch area.

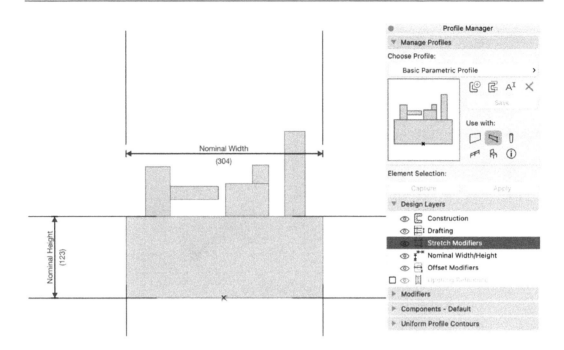

Figure 11.3 – Setting the Nominal Height and Width stretch modifiers in a profile

You can now apply the profile and make adjustments via the Pet Palette. Everything within the stretch area marked by these two modifiers gets stretched and everything outside remains at its fixed size.

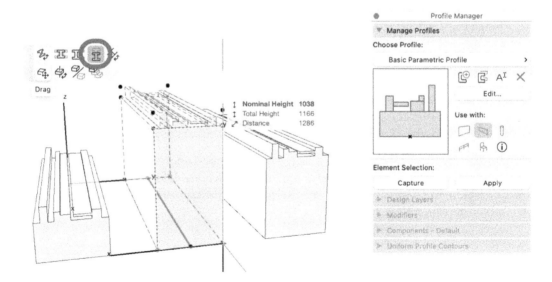

Figure 11.4 – Example of three stretch variants of a profile

But this does not work for every situation (e.g., when you also want to control other dimensions of the profile). Fortunately, Archicad allows you to insert additional modifiers into the profile. These are parameters that are displayed as editable dimensions: adjust the value and the geometry follows.

The following example shows how we replaced the series of random rectangles on top of the beam profile with an additional extension that is easier to control. We start by drawing a simple rectangle centered on top of the base rectangle and remove the random rectangles from the previous example.

From the **Profile Manager** palette, you can add a new modifier from the **Modifiers** section and set a name in **New Profile Modifier** or pick one from the list of already defined modifiers.

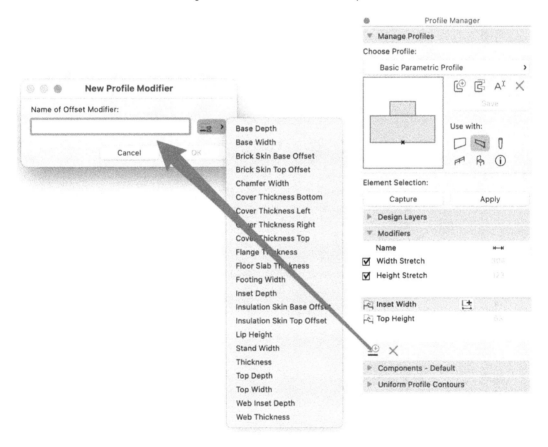

Figure 11.5 – New Profile Modifier and list of existing modifiers

Each modifier is known by its name. From the **Profile Manager** palette, you can start adding the modifier into the 2D view as a stretchable dimension. By picking edges or nodes, the fills get locked to the dimension and follow along when you change the modifier value.

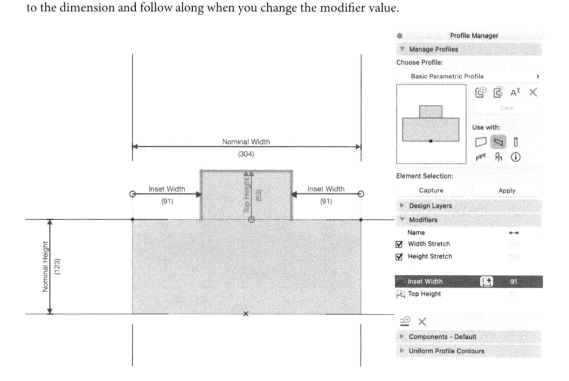

Figure 11.6 – Added two new modifiers and linked them to the geometry

The way you assign the dimensions influences their impact on the geometry:

- When you click on an edge, you get a two-directional modifier, indicated by two arrows. It can be stretched in two directions.

- When you click on a node, it becomes a single-directional modifier, remaining fixed at the start of the dimension.

After you save the profile and return to the model, your profile now displays the modifiers as thick, dark blue Reference Lines. The Pet Palette gives you the **Offset Edge(s)** option, which is tied to the modifier.

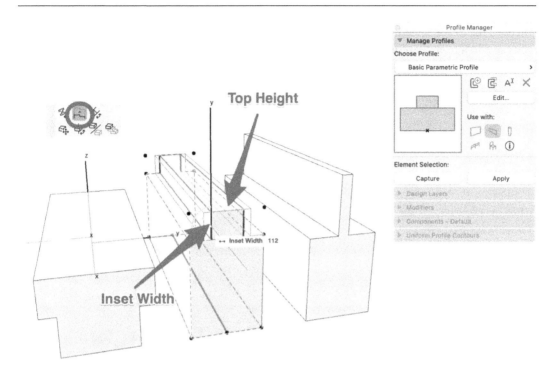

Figure 11.7 – Interactively editing the profile from one of the modifier edges

Notice how the preceding example allows you to set a negative inset width, making the top box extend over the edges of the lower part of the beam, with no additional effort. It doesn't work that way with a negative top height, as we still have two rectangles.

Challenge

Do you think you could modify the fills so the top height could also become negative?

If you open the **Beam Selection Settings** dialog, the modifier is also available as a numeric property, so you can edit it from there as well. The modifier has become an integral part of this **Beam** segment when the Parametric Profile is applied to it.

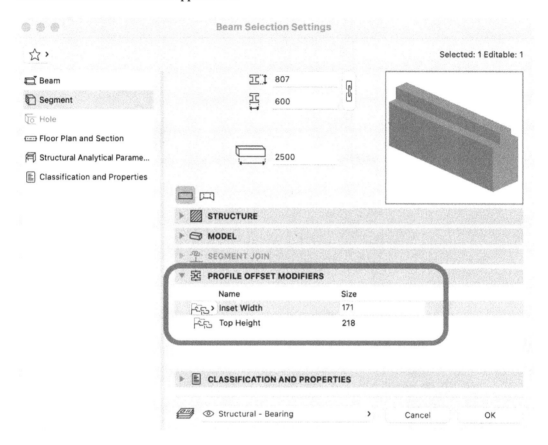

Figure 11.8 – Profile offset modifiers as properties of the element settings dialog

---

Pro tip

From the template, a series of modifiers is already defined. You are strongly advised to reuse existing modifiers in your Parametric Profiles, as this will limit the number of modifiers to maintain, but also makes it more convenient to collect them into schedules. But feel free to add new ones if they make more sense to you.

Applying profiles to improve models

This last section gives a few practical situations where a Complex Profile is applied and shows how to use parameters.

Beams and columns

Many beams and columns are not rectangular or round. Using a Complex Profile, you can recreate any common **Steel**, **Concrete**, or **Wood** profile, including combinations of materials, for example, for **Cross-Laminated Timber** (**CLT**) beams.

Since so many **Steel** profiles are standardized, Archicad provides a catalog of profiles from which you can pick and choose:

1. Open **Standard Steel Profile Database** from the **Options** > **Complex Profiles** > **Import Standard Steel Profile…** menu.

2. The dialog that opens is very straightforward to use: pick the right **Country Code** (**a**) option (we opt for the default **EuroCode** profiles), select the required **Geometry** (**b**) (**I/H**, **T**, or any other category), and this presents the **Available Profiles** list (**c**).

3. You can then select **Add Profile to the Project**.

4. If needed, adjust the **Assigned Building Material** selection.

5. Finally, press **Import** to bulk import the different selected profiles into the project.

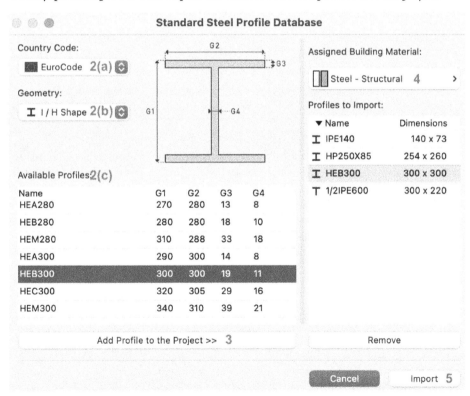

Figure 11.9 – Adding profiles to the project from the Profile Database

Foundations and precast beams

Sole foundations don't have a dedicated command in Archicad, but they can be created easily using walls or beams. For plain rectangular foundation soles, you could simply place a regular single-material wall or beam underneath the edge of the building. We recommend you place them first in a floor plan view to set their horizontal position and then switch to a section window to verify or adjust their vertical position.

However, with profiles applied to beams (or walls), you have more options and can control multiple beams with a single profile. Here are a few examples from predefined profiles in the template. Notice the blue modifier lines and the main profile Reference Line:

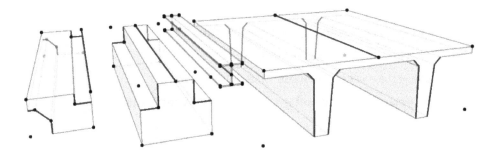

Figure 11.10 – Four examples of profiles included in the template (from left to right):
Concrete Slab Edge Footing, Precast Inverted Tee, HEA200, and Precast Double Tee

> **Pro tip**
>
> In some cases, even using a slab is possible, but we recommend you preserve slabs for foundation slabs rather than foundation beams, which behave more like line-based elements, especially when you need to take off quantities. Moreover, using a wall or beam allows you to keep each beam as a single element and still have proper connections with adjacent and perpendicular beams. Finally, the display options for these three element types are quite different and it is advised to choose your tool with this in mind (for example, a wall can show a cut fill in a plan view, but a slab or beam cannot do this).

You also have to decide whether you want their reference line to be centered or placed at one of the sides (inner or outer edge). Again, there is no wrong or right way to apply them, but rather a choice of the most convenient way to place them and to manage their intersections.

If you browse through the Archicad library, you may also find some foundation and beam library objects, as an alternative to profile-based elements. They both have their purpose: an object-based foundation can be fully parametric, but only exists on its own. A profile-based foundation also allows clean connections with other foundations, preferably using the same profile.

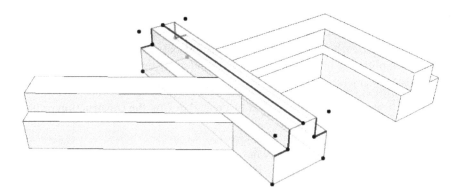

Figure 11.11 – Profile-based beams (or walls) can properly connect

From the preceding example, you will understand that for a foundation beam around the contour of the building, a profile-based foundation is recommended. You may still need to decide whether you see it more as a beam or as a wall, as they may both apply the same profile.

Using the technique with Complex Profiles, you are now able to model foundation elements, although there is no dedicated tool for this type of construction element. Next, we will further explore how we can use data in our model. As a first look into the *I* of BIM (*Information*), we will focus on Element Properties for renovation, function, and position, which we consider to be basic information in any model when applying BIM.

Element Properties for renovation, function, and position

Gradually, we dive deeper into the information management aspects of models, starting with the Element Properties covered in the following subsections. They allow you to add more in-depth characteristics to the elements and they may even have a direct impact on the display and visibility of elements or their components.

Renovation status

Each element in Archicad has a setting to control its **renovation status**. You can indicate this to set the element in three possible states and they all relate to the activities that are planned during the construction project:

- **Existing**: This element is part of the building already before the construction starts and is intended to remain after the project finishes.

- **To Be Demolished**: This state indicates that the element will be removed during the construction works. In renovation projects, this is very important, as it helps with estimating the demolition efforts of the project, which represents a significant cost. It also enables you to create and document demolition plans and it enables you to identify how your new elements interact with the existing elements.

- **New**: This indicates that the element will be created during the construction project, as an addition after the demolition works.

It is not possible for the renovation status to be not set, it cannot be *Undefined*, as is the case for many other properties. The default setting in the *Archicad 26* template we use is **New**. Renovation status is available for all 3D elements (walls, slabs, etc.) and for 2D elements such as lines, fills, and dimension lines, but not for the section line, elevation markers, or other viewpoint markers.

> Beware
>
> Note that the status is set for the element as a whole and that you cannot indicate individual components of an element with a dedicated renovation status. If only a part of a wall is demolished, for example, when only the facade bricks are to be removed or if only the beginning of the wall is impacted, you should split the wall into different elements and indicate their status separately.

How do you assign the renovation status?

There is a dedicated palette that you can leave open while modeling (**Window** > **Palettes** > **Renovation**). Whenever you select an element, you can alter its status using one of the three icons.

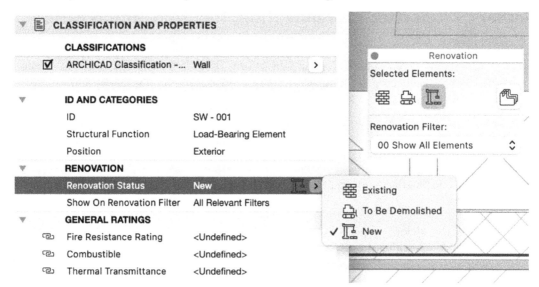

Figure 11.12 – Edit Renovation Status of an element via palette or element settings

As an alternative, you can open the element settings where you can find **Renovation Status** as a dedicated property in the **CLASSIFICATION AND PROPERTIES** pane (in the **RENOVATION** section at the top).

Renovation filter – impact on element display

Apart from enriching the model with this information, which is important for project budget estimations, **Renovation Status** can be used to toggle the display style of elements.

Renovation Filter is a series of presets you can define in a project, where the visibility (**Show**/**Hide**) and the display style (**Override**) of elements can be set based on their assigned renovation status. For each filter, you can configure a series of settings per state (**Existing Elements**, **Elements to be Demolished**, and **New Elements**).

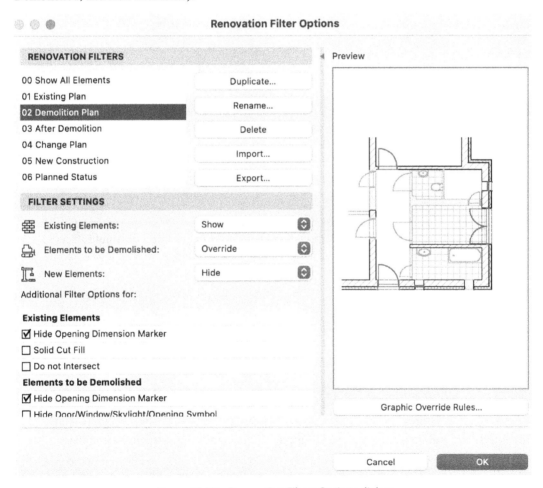

Figure 11.13 – Renovation Filters Options dialog

When you switch between the different configured renovation filters, you can effectively tune the model display according to these different project states. The default template already provides a few combinations: **Show All Elements**, **Existing Plan**, **After Demolition**, and so on. You also get a graphical preview to give an indication of the effect of each filter.

> **Pro tip**
>
> Rather than using complex combinations of layers or even using different models, this approach allows you to keep the project status before, during, and after the project execution in the same model.

Archicad is quite clever and tries to prevent you from contradictory situations. If you have a wall that is set to **New Construction**, you cannot set a window or door in this wall to **Existing** or **To Be Demolished**. The renovation state that is not allowed will be grayed out.

Show on all relevant filters

By default, each element is available for all renovation filters. This is indicated by setting the **Show on Renovation Filter** property to the **All Relevant Filters** value, which is the recommended setting.

Alternatively, you can also pin the element to only a specific filter by selecting the filter from the drop-down list. This can be used to limit the display of the element (e.g., when you want to include some temporary element during one particular renovation filter).

> **Pro tip**
>
> We don't often use that option, as elements may get lost a bit, but it is sometimes required when we need more control or if the display of an element across the filters does not suffice. In that case, we may sometimes add the object twice, each with a different renovation status and possibly limited to one renovation filter.

As **Renovation Status** is not available for section lines, elevation markers, and such, you will have to place viewpoint markers on a specific layer if you want to hide them in a certain renovation phase!

Graphic Override for Renovation Status

When you opt to override the display of elements, you are brought to the **Graphic Override Rules** dialog. The first three entries in this dialog are dedicated to defining how the display of these elements (based on their renovation status) is overridden, by optionally setting a fixed line type, pen, fill type with related pens, and the surface for the 3D display. Combine the pen overrides with different Pen Sets (discussed in *Part I*, *Chapter 6*, in the *Using Pens and Pen Sets to change the weight and color of lines and fill patterns* section), and you have a multitude of ways to show your renovation project at your fingertips.

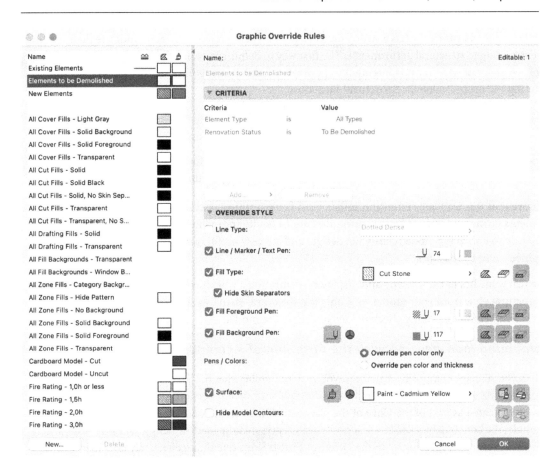

Figure 11.14 – Graphic Override Rules dialog to define the Renovation Status override style

We will return to this dialog in *Chapter 12*, when we will discuss **Graphic Overrides** in more detail, as this is a flexible way to display how information in the model can control their display style, regardless of assigned materials, fills, pens, and surfaces.

Structural Function (Load-Bearing)

As explained just before the introduction of the *Element properties for renovation, function, and position* section, we are focusing on basic information that is being applied to elements in a model. Besides the renovation status, there are two more properties we would like to dive into: Structural Function and Position.

While Archicad historically has a strong focus on architecture, architects need to indicate their design intent relating to structural performance. The first way to define structural information is by indicating for each element whether it is intended to perform a Load-Bearing function or not. You do this by entering the proper value for **Structural Function**:

- **Load-Bearing**: The element is intended to carry weights. In construction, we expect a continuous set of elements carrying weights from the roof to the foundation.

- **Non-Load Bearing**: The element is not taken into account for structural performance. This is true for many partitioning walls, finishes and coverings, railings, equipment, and furniture.

Beware that this indicates intention rather than facts: nothing prevents you from placing floating elements or making objects load-bearing that are in reality incapable of carrying such weights. But they help to enrich the model, clarify the design, and are very useful in communicating with structural engineers and contractors.

The real value lies in the filtering and display configurations: when properly assigned, you can derive a structural view from your model with only the structural elements showing. This improves insight but is also helpful as a sanity check before you share the model.

Combining load-bearing with the Core skin of a composite structure

When you manage composite structures, you may remember that the Skins in the structure also have a function. As shown in *Chapter 8*, in the *Creating a new composite* section, one or more adjacent Skins can be indicated as part of the **Core** of the element, which complements the structural function: for example, the concrete blocks Skin in a cavity wall represents the load-bearing part of the wall, when the wall is set as load-bearing, with the **Other** or **Finish** Skins remaining effectively non-load-bearing.

This is used in the **Display** settings of a view, where you can hide non-load-bearing elements, and for the remaining elements, you can also limit the Skins to be displayed to those indicated as **Core**.

Here is an example of the four possible display configurations showing the same model:

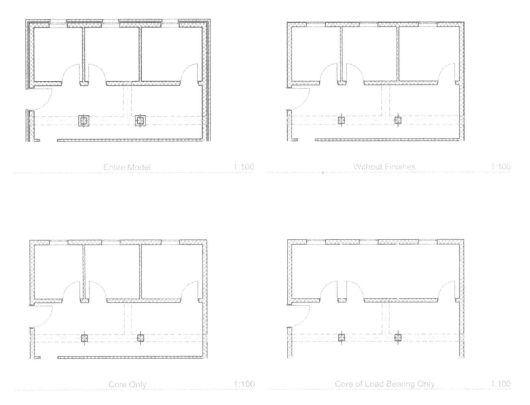

Figure 11.15 – Four options to display elements according to their structural function

This impacts the Skins of composites (for walls, slabs, roofs, and shells), and also columns and beams.

For rectangular or round columns, you can control the display of the column veneer by indicating **Veneer Type** as **Core**, **Finish**, or **Other**.

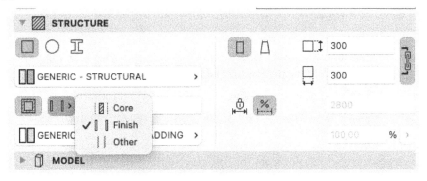

Figure 11.16 – Structural function for column veneer type

For profiles, this can be set for the fills inside **Profile Manager**, where each fill can also be defined as either **Core**, **Finish**, or **Other**.

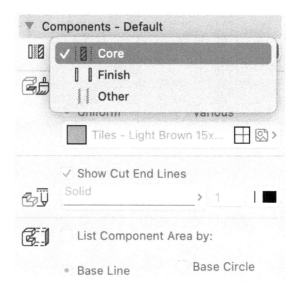

Figure 11.17 – Structural function for the Profile component

Position (exterior/interior)

In buildings, we protect people from the outdoor environment. The building shell or envelope consists of walls, roofs, exterior slabs, doors, windows, and curtain walls. This is information that is required in energy calculations.

To assist such analyses, Archicad provides the **Position** property, which is available for most elements and has three possible values:

- **Exterior**: This is set for elements that present a barrier against the environment. This is typically true for all exterior walls, roofs, exterior slabs (for example, flat roofs or cantilever floors), or foundation slabs.

- **Interior**: This is set for the opposite: elements that are entirely inside the building envelope and thus have no role to play in the protection against the outdoor environment.

- **Undefined**: This is the default when you haven't yet assigned a position. For many objects, this is just fine (for example, furniture, as you would probably not rely on it anyway).

As a property, Archicad simply relies on the user to indicate the correct position. It is not derived from thermal or other properties, nor from the thickness of the element. This implies that you are responsible for making the right interpretation yourself. There is nothing preventing you from setting an interior wall to be **Exterior**.

That said, there is an integrated **energy evaluation module** (which is not explained within this book), which performs a more thorough assessment of the model.

Use of the Position property

The **Position** property is available for most elements from its settings:

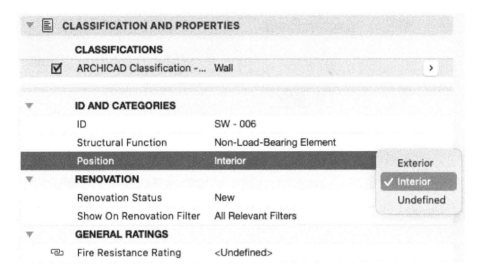

Figure 11.18 – Position property in the Settings dialog

Beware that the setting applies to the element as a whole, so a multilayered wall will have all of its Skins indicated as the exterior. This is not so strange, as the interior Skins are also part of the building envelope.

Pro tip

Unlike the renovation status, there is no special configuration of the display of elements based on their **Position** property. However, you can use the *Graphic Override Combinations* technique, as explained in *Chapter 12*, to help you visualize the elements indicated as **Exterior**, **Interior**, or **Undefined** if you want.

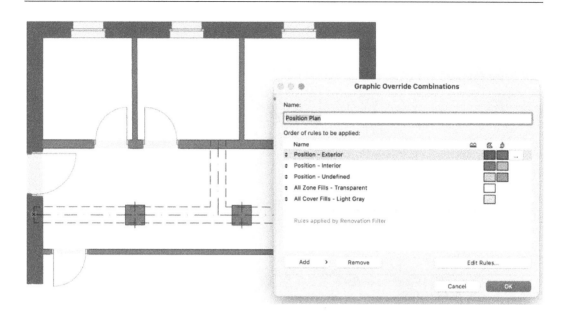

Figure 11.19 – Customised Graphic Override combination to visualize Position

After completing this section, you have now developed more insight into some of the basic types of information included in any Archicad element. As for first use, these open up a vast array of visualization options similar to the example shown in *Figure 11.19*, to which we will return in *Chapter 12*.

Summary

In this chapter, we started exploring complex profiles in more detail and learned how to make them parametric using modifiers, which helps us to cover more situations with fewer profiles.

The last sections discussed three element attributes that enrich our model for renovation and for structural design and are used to indicate the exterior and interior elements (position).

Next up, we return to creating construction drawings, with annotation and a few more views of the model.

12

2D Construction Drawings and 3D Views with Linked Annotations

In this chapter, with most of the model in place, we will learn about setting up our various 2D views to fully control our Annotation, including importing or creating our own Line Types, Fill Types, and other attributes. We also introduce new Views, such as Schedules, 3D Documents, and Detail windows.

Topics covered in this chapter are as follows:

- Creating and managing attributes used in the model and 2D Annotation
- Configuring Sections and Elevations in detail
- Other ways to display the 3D model by creating 3D projections and 3D Documents
- Construction detailing, based on the 3D model

Learning these skills is an essential step toward creating fully detailed construction plans and layouts, allowing you to take full control of how the model is represented.

Annotating 2D documents and creating attributes

As explained throughout *Chapter 7*, Archicad has a good set of tools for adding text, labels, and dimensions to a drawing. These tools can use the geometric and non-geometric (meta)data from model elements and display this information in any 2D view. In a design process, we often also need additional Annotation that is not necessarily linked to a model. For example, in an early design stage, we may not have added all the data into the model already, and thus, we cannot use a linked Label to show data (as demonstrated in *Chapter 7*, in the *Exploring labels* section). In such a case, we still need ample 2D graphical elements such as lines and fills to express our ideas. Furthermore, there are also many cases in which using a line or a fill is necessary to express something graphically, mark an area, or just make a quick 2D sketch before modeling a new idea in 3D. And of course, there is always the need to just add 2D detailing, or some extra adornment on our 2D output that cannot be created using one of the (many) objects in the parametric library. All the preceding examples show that there is still a need for 2D Annotation and the use of 2D elements such as lines and fills to this end. The following section will teach you more about attributes related to 2D lines and fills and how you can create and manage your own attributes.

Adjusting attributes

We introduced you to the main drafting tools and how they are specified using attributes in *Chapter 6*. While the template comes with a reasonable selection of configured fills and line types, you can adjust them extensively. This section will explain how to do this for 2D lines and fills and textured surfaces.

First, we'll begin by customizing line types.

Custom Line Types

To customize line types, open the **Lines** attribute through **Options | Element Attributes | Lines…** and choose a line type you want to edit through the dropdown, or create a **New…** type from scratch.

A **solid** line type has nothing to customize it. Remember that thickness and color of lines are completely defined by the pen in the active Pen Set that is used to draw them and not by the Line Type itself.

Customizing **Dashed Lines** types is also quite easy – you define a pattern of Dashes and Gaps.

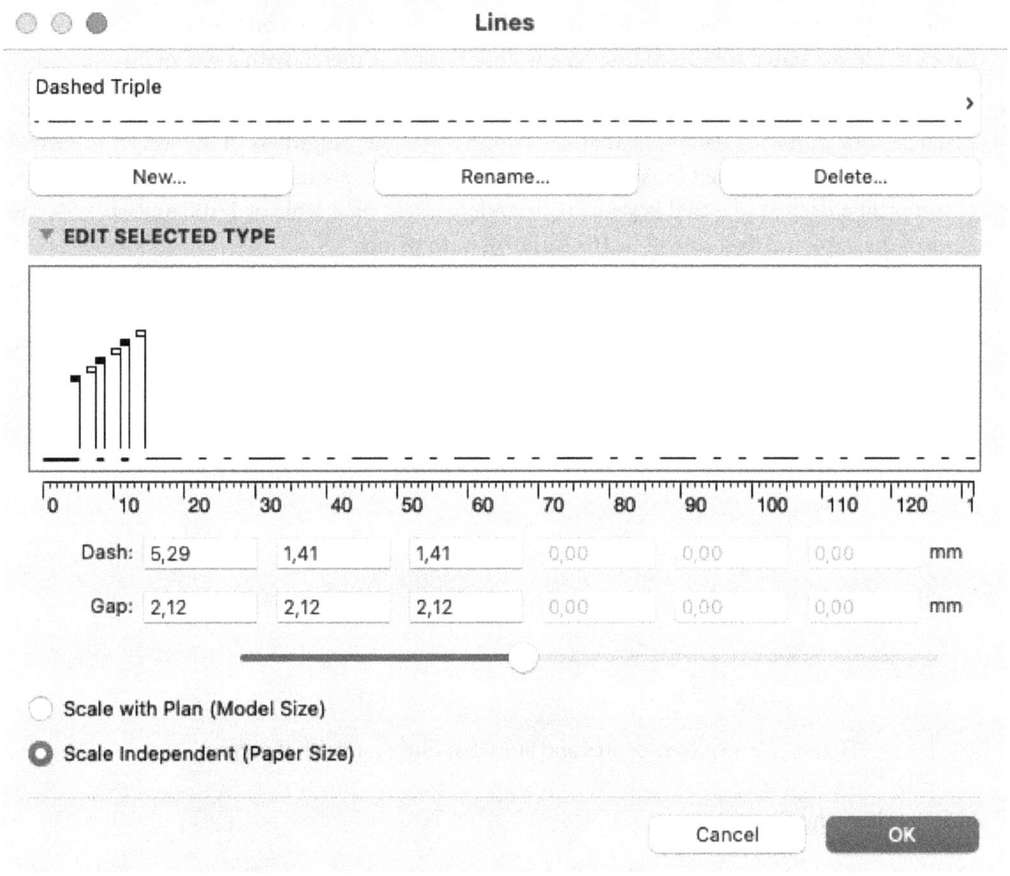

Figure 12.1 – A dashed line pattern adjustment

Up to six sequences can be set. Just set the length of each dash and gap, as needed. Using the slider, you can set how many sequences are used in this line type. It may not be obvious, but the preview is interactive – you can drag the handles to graphically define the pattern. However, if you want consistency between the patterns, it may be better to set this numerically.

If you want a dot, set the **Dash** length to 0. With this approach, you can define dashes, dash-dots, triple-dashes, and any other combination that is common in technical drawings.

You also must choose the scaling method, which impacts how the lengths of dashes and gaps are applied:

- **Scale with Plan**, where the pattern is defined in sizes that are based on the size of the model – for example, an actual (painted) line on a wall or floor of 1 meter, with a gap of 20 cm.

- **Scale Independent**, where the pattern is defined by the output on a drawing sheet. This means that a gap of 2 mm will (always) be that size (2 mm) on paper, regardless of the scale it is printed on. This type is mostly used for symbolic representations – for example, for an overhead edge. If you print a dashed line that is set as scale-independent on a scale of 1:100 and on 1:50, the sizes of the gaps and dashes will be the same on both prints.

Symbol lines may seem more complicated, but they are actually straightforward to create. Just draw what you need using regular arcs, lines, and hotspots and copy and paste them into the editor dialog.

We promised in *Chapter 6*, that we would draw a series of Christmas lights.

Figure 12.2 – A series of arcs and lines that form our symbol pattern

We'll show you how in this chapter:

1. Draw our pattern using lines and arcs. Here, we start from two arcs, which were adjusted so that they align and resemble a sine function. The pen color is irrelevant to the resulting symbol line, but it can be useful during drawing.

2. Draw a single symbol to represent a small lamp – again, only using arcs and lines.

3. Drag copies across the line by using guides.

4. Rotate the symbols one by one to get a pleasing orientation across the two arcs.

5. Select the pattern and copy it to the clipboard (*Cmd + C*/*Ctrl + C*).

6. Open the **Lines** attribute editor and create a new line by clicking **New…**, and then choose the **Symbol** line type.

7. Paste the pattern from the clipboard into the symbol line type.

8. Set **Dash** to a suitable size (e.g., 40 mm and **Scale Independent**), and set **Gap** to 0 so that the symbol line can repeat without gaps.

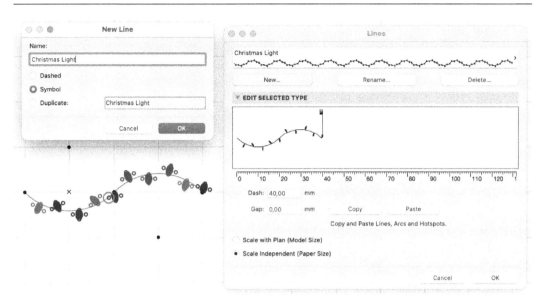

Figure 12.3 – Turning selected lines and arcs into a symbol line type using copy and paste

Now, you can draw lines, arcs, and splines and assign this symbol type to them. Don't overdo the use of symbol lines, as this can put a burden on a drawing, as opposed to dashed or solid lines, which use fewer resources when being rendered to your screen.

Custom fills

This is a very similar story as with the line types – fills use pens to set the color and line thickness. They can also be customized and you have, again, a few different fill categories, as explained in *Chapter 6*.

For plain **Solid** and **Vectorial** fills, the customization is done from the **Fill** editor dialog. While there are different vectorial fill patterns, you cannot create one from scratch. As long as a suitable pattern is available, you can make a copy and adjust its unit size and rotation.

The **screen-only pattern** is an interactive widget, where you can toggle pixels to be shown as an icon or miniature preview in the selection and pop-up dialogs, where fills can be chosen. It's a bit limited, but you can use this as a hint to a user on how the fill will look. This pattern is not used to draw the fill, only as a preview in the interface.

A **Symbol fill** can be created by copying and pasting a series of lines, arcs, and hotspots and adjusting the unit size, scale, and rotation of the pattern in the dialog. The **Strokes** setting is used to define the shift that can occur per row and column of the pattern. Create your own symbol fill by following these steps:

1. Use any technique you want to create a series of lines, arcs, and hotspots. You can use the Construction Grid or Guide Lines while drawing the pattern, but these drawing aids will not be retained in the **Symbol fill**, should you copy them along with the pattern you created.

2. Select them all and copy them to the clipboard.

3. Open the **Fills** attributes dialog and create a **New...** fill. Select the **Symbol fill** and give it a suitable name.

4. Back in the Fills editor, press **Paste** to bring the arcs and lines from the clipboard into the fill definition.

5. Adjust the **Scale**, **Rotation**, and **Stroke** values of the pattern. In the example shown in *Figure 12.4*, some circles and rectangles have been placed across our 2,000 mm by 2,000 mm Construction Grid (the orange dashed line in *Figure 12.4* was created in the first step), so Archicad picks up a pattern unit size that is slightly wider. When we set **Stroke** back to our intended 2,000 mm horizontal and vertical sizes, the original overlap for the pattern is recovered. Otherwise, there would be a larger gap between the units. This will not be the case if the **Original Arrangement** option is ticked, as this enforces all the drawn elements into their outer bounding rectangle.

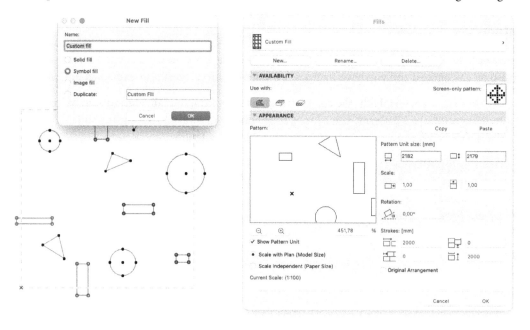

Figure 12.4 – Creating a custom fill pattern by drawing lines and pasting them as a Symbol fill

Finally, the **Image fill** can be created from any bitmap picture available in the Archicad library. The rest of the customization is defined in the **Fill** dialog, so there is not much more to it, to be honest.

Organizing attributes

Since the release of Archicad 26, it is possible to organize the different attributes into a folder structure, allowing you to group related attributes in a logical tree-like structure. Instead of inventing complex names to help with sorting, each attribute can be placed in a relevant folder. The default template already hints at a possible organizing method, but you can adapt this to your liking.

Using the Attribute Manager

As part of the organization of attributes, you can use the **Attribute Manager**, which helps you not only get a complete overview of all attributes but also transfer attributes across files, so you can refine or reorganize an existing project and import attributes from other projects.

You can open **Attribute Manager** from the menu (**Options** > **Element Attributes** > **Attribute Manager…**), which gives you an extensive two-panel dialog, as shown in *Figure 12.5*. On the left side, you get a full list of all attributes as defined in your currently open project file. On the right side, you have a temporary area for attributes.

Each attribute is listed with its **# index** (the hash column) and its name. While the name is what you typically have in mind when working with attributes, it is the index that is the main organizing method. Names can change as you see fit, but if the index changes, it may have a big impact on the way the attribute is used in a project.

> Tip
>
> Always be mindful of the index when performing attribute management operations! It is the basis for how attributes are referred to and linked to each other in Archicad and, therefore, essential for achieving the results you expect.

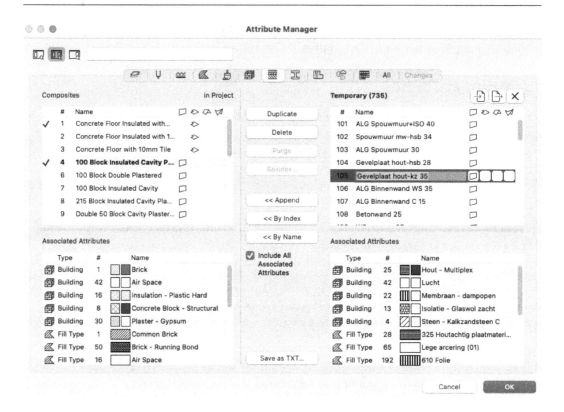

Figure 12.5 – Archicad's Attribute Manager lets you control the attributes of your elements

You can use Attribute Manager in the following ways:

- Use the right side as a **Temporary** area, into which you can copy attributes from the left side of the panel, adapt them on the right side, and copy them back into the left panel

- Use the right side to export a group of attributes into an external file (XML) – for example, to be used as an attribute template file for your office

- Use the right side to import attributes from an attribute file (the XML format from the previous step or the older AAT format) or another Archicad project file (PLN) – for example, if you have a project or template that is already set up as you want

Main operations in Attribute Manager

At the top of Attribute Manager is a **Search** field, which you can use to search the left panel, the right panel, or both panels, using the toggle button.

Underneath is a series of **tabs**, which act as filters; you can either show all attributes (the second-to-last tab) or focus on a specific attribute category (from left to right) – **layer sets**, **pens**, **line types**, **fills**, **surfaces**, **composite structures**, **profiles**, **zone categories**, MEP systems, and **operation profiles**.

The last tab is used to display the planned changes, allowing you to either apply them all or revert (some or all) the changes.

Depending on the chosen attribute category and whether you select an attribute in the left or the right panel of the dialog in *Figure 12.5*, you can perform a series of attribute operations using the buttons in the middle.

> Caution
>
> Beware that attribute operations can be complex and may have interdependencies, which is why Archicad presents this as a transaction or sequence of operations – you prepare all operations but only make them final when all modifications are staged.

Let's go through the different available operations. Please keep in mind how these operations act based on the index number of the attribute, as explained in the introduction of *Using the Attribute Manager* section:

- **Duplicate** copies the currently selected attribute and adds it to the end of the list. It becomes a new, independent attribute. Beware that associated attributes are *not* duplicated, but still reference their original attribute.

- **Delete** removes the attribute from the list. This does not remove associated attributes.

- **Purge** removes all attributes in the list that are not referenced (used) anywhere in the project. This is a dangerous operation, as it may turn your model into a state that makes it very difficult to change afterward, since you will be limited to attributes that were in use at the time of the purge, and you will no longer have access to all the attributes provided in your template.

- **Reindex…** is used to change the index of an attribute, provided you select one of the available indices. This change is applied everywhere the attribute is referenced. This will open a small pop-up dialog in which you will be able to see whether the chosen new index is available or not.

- **Append** makes a copy of the attribute on the selected side and adds it at the end of the list on the other side. Note how the button displays << or >> to indicate the direction that the operation will follow (from left to right or from right to left).

- **By Index** overwrites the attribute on the other side, based on the index. This works well if you have files that started from the same template but you want to update the appearance of the attribute, without changing the index structure.

- **By Name** overwrites the attribute that uses the same name, regardless of index.

Note that **Append** and **By Index** (between the two columns, at the center of the dialog in *Figure 12.5*) never result in attributes with identical names. When a duplicate of an attribute is made using these methods, (1), (2), and so on are automatically added as a suffix at the end of the name of the respective first, second, and so on copies.

The complexity of Attribute Manager

One of the biggest challenges when extensively editing with attributes is ensuring that related attributes remain related! The key (no pun intended) is in the *key index number*. While each attribute has a name and, in many cases, a small preview, it is the key index that is used in the model. When a wall references a composite structure attribute, it does so by keeping the key index number, so the full attribute definition does not have to be copied and maintained as part of the wall. Choosing another attribute is merely switching the key. This allows attributes to be shared between elements, but it also adds some complexity, with attributes that reference other attributes.

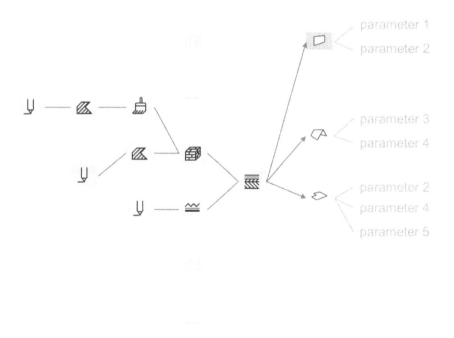

Figure 12.6 – An example taken from Chapter 6 – illustrating attribute dependencies

A composite structure references potentially multiple building materials, which in turn reference surfaces, fills, and pens, as shown on the left side of the diagram in *Figure 12.6* (a copy of *Figure 6.16* in the *Learning to use line types as an introduction to Archicad attributes* section in *Chapter 6*). You can copy attributes back and forth, but it is the index that links them together. If you copy an attribute by name, it may receive another index or overwrite an existing attribute that happens to have the same index.

This is especially challenging if you try to manage attributes across files that use a completely different template, as all indexes may be used completely differently. So, we advise you to be very cautious in working with Attribute Manager and suggest you work in a specific order – composites, then building material over surfaces and fills, and then lines and pens. Don't do this the other way around, unless you know exactly what you are doing.

This concludes the discussion on annotating 2D views and creating your own line and fill attributes to do so. With the knowledge of Attribute Manager, you can manage your self-made attributes and even transfer them to other Archicad projects. In the following section, we will further explore the settings available for sections and elevations.

Configuring sections and elevations in detail

In this section, we will go back to creating **viewpoints** for sections and elevations and see how we can use the settings to get exactly the 2D representations our project needs. Creating a viewpoint involves just two clicks, but setting one up in detail does need more explanation! You can use the model you have created throughout this book (or download the result from the *Chapter 7* folder on GitHub: `https://github.com/PacktPublishing/A-BIM-Professionals-Guide-to-Learning-Archicad/blob/main/CH12_Result.pln`).

Fully configuring a section

In the first section of *Chapter 6*, *Using basic section, elevation, and independent viewpoints*, we learned how to create a section (viewpoint) and how to edit it in a basic way. However, there is much more to a section than drawing a line on a floor plan to create it and merely giving it an ID and a name. In this section, you will learn how you can use the detailed settings of the **Section** tool to fully configure how the 3D elements of the model are visually represented in the 2D drawing that is created.

When you select the section you created in the project for the first module on a floor plan and open its settings (*Cmd +T/Ctrl +T*), you get a long list of options. In *Chapter 6*, we only focused on the **GENERAL** tab. Right beneath this tab, you can open the **MODEL APPEARANCE** panel. There, you have a series of settings that define how the 3D geometry is projected onto the 2D section view and how it will appear.

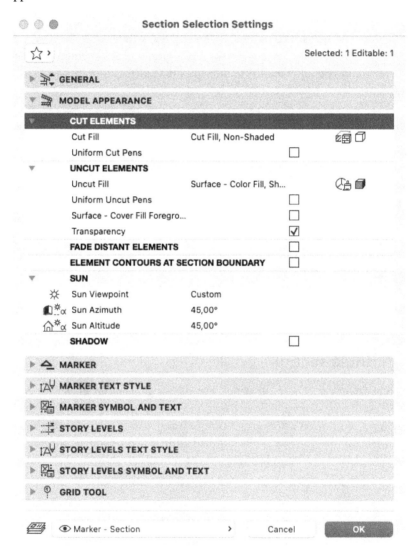

Figure 12.7 – Section Selection Settings – MODEL APPEARANCE

There are many options here, which allow you to go from the default, somewhat random colored view, via a clean, black-and-white technical drawing, all the way to a colorized rendition of the view, including shadows, textures, and transparency.

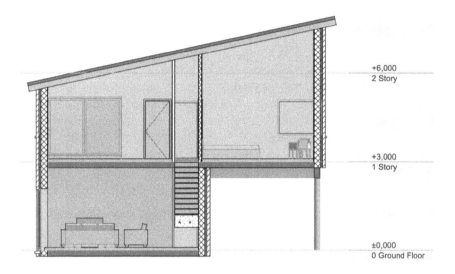

Figure 12.8 – Default settings result in a colored, slightly cartoonish rendering

There are a few subsections in this settings dialog:

- **CUT ELEMENTS** allows you to define how to render all elements that are cut by the section plan:
  - **Cut Fill** defines the fill or color used. By default, it is derived from the element's composition, displaying the fills from the different Skins (**Cut Fill, Non-Shaded**). You can switch to display the **Surface - Color Fill, Shaded or Non-Shaded** fill of the element instead, which uses a monochromatic representation of the used surfaces of the materials. Remember how a building material attribute depends on surfaces for its 3D representation. This results in a less technical look for the section, as it doesn't use complex fill patterns, based on (local) drawing conventions. With **Uniform Surface - Color Fill, Non Shaded**, you can overwrite all section fills to have the same surface, using one of the available surfaces from the attributes. In addition, due to the way Archicad merges fills, you get a very clean-cut drawing, with no intermediate lines where two objects touch. This is very useful for preliminary design presentation drawings or when your section is displayed very small, such as those you would see in an architectural magazine. To achieve the result shown in *Figure 12.10*, choose this setting and combine it with **Paint - Ivory Black** as the **Surface - Color Fill** option.

> **Note**
> The same "effect" can be created by using **Graphic Overrides**, as described in *Chapter 13*.

- Additionally, you can override the cut pens so that every element that is cut uses the same uniform cut pen. A black, medium-thickness pen is an obvious choice here.

- **UNCUT ELEMENTS** configures the projected geometry, with quite similar options:

  - **Uniform Pen - Color Fill, Non-Shaded** gives you the option to override all pens with a uniform cut pen and apply **Surface - Cover Fill Foreground** for a more abstracted display of everything at the back of the section. When you opt for a surface fill, this space can be combined with a color or texture fill, with or without shading. With these variants, you can make the section appear more photorealistic, without having to resort to manually coloring the section using symbol or image fills. Let's set this pen value to 121 to get a white background, as shown in *Figure 12.10*.

  - Combine this setting with **Uniform Uncut Pens** activated and set to a gray pen, such as 102, to achieve the desired pre-design rendering.

You can see the preceding settings in *Figure 12.9*.

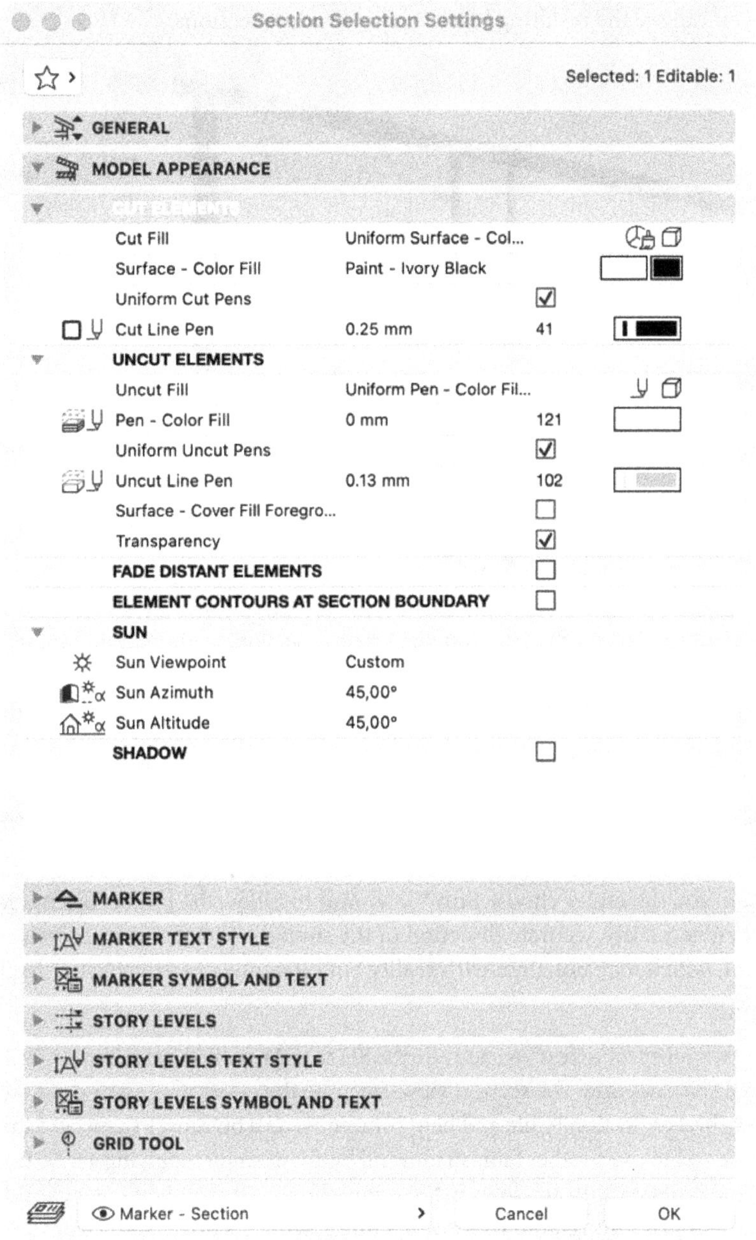

Section Selection Settings

☆ ›   Selected: 1 Editable: 1

▶ GENERAL

▼ MODEL APPEARANCE

▼

	Cut Fill	Uniform Surface - Col...		
	Surface - Color Fill	Paint - Ivory Black		
	Uniform Cut Pens		☑	
Cut Line Pen	0.25 mm	41	

UNCUT ELEMENTS

	Uncut Fill	Uniform Pen - Color Fil...		
	Pen - Color Fill	0 mm	121	
	Uniform Uncut Pens		☑	
	Uncut Line Pen	0.13 mm	102	
	Surface - Cover Fill Foregro...		☐	
	Transparency		☑	

FADE DISTANT ELEMENTS   ☐

ELEMENT CONTOURS AT SECTION BOUNDARY   ☐

▼ **SUN**

☼ Sun Viewpoint   Custom

Sun Azimuth   45,00°

Sun Altitude   45,00°

SHADOW   ☐

▶ MARKER

▶ MARKER TEXT STYLE

▶ MARKER SYMBOL AND TEXT

▶ STORY LEVELS

▶ STORY LEVELS TEXT STYLE

▶ STORY LEVELS SYMBOL AND TEXT

▶ GRID TOOL

◉ Marker - Section   ›   Cancel   OK

Figure 12.9 – The settings for a simplified black-and-white representation

In *Figure 12.10*, you can see the resulting black, gray, and white sections.

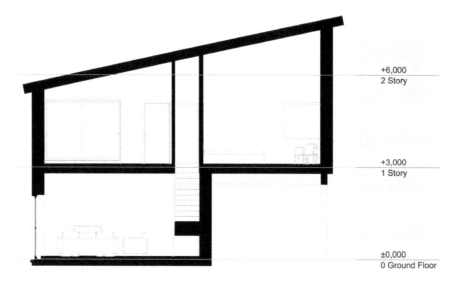

Figure 12.10 – The result of the settings – a clean section, ready for publication

To give more depth to the section – for example, to give a hint of the surroundings – without making the section drawing too busy or pronounced, you can activate the **FADE DISTANT ELEMENTS** option. This basically gives you a second set of options, like **UNCUT ELEMENTS**, to configure elements that lay further away from the section plane. We will use this setting in an example of the elevation, shown in *Figure 12.18*, for which it is more commonly used.

In the **SUN** section, you can either choose **Sun Viewpoint** to follow the project orientation as it is in the 3D window to give a fairly accurate direction of the sunlight and shadows, or you can opt for a **Custom** viewpoint, which may not align with reality but often gives sections better legibility, albeit looking a bit fake.

The **SUN** settings take no effect unless you activate the **SHADOW** section, which uses the 3D geometry to project vectorial shadows onto the section view. Imagine the effort needed to draw such shadows manually! Shadows bring a section to life and help communicate a building's shape and the position of elements. We advise you to use a rather faint shadow fill by, for example, applying a 50% transparent fill and a gray pen. One problem with shadows in a section, though, is that light enters as if the building were cut open and not just through the window openings, which is not always clear to the viewer.

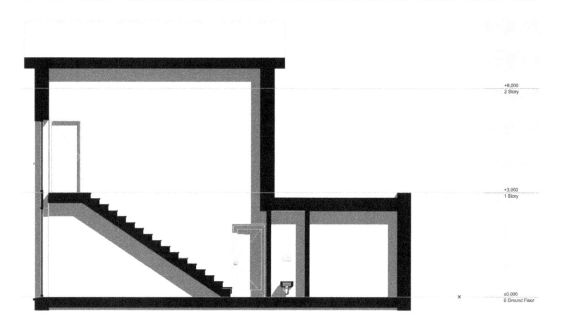

Figure 12.11 – A cut-open 3D model created with the Section tool and using a shadow fill

> **Note**
>
> Beware that the settings in the previous part are *not* defined *for the view* but for the Section *as a viewpoint* – an object represented as a linear symbol on the floor plan. So, if you have multiple Views referring to the same Section, they will look the same. This has the benefit that you can copy settings easily when creating a new Section, by picking up parameters – *Alt + LMB* – as shown in *Figure 12.12*. If you want to create different versions of a Section, you have only two options – creating two overlapping Sections while applying different Section Settings, or creating just one Section and seeing whether you can attain the desired variation using Graphic Overrides. How these work is explained in *Chapter 13*, in the *Graphic Overrides* section.

Let's apply the preceding remark to our project. Start by picking up parameters from the Section you have just set up in the previous subsection.

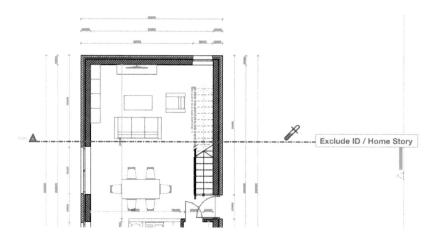

Figure 12.12 – Pick up parameters (settings) from the first section object in the floor plan

Activate the **Section** tool to create a second Section.

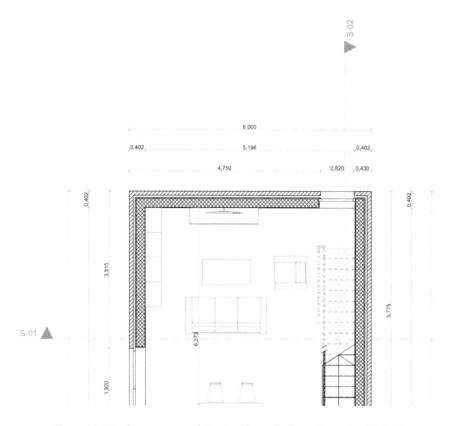

Figure 12.13 – Create a second Section through the stairs and call it S-02

Open the Section by double-clicking it in the Navigator or *RMB* on the section object you just placed and choose **Open in New Tab…**.

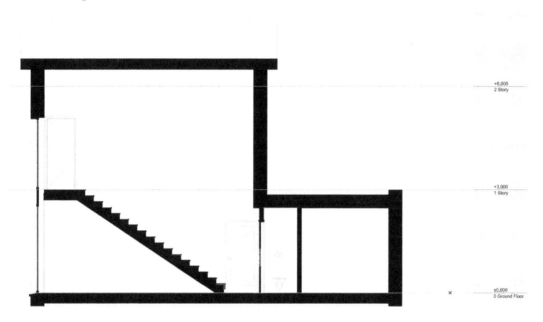

Figure 12.14 – The second section reflects the settings made in the original section

By now, you should have a clear understanding of the settings for sections and how these influence the way these viewpoints are represented. In the next part, we will do the same for elevations. We will see a lot of similarities, but we will also explore a few options that are more common for elevations.

What about Elevations?

Once you understand how the settings work for a Section, you'll have no issue configuring the model appearance of elevations, as they have the same options. You can play with surface colors, fills, and shadows to turn your flat elevations into attractive, colorized, and more realistic façades. For elevations, adding shadow and textures is often a good way of showing some depth and realism. Of course, sometimes a technical drawing is also needed.

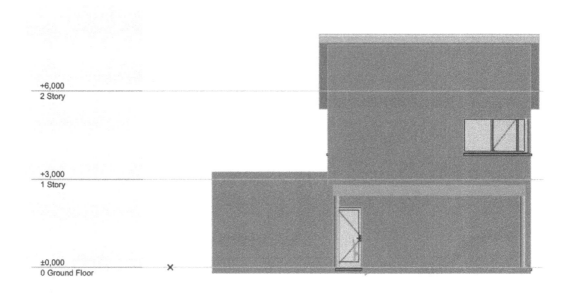

Figure 12.15 – East Elevation with default settings

By using what we have learned in the *Fully configuring a section* section, we can easily set up an elevation for the desired result. The available options are the same, but for an elevation, we often make other choices. This time, adding shadow is crucial, and for more depth, we can experiment with fading the more distant elements. Use the settings in *Figure 12.16* to achieve a similar result and experiment a bit.

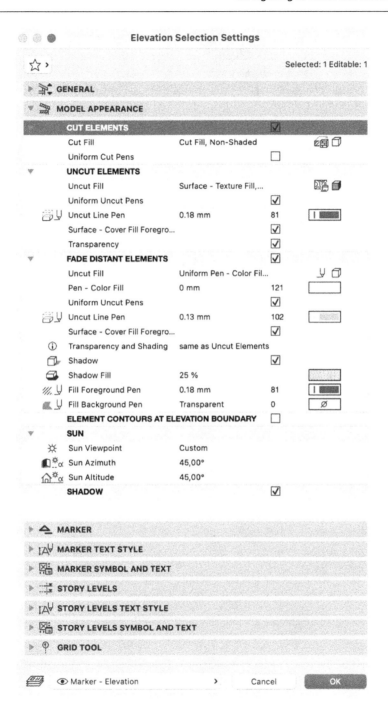

Figure 12.16 – Settings used for the result in Figure 12.18

To fade distant elements, you must set a threshold. When **FADE DISTANT ELEMENTS** is activated, a second line appears in the floor plan at a distance from the elevation object itself. Everything beyond this mark will be faded according to the elevation settings. Using the pet palette, you can move this line to the desired position (*Figure 12.17*) – in our example, just in front of the eastbound, outside wall on the ground floor.

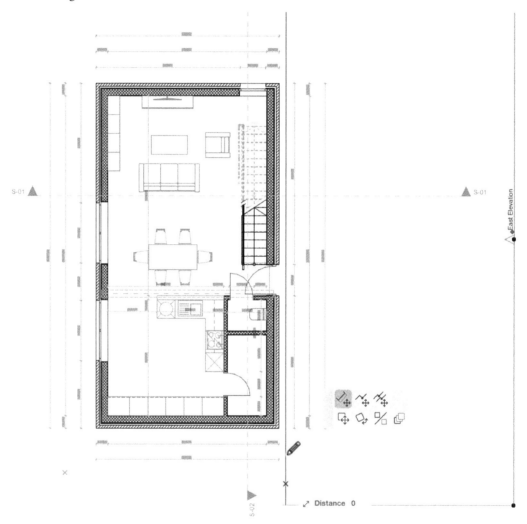

Figure 12.17 – Using the pet palette to adjust the threshold for fading distant elements

After positioning the threshold, open your elevation view and watch the result.

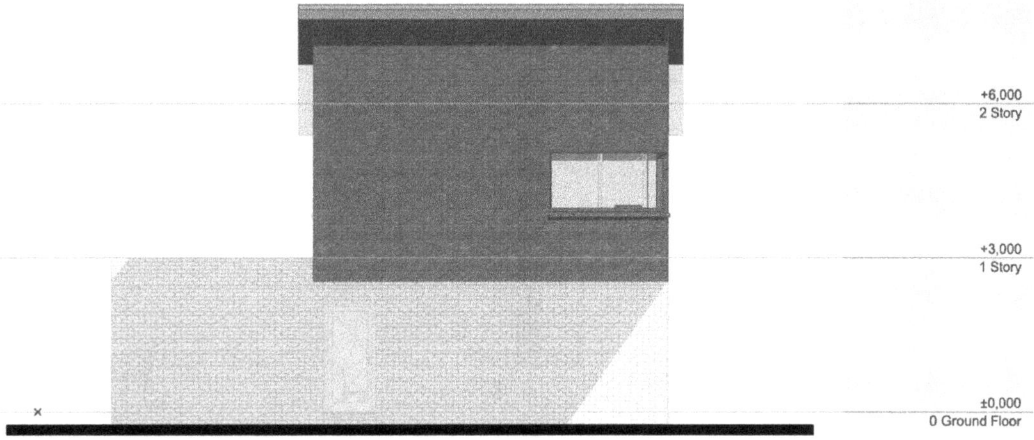

Figure 12.18 – The final result for the east elevation with a black fill added at the bottom

This concludes our exploration of Section and Eswswlevation settings. Now that you know how to set these up according to your wishes, let's have a look at how we can keep track of the design in the different views you will have created by now.

Combining elevations and sections with other views using Trace and Reference

In traditional 2D drawing methods, you often refer to other drawings to ensure coherence between sections, elevations, and floor plans. While less of a concern with BIM software, Archicad provides the **Trace and Reference** tool to help align different parts of a building. With this tool, you can reference a façade and align it with the floor plan to check the alignment of windows across stories, or you can reference a section in an elevation view to see interior and exterior views at the same time.

Just like we did in the floor plans in *Chapter 3*, you can reference any other Viewpoint inside one of the Section or Elevation Viewpoints and play with colorization, transparency, and alignment, with full support for Guidelines and snapping.

In *Figure 12.19*, you can see the Trace Reference in use.

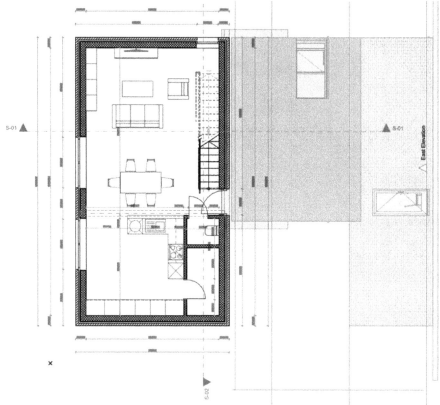

Figure 12.19 – East elevation shown as a trace reference in the ground floor plan view

This concludes the exploration of the **Trace and Reference** tool. By now, you have a good understanding of how the most common viewpoints in architectural drawing can be created, adjusted, and coordinated within Archicad. There are some more uncommon options as well, of which we will explain interior elevation next.

The Interior Elevation tool

The **Interior Elevation** tool is not an entirely different tool compared to the **Section** or **Elevation** tool but, rather, a wizard to generate a series of aligned and linked elevations for a selected zone. In fact, you could emulate this manually with sections and elevations, but using an Interior Elevation is more convenient and can save considerable time.

You can insert an Interior Elevation from the menu (**Document** > **Documenting Tools** > **Interior Elevation**) or from the Toolbox.

Switch to a suitable geometry method (using the *G* shortcut). You can toggle between **Singular**, **Polygonal**, **Rectangular**, and **Rotated Rectangular**. We will stick with **Rectangular**.

The first two clicks define the extent of the Elevations, whereas the next clicks define an offset, indicating the position from which the sections are positioned. Remember that the Status Bar at the bottom left guides you through the steps of any tool in Archicad (see the *Understanding on-screen feedback* section in *Chapter 2*)!

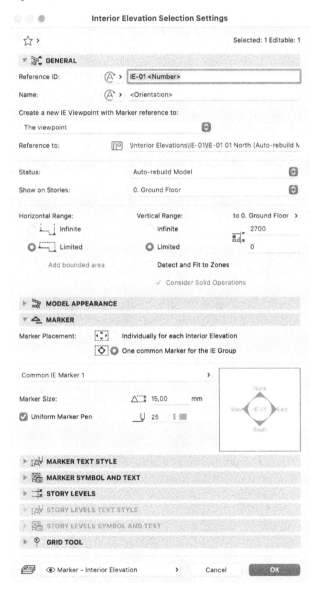

Figure 12.20 – Interior Elevation settings

The settings for the model appearance in an Interior Elevation are the same as what we saw for sections and elevations. With the knowledge from those sections, you should be able to set up Interior Elevations on your own.

This concludes our journey of setting up 2D views derived from the 3D model. Next, we are going to explore how we can represent our 3D model in a bit more in detail. After that, we will finish this chapter with a part on 2D construction detailing in Archicad.

Other ways to display a 3D model

Obviously, creating a model in 3D has the benefit of automatically coordinating changes across the different viewpoints (plan, section, elevation, etc.), but why limit the graphical representation to just 2D drawings when you can benefit from the clear communication inherent to a 3D perspective, or *axonometric* (parallel) projection? This section will tell you how to create specific 3D views, helping you to maximize the use of your 3D model.

Showing specific parts of a project using the Marquee tool

When you are developing a 3D model of a building, it is often difficult to see what you are doing, especially when you are looking at fragments inside the building. Even when you move your view position to the interior, objects may be covered by other objects. You can switch off some layers to hide objects, but you may need to see everything to see how it all works out.

The **Marquee** tool is useful for picking out fragments of a floor plan window and temporarily isolating them in the 3D window, letting you focus on a part and/or look inside your virtual building. It is inspired by the selection tool in graphics software such as *Adobe Photoshop* and uses the same *marching ants* animation.

Here's how you can use the Marquee tool (*Figure 12.21*):

1. Activate the **Marquee** command from the main Toolbox.

2. Now, you should indicate a rectangular area in the 2D window. We will pick an area around the staircase as our region of interest.

3. After you've picked an area, you can show the marquee in 3D (**View** > **Elements in 3D View** > **Show Marquee in 3D**) or use the *F4* shortcut. This will display only the parts of the 3D model that fall within the **Marquee** area in 3D; the rest is sliced away. You can see this in *Figure 12.21*.

Figure 12.21 – Marquee to display a model fragment in 3D

This is a very interesting visualization method, as it allows you to focus on a small section of the building. Moreover, especially when working with large models, this speeds up the display, as Archicad can ignore all parts outside the marquee. It is very helpful when working on corners or intersections because it gives us a closer look at the area of interest.

By default, the **Marquee** tool is shown with a *thin* line. This limits the marquee to only consider the elements of the current story, as displayed in *Figure 12.21*. When you switch the selection method to a *thick* line, all stories are considered.

> **Tip**
>
> If you want a more complex cutout, use the **Polygonal** geometry method. This allows you to draw a precise contour to use as the Marquee area. Note that the **Marquee** tool has similar options as other drawing tools, such as the Geometry Methods.

Figure 12.22 – A polygonal Marquee across all stories

A 3D Cutaway using Cutting Planes

A **3D Cutaway** allows you to add one or more so-called **Cutting Planes**, which allow you to create a *live section* through a model. Such a Cutting Plane will not only clip elements but also show their composite structure. They work very similarly to regular Sections and Elevations but use a 3D viewpoint instead, which means that you can still navigate around the sectioned model.

This gives you a whole range of possibilities to explain and inspect a model and is a wonderful option to have to document your design.

Here is how you can use the 3D Cutaway:

1. From the standard toolbar, select the **Cutaway** tool (*Cmd*/*Ctrl* + *Y*).

2. At the edges of the viewport, a *scissor* icon appears. Click on the left one to start adding a vertical Cutting Plane.

3. Drag the plane across your building. You get an interactive preview of the effect of the Cutting Plane. Click the **Finalize** button that appears when you have completed the positioning of the cutting plane. Note how the view is recalculated and the hollow representation of the elements is now replaced with a fill, according to their composite structure or building material.

4. Add another plane using the scissor icon at the top of the screen to insert a horizontal cutting plane. Drag it downward. Press **Finalize** again to complete the cutting plane placement.

In the following figure, you can see the resulting 3D view after placing two Cutting Planes:

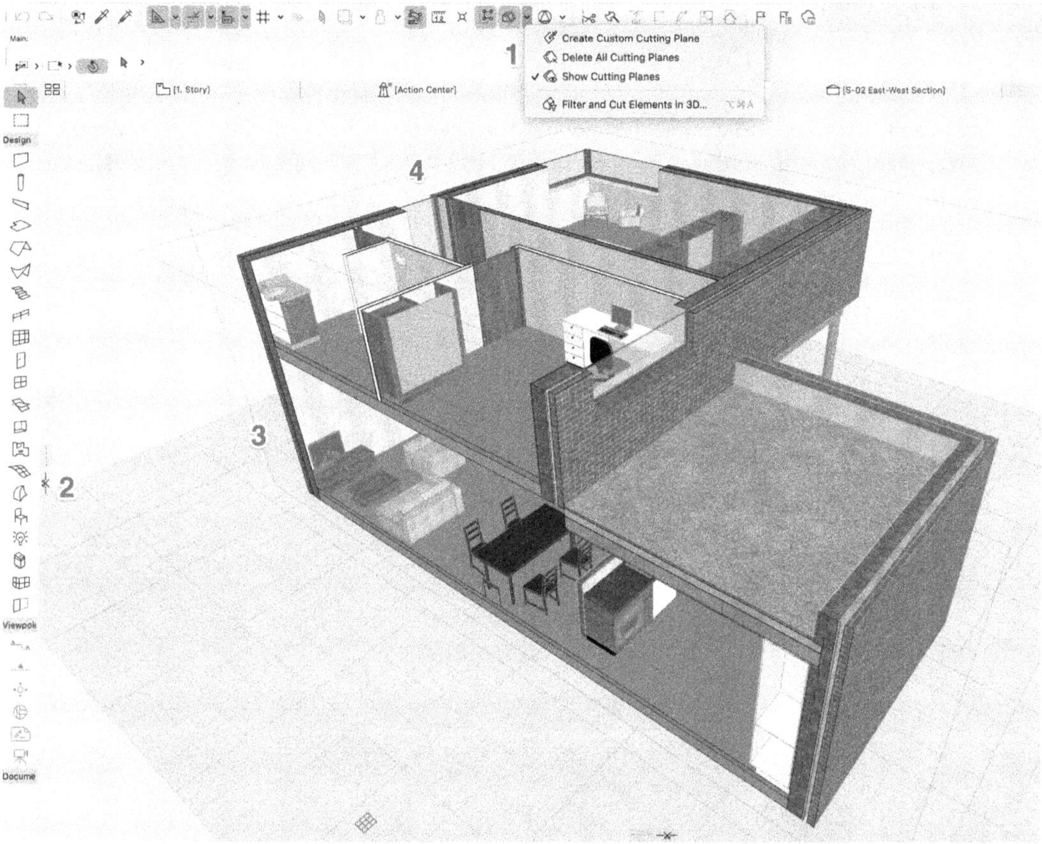

Figure 12.23 – A 3D Cutaway after placing two Cutting Planes

You can still edit the Cutting Planes by clicking on them. From the pop-up menu, you can reverse, rotate, or delete them. This menu also allows you to show or hide Cutting Planes.

To deactivate but not remove Cutting Planes, toggle the shortcut again (*Cmd/Ctrl* + *Y*). This is very useful when modeling the inside of the building from within the 3D window. The outside walls are still visible, and you have all the context you need. This will not be the case when hiding layers, although you can combine Cutting Planes and layers to view the inside of a project more easily. And since this is a fully active model view, you can edit your design as needed, from different viewing directions.

Remark

When you move a Cutting Plane across an area that was already clipped, you won't immediately see the clipped geometry appear again. You have to finalize the position (*step 4* in the previously described workflow) before the geometry is recalculated.

Toggle element categories in the 3D window

From the menu, when you follow the **View** > **Elements in 3D View** > **Filter and Cut Elements in 3D** path, or use the contextual menu on the active 3D Viewpoint in the Navigator, you get a dialog as shown in *Figure 12.24*, where you can adjust the **Element Types to Show in 3D** settings (zones are off by default). You can also change the way the Marquee impacts the view (**Inside Marquee** or **Outside Marquee**), limit the stories to display (**Infinite** or **Limited**), and adjust how cut surfaces are displayed (**Use Element Attributes** or using a **Custom** pen and surface).

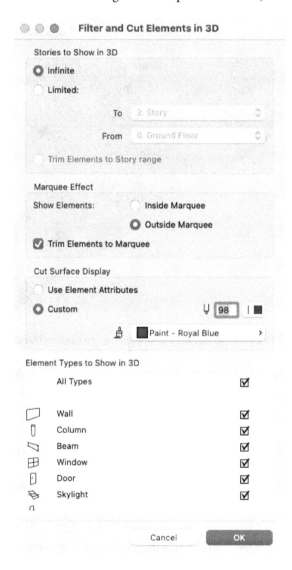

Figure 12.24 – Filter and Cut Elements in 3D

Note that you can even combine the Marquee and the Cutting Planes to achieve a result like the one shown in *Figure 12.25* – blue-colored section fills and the Marquee used to define an L-shaped building fragment.

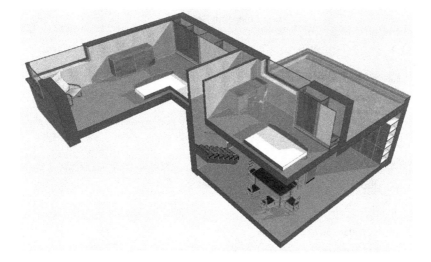

Figure 12.25 – Combining Marquee, Cutting Planes, and the elements in 3D options

Throughout this section, we focused on the visible 3D geometry (walls, floor slabs, furniture, etc.), but we also have tools and options available in Archicad to display the invisible spaces created with the **Zone** tool. Let's see how this works in the next section.

Zones in 3D

When we discussed zones in *Chapter 4*, they were mainly added in the floor plan to indicate the rooms of our model. However, as indicated by their presence in between the design commands, zones are effectively 3D objects; they have a height, but they can also be trimmed or cut off by roofs or floors. This allows you to determine not only the area of a zone but also its exact volume.

Zones can be displayed in other views too, confirming their 3D volumetric shape.

In sections and elevations, they are shown as fills, based on the surface assigned to the zone in its **MODEL** settings (*Figure 12.26*). They typically use a transparent surface, like glass, but you can create dedicated zone surface materials to help their display.

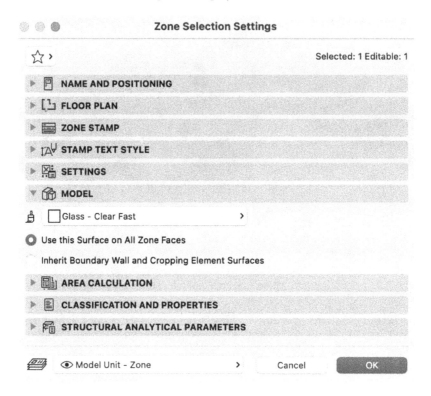

Figure 12.26 – The surface used for the display of the 3D zone volume

You can show or hide zones in a section by toggling their host layer on or off.

In a 3D window, their visibility not only depends on their layer being toggled on but also on **Element Types to Show in 3D** (**View** > **Elements in 3D View** > **Filter and Cut Elements in 3D**). Ensure that the **Zone** element type is toggled on, as it is off by default.

To help make Zones stand out clearly against other elements, we use the **10 Show 3D Zones as Solid** layer combination. This ensures that all other Layers are shown in the Wireframe Display mode. Now, it becomes obvious that our Zones still need to be trimmed against the Roof and Floors of our model.

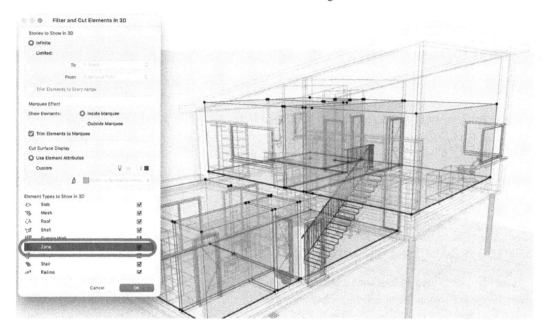

Figure 12.27 – Toggle the display of the Zone element type in a 3D window

You should be familiar by now with **Trim Elements to Roof/Shell** and other **Solid Element Operation** commands, which can both be used to ensure that Zones are properly attached to their floor, ceiling, or roof above. Beware that you must increase the zone height first to ensure that the sloped roof above is touched in this model.

This concludes the part on how you can manipulate the 3D representation of your BIM in general and the options for Zones specifically. In the last section of this chapter, we will look at creating 2D construction details within Archicad.

Construction detailing in Archicad

Up until now, the models we have developed were rather rough in terms of construction detailing. Overall, most of the 3D model will probably not be developed further than necessary for a scale of 1:50, perhaps with some minor aspects on a higher scale. However, to be able to construct a building on site as it was designed, the contractor will often need additional construction detail drawings to explain the configuration of masonry, concrete columns and beams, steel and wood frames, cross-laminated timber, and so on that is used in the design. Instead of adding every small detail for the whole model in 3D, Archicad provides several methods to refine parts of the model at a more detailed scale.

We discussed *Parametric Profiles* as a good way to add more detail to a 3D model in *Chapter 11*, in the *Creating Parametric Complex Profiles and using Geometry Modifiers* section. These profiles can incorporate more detail than plain walls or beams and are very useful up to a scale of about 1:50.

To add even more detail, we will explore the possibilities of 2D construction detailing in Archicad, for which we combine **detail viewpoints** with specific parametric objects and some 2D drafting techniques that we've already shown in *Chapter 6*, in the *Getting started with drafting tools* section.

Creating a detail viewpoint

Archicad has a dedicated **Detail** tool that creates a *detail viewpoint*. This viewpoint can be linked to the **Source View**, which is the original view where the detail was created. The benefit of linking is that if we update the contents of the detail viewpoint, the changes made in the original floor plan or section will be reflected in the details. Although the preceding workflow is the preferred method, it has some drawbacks – it is difficult to make adjustments to graphics that are generated from the model, and you don't have the full flexibility of a pure 2D detail drawing approach. These limitations can be avoided by using Worksheets creatively. The method we will explain further also has the added benefit of giving a user a better overview of details in a project.

Using the Detail tool

Open or create a section in the project you have been creating throughout this book. Make sure your section shows building material fills by setting **Section Settings** correctly, as explained in the *Configuring sections and elevations in detail* section earlier in this chapter. From the Toolbox in the viewpoint section, we can activate the **Detail** tool, as shown in *Figure 12.28*, to create a new viewpoint and, at the same time, define its borders. This can be achieved following these steps:

1. Activate the **Detail** tool.

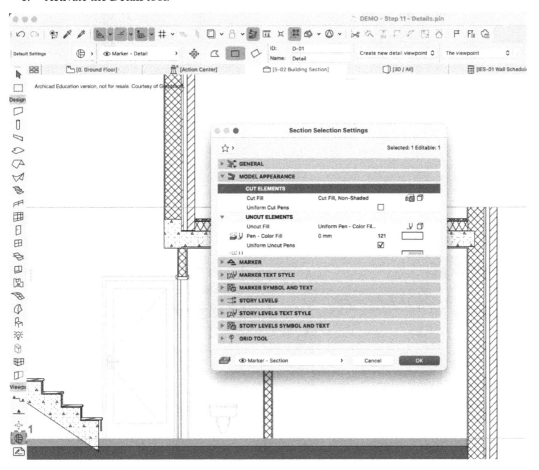

Figure 12.28 – The location of the Detail tool and the correct section settings

2. Click the opposite corners (**2a** and **2b** in *Figure 12.29*) when using the Rectangle geometry method available in the Info Box.

3.   Place the marker using the hammer-shaped cursor.

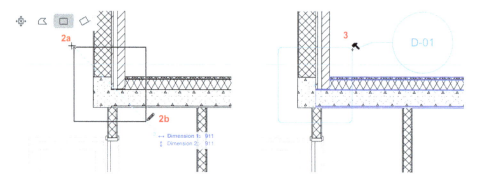

Figure 12.29 – Creating the detail marker

4.   You can find the resulting **D-01 Detail (Drawing)** viewpoint in the **Project Map** of the Navigator.

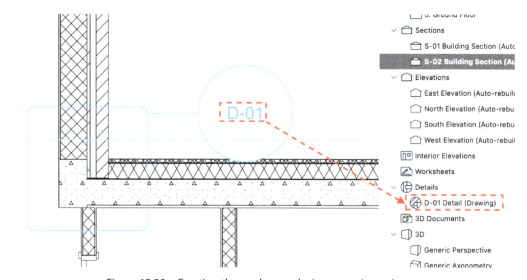

Figure 12.30 – Creating the marker results in a new viewpoint

5.   Open it by double-clicking on it. You will see that Archicad has copied all construction elements from the original section, cutting the resulting (2D) fills with a dashed line.

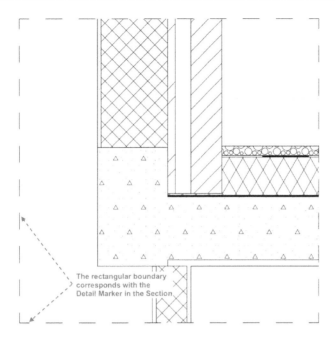

The rectangular boundary
corresponds with the
Detail Marker in the Section

Figure 12.31 – The resulting detail drawing

The preceding steps show you how to derive a 2D detail drawing from the model. As you can see in *Figure 12.31*, this is more like a first sketch instead of a finished drawing. In the next section, we will improve the drawing by adding and editing 2D elements.

Editing and adding 2D elements in a detail drawing

The detail viewpoint you just created shows nothing but a 2D copy of the area that is marked with the tool Tool. Hence, we can safely say that it does not have more detail than the original source view. To develop your details further, you must add extra 2D drafting elements, such as Lines and Fills, or you will have to edit what was generated in the detail by Archicad (*Figure 12.31*). In general, elaborating a detail works just like general 2D drafting, as explained in *Chapter 6*. There are some points to note and tips though, which we will go through using the newly created detail viewpoint.

To start, we will edit some existing elements. In the example used, the alignment of the concrete beam and the plasterwork finish is not perfect, and we also want to change the height of the concrete beam. To do that, follow these steps:

1. Select the fill of the concrete beam.
2. Use the Pet Palette to change the height.

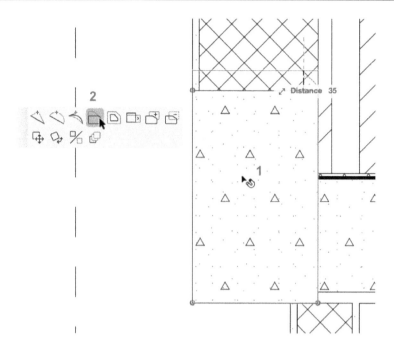

Figure 12.32 – Editing a detail is mainly like editing fills

Note how only the fill was edited and the line representing the edge of the beam stays at the original position. You could, of course, move this line as well, but there is a more efficient technique to change a fill and line at the same time!

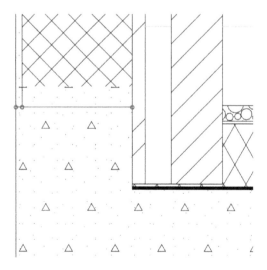

Figure 12.33 – The fill boundary wasn't moved during the edit in the second step

> **Note**
>
> The **Detail** tool marks the part of your model to be shown in a detail viewpoint, with the correct Building Materials and using (Building Material) Fills. In this detailing process, linework and fills are separated from each other. The benefit is that you have more freedom in drafting a detail, but the drawback is that it makes editing more tedious.

Undo any edits you made and clear the selection in the detail viewpoint. Let's see how we can stretch multiple elements more easily:

1. Activate the **Marquee** tool.

2. Place a rectangle around the edge(s) of the element(s) you want to stretch.

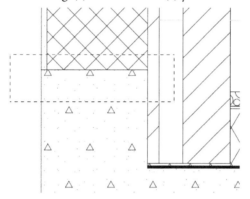

Figure 12.34 – Marking the area you want to stretch

3. Click inside the marked area and move in the desired direction, using the Tracker for exact input.

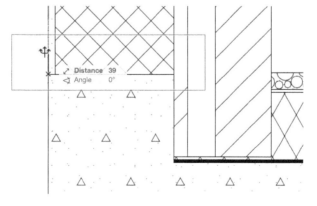

Figure 12.35 – Stretching the "contents" of the Marquee

4. Click outside the marked area to finish the operation.

To finish our edits, we could, for example, add plasterwork onto the beam and add a load-bearing *insulation block* to avoid a thermal bridge at the connection with the roof. Use the following techniques to achieve these edits:

1. Move the edges of the concrete beam (fills and lines) using the same stretching technique described in *Figure 12.35*.

2. Draw a fill with the gypsum material. You can use *Alt + LMB* to pick up parameters from adjoining fills. Make sure you activate the boundary in the Info Box.

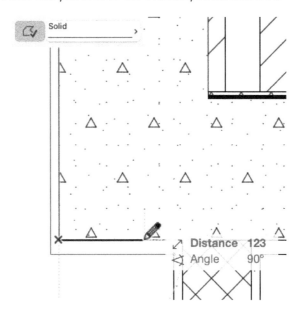

Figure 12.36 – When adding new elements, you can use fills that do have a boundary!

3. Correct the masonry and gypsum fills at the connection of the lower wall.

4. Draw a rectangular fill with the **Insulation - Fiber Hard** insulation material – make sure to change the display order to see the fill by opening the contextual menu.

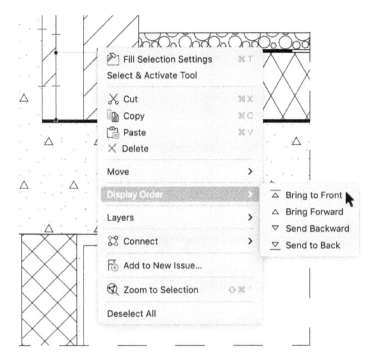

Figure 12.37 - Display Order lets you rearrange the "stack" of elements

Display Order

In most circumstances, the display order of elements and objects in Archicad will be correct by default. **Display Order** is regulated in Archicad by using a mechanism of 14 stacking levels and 7 element classes (e.g., Annotation, Library Parts, 3D structures, etc.). These classes have a hierarchy – Annotation is placed in front of Library Parts. Within the same class, the display order is determined by the order in which elements have been placed.

This seems a little complex at first. You can read more at `https://help.graphisoft.com/AC/26/INT/_AC26_Help/030_Interaction/030_Interaction-102.htm`, but for now, just remember that you can create a custom order using the contextual menu!

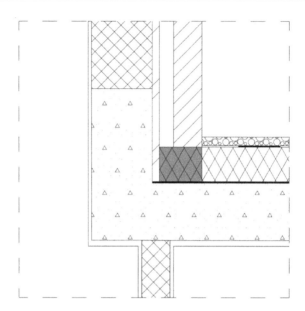

Figure 12.38 – The result of the edits we made

If you followed the preceding steps and have applied your knowledge of how 2D drafting works in Archicad, you should be able to achieve a result similar to the one shown in *Figure 12.38*.

Evolving toward an efficient detailing workflow

The workflow we described in the previous part is rather linear; you may get the feeling details are only created at the end of the design process. Although this is common practice, developing a detail can also be useful earlier on during a project. Imagine, for example, how the edge of a roof determines the overall look and must be aligned with an adjacent wall edge. Remember that the fills from these 2D details can also be reused by copying them into the definition of a parametric profile, as described in *Chapter 11*. This turns a 2D detail into a 3D element of the model.

Knowing this, you might wonder why we still need 2D detailing. Once you have some more experience in modeling, or if you work on larger and more complex models, you will understand that developing everything in 3D is not always the correct or only answer. Creating overly complex 3D geometry tends to slow down your modeling process, as it not only presents a lot of work but also needs a lot of computer resources to calculate the geometric representation, especially with lots of highly detailed or rounded objects.

The key to attaining a fluent workflow is to find the right balance between 2D and 3D detailing. It's a mix of interactive 3D parametric profiles and adding 2D detail elements. After we edited the detail shown in *Figure 12.38*, the result could be copied in its fairly rough state as a Parametric Profile. This way, the updated Section could be repurposed for a more detailed version of the construction detail.

Luckily, such a process does not require redrawing the 2D detail entirely. In the next part, we will learn that Archicad has a built-in update option for details, albeit with some disadvantages.

Updating from a model

During a non-linear design process with multiple iterations, you might imagine that the basis for a construction detail (the source view pointing to a model) might change after the detail was designed. Let's emulate this process in the section we used and with the newly created detail (*Figure 12.38*):

1. Open the section you used.

2. Change one of its elements – in our example, we are going to adjust the height of the concrete beam, maybe as a result of reflecting on the detail (see *Figure 12.38*).

3. Go back to the detail viewpoint; your detail will not update automatically.

4. In the Navigator (**Project Map**), right-click on the detail viewpoint and choose **Rebuild from Source View**.

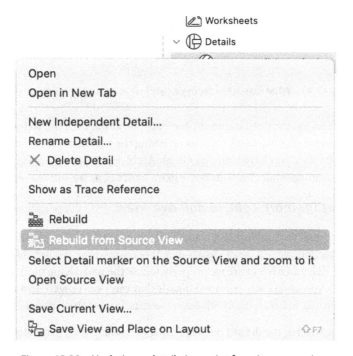

Figure 12.39 – Updating a detail viewpoint from its source view

5. The result will be a mixture of updates from the model and the *added* elements in the detail viewpoint. Unfortunately, all *edited* elements (lines, fills, etc.) will be restored to correspond with the source view.

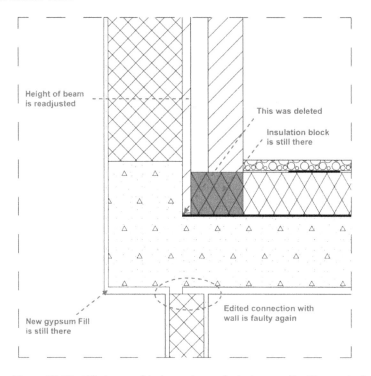

Figure 12.40 – What was added remains and what was edited is reverted

Although the **Detail** tool in Archicad has many benefits out of the box, the **Rebuild from Source View** option presents some disadvantages, such as reversing the modifications we made, as shown in *Figure 12.40*. Luckily, by now, you have learned enough Archicad tools to use a workaround that not only solves these disadvantages but also adds some extra benefits, as we will see next.

Using a Worksheet for more control and overview

This method uses the **Detail** tool, a Worksheet, and Trace and Reference. Instead of placing the detail markers directly on a section, we will create details on a Worksheet using regular 2D drafting methods and rely on Trace and Reference to ensure we properly follow the underlying model. The **Detail** tool will be used to create clean and neatly trimmed details that can be arranged on a layout. The detail markers also create the link between the Worksheet, the detail viewpoints, and the source view.

To use this workflow instead of the default one explained previously, it is best to delete the detail you have created via the **Project Map** in the Navigator; select **Delete** from the contextual menu (*RMB*) and confirm.

This alternative workflow goes as follows:

1. Create an empty Worksheet, give it a suitable name, and open it.

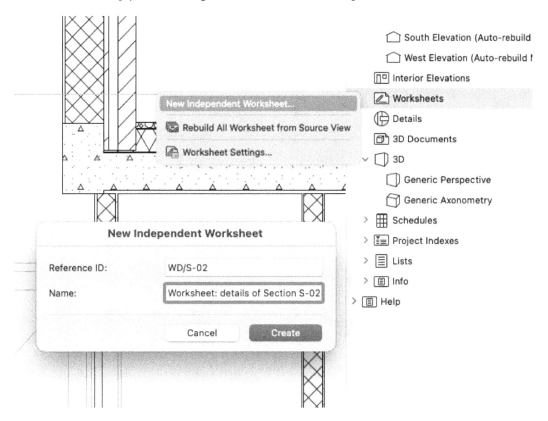

Figure 12.41 – Create a new Worksheet

2. Activate the **Trace and Reference** tool and choose a viewpoint or view to see as a reference (see *Chapter 3*, in the *Trace and reference* section). In our example, we will use the section in which we created the first detail.

3. Using regular 2D drafting techniques, develop the detail with fills and lines. Optionally, you can copy and paste elements from the section viewpoint into the Worksheet. To do this, open the section you are using as a reference and select the elements you need in this viewpoint. Copy the elements to the clipboard by pressing *Cmd/Ctrl + C*. Navigate back to your newly created Worksheet and hit *Cmd/Ctrl + V* to paste the contents of the clipboard. Confirm the position by clicking outside the marked area, or move elements to the correct position by clicking inside the marked area and using the pet palette.

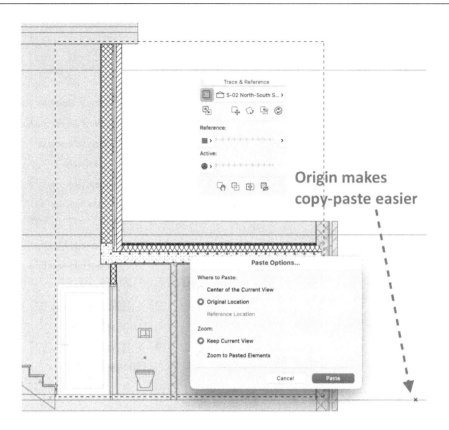

Figure 12.42 – Trace and Reference is set up and elements from the source have been pasted

---

Tip

When copying and pasting, Archicad always pastes elements to their original location. Depending on whether this location is visible in the target view (you are zoomed out far enough or have panned to this location), you are presented with a dialog with two options – **Paste at original location** or **Paste in center of current view**. If the original location is visible, Archicad automatically places the copied elements at this location. In most cases, using the original location is the preferred option, as it makes it easier to align sections, details, and other views.

---

4. Use the **Detail** tool, as explained in the *Using the Detail tool* section, to create a detail viewpoint.

5. Select the detail marker and copy (**a**) and paste (**b**) it into the original section as well. A detail marker is an object that creates a link between viewpoints. A copy of this marker will refer to the same viewpoint as the original marker. See for yourself – right-click on the marker in the Worksheet or the section and choose **Open with Current View Settings** (**c**), and you will end up in the same detail.

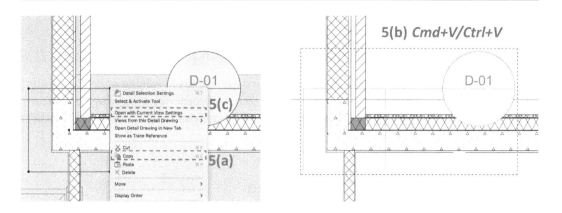

Figure 12.43 – Copy and paste a marker to connect the section and Worksheet with the detail viewpoint

When you edit something in your model now, rebuilding the detail(s) from the model will no longer overwrite the changes you have made to the elements in the Worksheet. To update your details, use the section as a reference and edit the drawing(s) in the Worksheet, and then use the **Rebuild from Model** command.

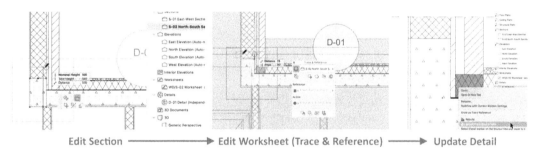

Figure 12.44 – Rebuild the detail without unwanted side effects with this workflow

Now that we have learned an efficient method of creating and updating details, let's see how the detail drawings themselves can be further improved.

Using 2D drafting for construction details

How you can use 2D drafting to edit details is mostly explained in the *Editing and adding 2D elements in a detail* section. However, there are some extra techniques, such as using lines and fills and adding parametric objects, which can come in handy.

Using lines and fills

In this section, we will talk about using lines and fills for construction details.

- Trimming lines by using **Trim** or *Cmd/Ctrl + LMB* is a quick way to edit linework. The cursor will show a pair of scissors to indicate that you can trim an element when you press the shortcut key. This tool is also available through **Edit** > **Reshape** > **Trim** or by using the icon in the Toolbar (as shown in *Figure 12.45*).

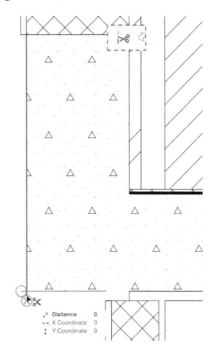

Figure 12.45 – Trimming elements

- You can use **Linework Consolidation** to quickly get rid of duplicate lines in detail viewports and Worksheets. Duplicates can either come from creating the viewpoint or after copying and pasting elements. Select the elements you want to have analyzed and edited, access the linework consolidation command through **Edit** > **Reshape** > **Linework Consolidation…**, and go through the steps in the pop-up dialog. In most cases, the default settings will suffice. In the end, you are presented with a summary, showing how many elements were adjusted and in what way.

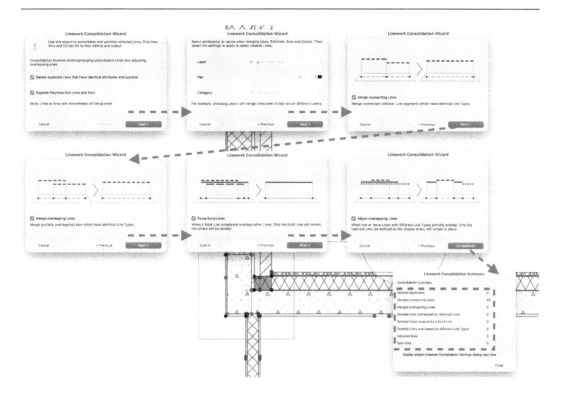

Figure 12.46 – Select the elements you want to consolidate and go through the steps

There is also a **Fill Consolidation** command available that works in a similar way, albeit with only one step – try it!

Adding parametric 2D objects

Aside from using 2D drafting techniques with lines and fills, you can also add parametric 2D objects to your details. They can save a lot of time, especially when adjusting a detail. These objects are typically available when using a Subscription license, as explained in *Chapter 1*, in the *Licensing* section. In *Figure 12.47,* you can see an example of a detail, with the used objects selected/highlighted in the image. Examples of such objects are insulation, masonry with configurable joins and bricks, and prefabricated concrete elements.

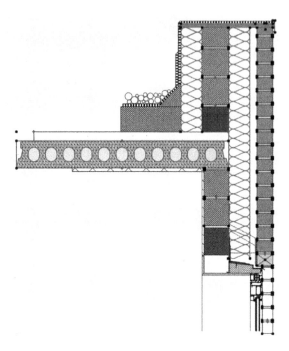

Figure 12.47 – Using 2D detailing objects included in a (Belgian) Subscription license

The image in *Figure 12.47* shows several 2D objects, with handles (interactive hotspots) to adjust them to your project. It clearly shows how details benefit from mastering 2D techniques in Archicad, which brings us to the end of this chapter.

Summary

In this chapter, we further explored the sections and elevations of our model. We added various ways of representing our model in 3D with Cutaways and the **Marquee** tool, allowing us to present our design in various ways. We also explored how we can add detailed 2D construction documents to our model, as a means of creating a complete set of construction documents. In the next chapter, we will explain how we can further use the information in our virtual building for data extraction and visualization.

13
Data Extraction and Visualization

In this chapter, we will finally explore the meaning of *I* in *Building Information Model (BIM)* a bit further and mainly focus on the information in our **BIM**, using the project we have been modeling in previous chapters. We will learn how to create and use **Schedules** to check the information in our model and will create Schedules to extract geometrical and non-geometrical information from the model. We will also further develop drawing styles with Graphic Overrides, visualizing the data inside the model in a graphical way.

The topics covered in this chapter are as follows:

- Using Schedules for information extraction
- Creating and using Graphic Overrides to change the way the model is represented in various views or visualize non-geometric information

These skills allow you to take full control of how you derive data from the model and how you can visualize this data in your 2D and 3D output.

Using Schedules for information extraction

A BIM is more than a collection of 2D drawings and 3D elements. In the database of the model, there is also all the information about the quantities, dimensions, names, renovation status, load-bearing functions, and so on of all elements that are added to it. The output of our model can, therefore, also consist of Schedules and preparations for a bill of quantities (BOQ). In a well-organized model, this output can even exceed the input (and this should always be the aim, as attaining an improved overall workflow should be one of the reasons to start using a BIM). On the other hand, sufficient and correct input is a prerequisite for generating useful output. In this chapter, we will look at how Schedules

can help us improve the model and how the quantities of the materials used can be retrieved from the model. For this, we will start from the examples that are available in the Archicad 25 Template and try to adapt them to our wishes.

Using Interactive Schedules in Archicad

In Archicad, we use **Interactive Schedules** to list elements and their properties. Several such Schedules are included in the Navigator's Project Map. Like floor plans, façades, and sections, these Schedules are also a view of the model. This means, among other things, that **Filters (Quick Options)** also affect these lists. Interactive Schedules can be used for the following:

- **Checking the model**: A list of the elements and the layer in which they are placed can quickly reveal errors.

- **Changing element properties**: The Schedules

- are *interactive*; some properties/parameters can be changed from within the lists – for example, the *ID* of the objects. This often works faster and more clearly than through the selection settings.

- **Selecting elements in the project**: Through Schedules, you can look up and visualize selected elements in the floor plans of a project.

- **Searching for and checking Industry Foundation Classes properties**: Industry Foundation Classes (IFC) (`https://www.buildingsmart.org/standards/bsi-standards/industry-foundation-classes/`) is an open international standard (ISO-certified since 2013) created by *BuildingSmart* to allow a vendor-neutral way of exchanging 3D geometry and building information across multiple native software. For the purpose of exchanging geometry and data/information with other software (by means of IFC), the IFC categories can be checked with a Schedule.

- **Listing quantities/preparing a BOQ**: Since (almost) every element is modeled in three dimensions; quantities can be listed very accurately. Of course, the precision depends on the detail with which the model is made (what is not modeled or cannot be listed). This is not only useful for surfaces and volumes of, say, walls and floors but also for making window and door lists.

Note

In Archicad, there are also "lists," which can also be used to list elements and their properties. This is an older technique and is not used much anymore. Limitations of this technique with respect to Interactive Schedules are as follows:

- It is a snapshot (and there is no live view of the BIM)

- No project changes can be made

- Setting up lists is much more complex

- The formatting possibilities are very limited

Understanding the different kinds of Interactive Schedules in Archicad

Archicad has three different types of Interactive Schedules:

- **Element Schedules**: With these Schedules

- all properties of any basic elements (walls, floors, objects, windows, doors, etc.) can be listed. The information on the individual layers (Skins) of a composite *cannot* be listed.

- **Component Schedules**: Specifically for listing properties of each component of composites (walls, floors, roofs, etc.). With these, you can request the same information as with Element Schedules but for each individual Skin.

- **Surface Schedules**: Specifically intended for listing the properties of element surfaces (walls, floors, roofs, etc.). With this, you can, for example, query the number of square meters of a certain finish.

Creating a Schedule

It is usually easier and quicker to convert an existing Schedule from a template into one that better suits your purpose, rather than creating it from scratch. However, to do this with sufficient insight, we must first look at how a Schedule is constructed.

To create a new Schedule, right-click in the Navigator on **Schedules**, and from the context menu, choose **New Schedule...**.

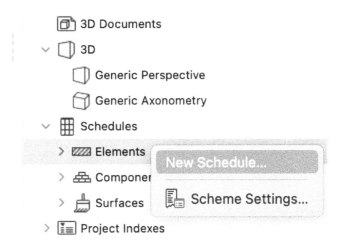

Figure 13.1 – Accessing New Schedule… scheme settings through right-clicking in the Navigator

In the dialog that pops up, you can enter an ID (**1**) and give your Schedule a clear name (**2**) (these can be changed afterward as well). Don't forget to specify which type of Schedule you want to create (**Elements**, **Components**, or **Surfaces**) (**3**). Optionally, you can duplicate an existing Schedule (**4**) – useful if you want to adopt the layout – instead of creating a new scheme (**5**).

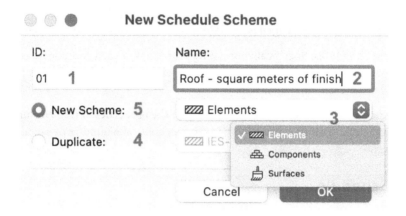

Figure 13.2 – The New Schedule Scheme settings

To add content to the Schedule, the scheme settings (accessed through the context menu of the corresponding list in the Navigator – see *Figure 13.1*) must be adjusted (see *Figure 13.3*) as follows:

- The content is determined by the following:

    - **CRITERIA** (**1**): Which elements are selected to be listed
    - **FIELDS** (**2**): Which properties of the selected elements are shown

- In this dialog, you can also add create new Schedules (**3**).

- Optionally, you can import (**4**) and export (**5**) Schedules. Exported Schedules can be saved in an XML format.

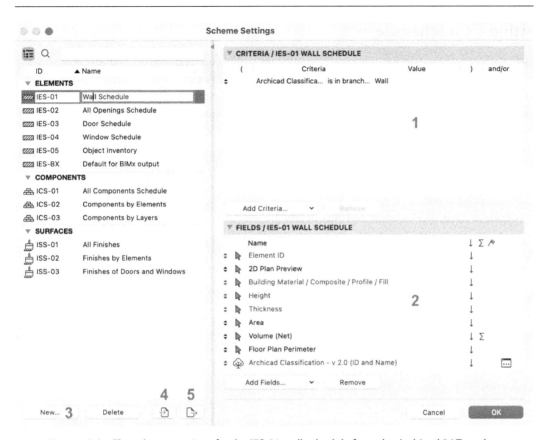

Figure 13.3 – The scheme settings for the IES-01 wall schedule from the Archicad 26 Template

Now that you know the most important definitions of Schedules, where to find them, and how to create or duplicate them, it is time to have a closer look at **Scheme Settings** and learn how you can use **CRITERIA** and **FIELDS** to filter out the data you want to collect from the elements in your project.

Setting up a schedule

To set up your first Schedule, we recommend opening the model you created in the first Part of the book (see *Chapters 3, 4, 5, 6, 7*, and *12*). Open the final result (see or download it from GitHub: `https://github.com/PacktPublishing/A-BIM-Professionals-Guide-to-Learning-Archicad/blob/main/CH12_Result.pln`) and navigate to the Element Schedule called **IES-01 Wall Schedule**. This Schedule lists all the walls in the project and shows a lot of the properties of these walls, such as the following:

- Graphical data such as **2D Plan Preview** (**1**)

- Geometrical data such as the dimensions (e.g., **Height** and **Thickness**) and derived quantities such as **Area**, **Net Volume**, and **Perimeter** (**2**)

- Non-geometrical data such as **Wall Type**, **Element ID**, and **Classification** (**3**)

- A lot of properties that are not yet defined (**<Undefined>**) (**4**)

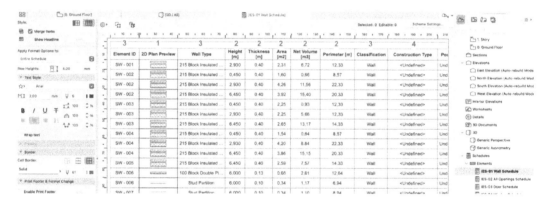

Figure 13.4 – A Schedule from the template, showing the different types of data in the model

> **Note**
>
> Some or a lot of the data in the Schedules might look new for now, but we will gradually get to know more of these details while exploring how Schedules work. There is a vast amount of data available in any BIM and certainly within the highly parametric Archicad software, so showing and explaining every Parameter, Property, or Classification is simply impossible. Do not worry, though. By the end of this section, you will have a clear understanding of how and where you can find data within Archicad!

The graphical and geometrical data is obviously directly derived from our (3D) model, as **2D Plan Preview** clearly shows the fills that are defined in the different wall composites we used in the model, and the dimensions just show the sizes of the different walls we modeled. Remember, as described in the *Using Interactive Schedules in Archicad* section, a Schedule is just another view of the model (just like a floor plan, section, or elevation). So, all the other data can also be found in the model (e.g., the **Wall Type** column in the Schedule actually shows the composite used for that element). To experience this connection between Schedules and the model fully, please do as follows:

1. Select any item in the Schedule by clicking the corresponding row.

2. Choose **Select in 3D** (**a**) by clicking the appropriate icon above the Schedule (you can also choose **Select in 2D** (**b**) if you want to see the element on its home story floor plan).

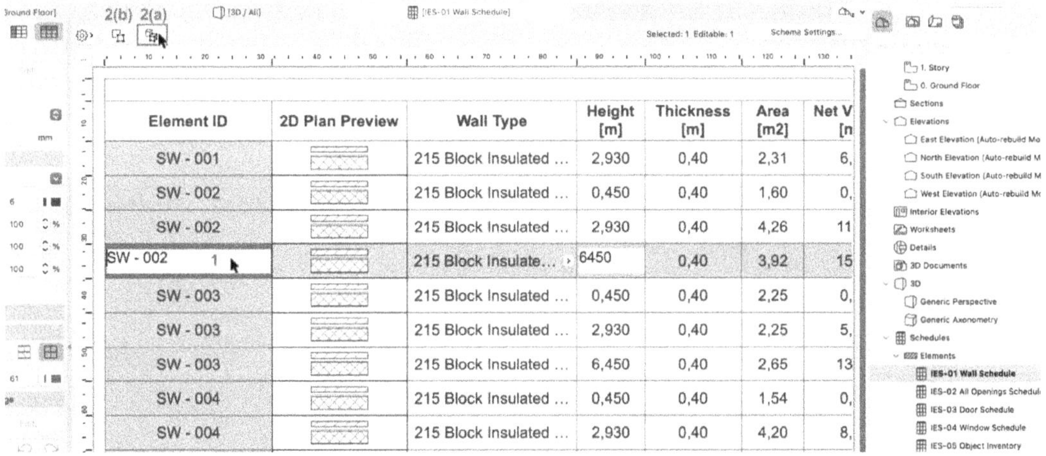

Figure 13.5 – Selecting an element in a Schedule and showing it in 2D or 3D

3. The element selected in the Schedule shows as selected as the 3D window opens.

4. Open **Wall Selection Settings** and scroll down, and in the last tab (**CLASSIFICATION AND PROPERTIES**), change the ID into something distinctly different from the generic IDs that are used in this project.

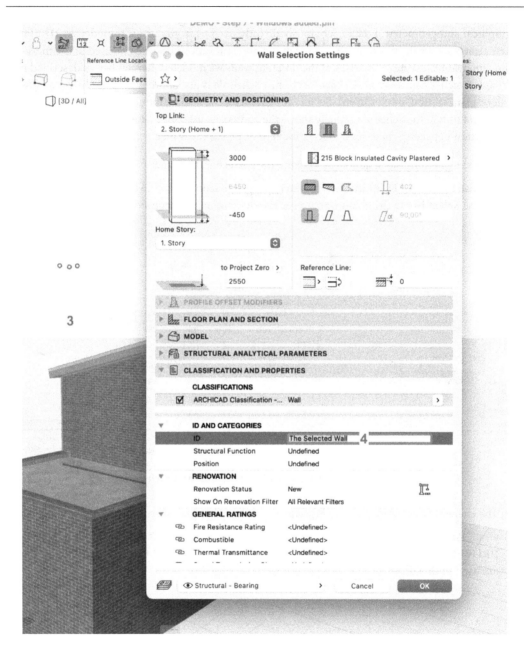

Figure 13.6 – The selection from the Schedule is shown in 3D, ready to change its properties

5. Go back to the Schedule (use the **View** tab) and scroll up or down until you find your renamed element (the list is currently ordered alphabetically). The name has changed in this **View** as well.

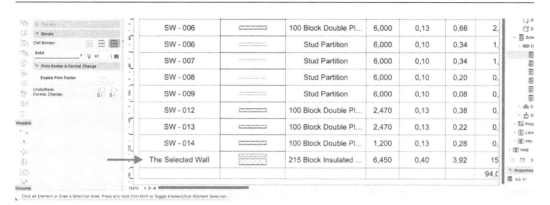

Figure 13.7 – Schedules are a view on the model, so changes work both ways

6. Try changing the name again, but now by clicking the cell inside the Schedule. That's right – some of the values can be changed in this view directly! Of course, this also changes the value in the 3D model (and it is derived 2D representations).

> **Note**
>
> Although we did not add any IDs when creating this model, every element seems to have one (as we can see in the Schedule). Archicad provides a semi-automatic way of creating unique IDs (and this method is applied in the Archicad 26 Template we use). If the first element of any tool gets an ID ending with a number (most logically, 1, 01, or 001), every next element will get an ID with 1 added to this counter. When an element is copied though, the ID is also copied. This explains why, in *Figure 13.5*, you can, for example, see multiple elements with **SW - 002** as an ID. How you can manage IDs or automatically change them in a systematic way is clearly explained in the Graphisoft Help Center (https://help.graphisoft.com/ AC/26/INT/index.htm?#t=_AC26_Help%2F070_Documentation%2F070_ Documentation-128.htm).

Now that we understand that a Schedule is just another view of the model (more focused on data than on geometry), let's see why this specific Schedule shows these elements, with this specific set of properties.

Using selection Criteria

First off, let's see how we can define what elements should show in our Schedule. In other words, we are going to make a selection of elements for this specific Schedule. Follow these steps:

1. In the Schedule View, you can access **Scheme Settings** in the top-right corner.
2. As shown in *Figure 13.8*, the criteria for these Scheme Settings are in the top-right part. This makes up our selection. For this Schedule, the selection is, thus, all elements of which **ARCHICAD Classification - v 2.0** (**a**) has a value, which is in the branch (**b**) of the wall (**c**). So, for example, elements classified as **Roof** will not show in the Schedule.

> **Note**
>
> Classifications are an important concept within BIM. Using these, you can organize the geometry of your model and exchange it in a clear and readable way with other users and other software (e.g., using IFC). For the purposes of this book, it is enough to know two things about Classifications in Archicad – you can freely classify any element, object, or space as anything, and Classifications are *not* automatically linked to an element or tool (although by default, a wall element will be classified in the **Wall** branch, a roof in the **Roof** branch, etc.).

3. As we do not yet know how Classifications exactly work, and so we are not sure that classification values are assigned correctly in our model, let's add some criterion that we do know. Click the **Add Criteria…** button (**a**) to do this (using the drop-down arrow (**b**) to the right provides more and more complex options).

4. In the pop-up list, available Criteria are arranged in groups, such as **Geometry** and **Surface and Materials**. Expand the tree for **Main** and **Model View**. You can see that it is not an easy task for a user to know where to look for the right criteria. For example, do you expect the layer to be under **Model View**?

Figure 13.8 – Opening Scheme Settings and adding Criteria

5. Luckily, you can also use the search window at the top of this list (*Figure 13.9*).

6. Add an **Element** type by selecting it from the list (or search results) and clicking **Add**. Simply double-clicking also works (*Figure 13.9*).

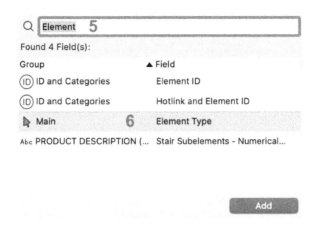

Figure 13.9 – Searching for criteria and adding one

7. We can now delete the classification criterium by selecting it (**a**) and clicking **Remove** (**b**) (see *Figure 13.10*).

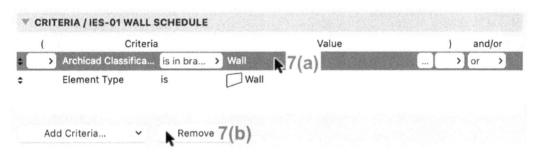

Figure 13.10 – Removing a criterium

Now that we know how to search, add, and remove Criteria, let us take a look at how we can narrow our selection. Follow these steps (see *Figure 13.11*):

1. Using the technique described in the preceding steps, add **Layer Name** as a criterium. As opposed to **Layer**, this criterium has the benefit that we do not have to select just one Layer and can use different statements. These are available in the dropdown when you click the right-hand side of the criterium:

    ▪ **is**: Select elements that are in one exact Layer (of which you know the exact name)

    ▪ **is not**: Select elements that are in any Layer except on this one (of which you know the exact name)

    ▪ **contains**: Select elements that are in all Layers with names containing a specific (set of) character(s)

- **does not contain**: Select elements that are in any Layer, except the Layers with names containing a specific (set of) character(s)

- **starts with**: Select elements that are in all Layers with names starting with a specific (set of) character(s)

- **ends with**: Select elements that are in all Layers with names ending with a specific (set of) character(s)

> **Tip**
> The values you fill in for the **Layer** criterium are *not* case-sensitive (`Structural - Bearing` as a layer name is exact, but `structural - bearing` will also work). Using any of the statements (**is**, **is not**, **contains**, etc.) does not influence this behavior. Also, so-called wildcards (`***`, ?, etc.) that you may have used in other tools (outside Archicad) are *not* recognized as such!

2. In order to finalize our Criteria, we should check the Boolean Operator at the end of the first line in *Figure 13.11*. There are two options:

- **and**: Both consecutive Criteria have to be met in order for an element to be listed in the Schedule. In this example, the elements in the list are of the **Element** type called **Wall**, and they are modeled in a Layer with a Layer Name containing the word `Bearing`. This means that, for example, **Wall** elements the **Interior - Partition** layer will not show, nor will, for example, **Roof** elements in the **Structural - Bearing** layer be added to the list, as these last two only meet one of the set criteria, and both have to be met!

- **or**: Only one of the two consecutive Criteria has to be met in order for an element to be listed in the Schedule. In this example, the elements in the list are either of the **Element** type called **Wall** or they are modeled in a Layer with a Layer Name containing the word `Bearing`. This means that if we use "or" in the example, **Wall** elements in the **Interior - Partition** Layer will show but also, for example, **Roof** elements in the **Structural - Bearing** Layer will be added to the list (and probably a bunch of other elements), as only one of the two Criteria has to be met for an element to be part of the selection in the Schedule.

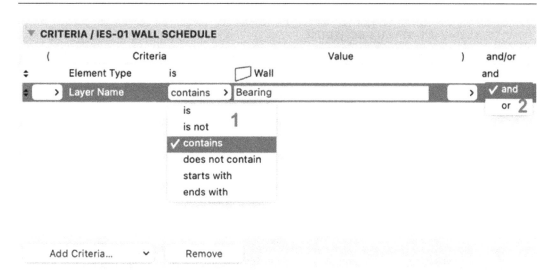

Figure 13.11 – Adding a layer name as a criterium to narrow down our selection

3.   Close **Scheme Settings** and return to the Schedule by clicking **OK**.

> **Note**
>
> When using or as a Boolean Operator, you should pay attention to the use of brackets in setting up your Criteria. This is quite comparable to the use of brackets in mathematical equations, where $4+5*7$ does not equal $(4+5)*7$ at all! The same goes for Criteria – Element Type is Wall and Layer Name contains Bearing or Layer Name contains Interior is not the same as Element Type is Wall and (Layer Name contains Bearing or Layer Name contains Interior). The first will result in a listing of walls on Layers with Bearing in the name and *any* elements in Layers containing Interior in the name, while the latter *only* lists **Wall** elements in any of these two types of Layers.

▼ CRITERIA / IES-01 WALL SCHEDULE

	(	Criteria		Value	)	and/or
↕		Element Type	is	⬜ Wall		and
↕		Layer Name	contains	bearing		or
↕		Layer Name	contains	interior		

▼ CRITERIA / IES-01 WALL SCHEDULE

	(	Criteria		Value	)	and/or
↕		Element Type	is	⬜ Wall		and
↕	(	Layer Name	contains	bearing		or
↕		Layer Name	contains	interior	)	

Figure 13.12 – Using brackets in Schedule criteria

Depending on how well you modeled this project, the resulting Schedule should only contain structural walls. This (probably) being your first Archicad model, it will also contain walls that are not load-bearing, mainly because when modeling them, you forgot to set the layer correctly.

Wall Schedule

Element ID	2D Plan Preview	Wall Type	Height [m]	Thickness [m]	Area [m2]	Net Volume [m3]	Perimeter [m]	Classification
SW - 001		215 Block Insulated ...	2,930	0,40	2,31	6,72	12,33	Wall
SW - 002		215 Block Insulated ...	0,450	0,40	1,60	0,66	8,57	Wall
SW - 002		215 Block Insulated ...	2,930	0,40	4,26	11,56	22,33	Wall
SW - 002		215 Block Insulated ...	6,450	0,40	3,92	15,40	20,33	Wall
SW - 003		215 Block Insulated ...	0,450	0,40	2,25	0,93	12,33	Wall
SW - 003		215 Block Insulated ...	2,930	0,40	2,25	5,66	12,33	Wall
SW - 003		215 Block Insulated ...	6,450	0,40	2,65	13,17	14,33	Wall
SW - 004		215 Block Insulated ...	0,450	0,40	1,54	0,64	8,57	Wall
SW - 004		215 Block Insulated ...	2,930	0,40	4,20	8,84	22,33	Wall
SW - 004		215 Block Insulated ...	6,450	0,40	3,86	15,15	20,33	Wall
SW - 005		215 Block Insulated ...	6,450	0,40	2,59	7,57	14,33	Wall
SW - 006		100 Block Double Pl...	6,000	0,13	0,66	2,61	12,64	Wall
SW - 006		Stud Partition	6,000	0,10	0,34	1,17	6,94	Wall
SW - 007		Stud Partition	6,000	0,10	0,34	1,10	8,94	Wall
SW - 008		Stud Partition	6,000	0,10	0,20	0,73	5,80	Wall
SW - 009		Stud Partition	6,000	0,10	0,08	0,26	1,85	Wall

Figure 13.13 – The resulting Schedule contains some Walls that are probably not load-bearing

Let's see how we can actually use Schedules to check for this common error and improve our model.

Choosing Fields

Fields determine which properties of the selected elements are displayed in the Schedule. In other words, Fields allow us to choose what information we want to see from the selection we made using the criteria. The list of available Fields is even more extensive than the list of Criteria, so we will not go through them all. To get to know the main ones, you should study the Schedules from the Archicad 26 Template carefully. Which parameters are available in the fields is determined by the selection made according to the Criteria. In the Schedule that we started from (**IES-01 Wall Schedule**), a lot of different parameters are listed. Let's take a look at how we can use these fields to check and improve our model:

1. Just like we learned in the previous subsection, *Using selection Criteria*, we can add Fields **1(a)** and remove **1(b)** them using the buttons below the list in **Scheme Settings**. Again, there is a dropdown **1(c)** available with more options to add Fields. Start by deleting the marked items from this long list (you can *Cmd/Ctrl + click* to select multiple elements).

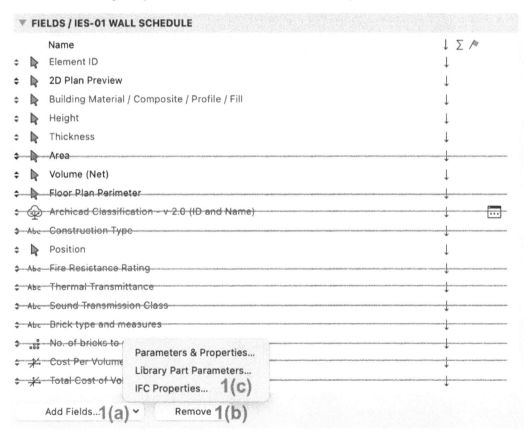

Figure 13.14 – Shortening the list of Fields

2. Click **Add Fields... 2(a)** and use the search bar **2(b)** to look for Layer. Double-click to add this Field.

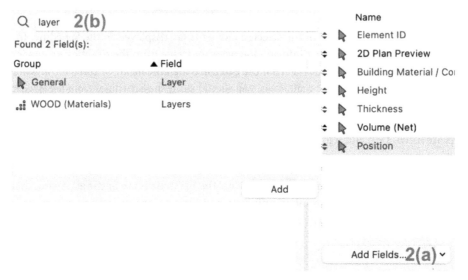

Figure 13.15 – Adding the Layer field

3. Use the small arrows to the left to move the **Layer** field to the fourth position (from the top). This is the order in which the columns are shown (from left to right) in the resulting Schedule.

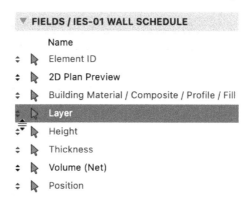

Figure 13.16 – Changing the order of Fields and, thus, the order of columns

---

Note

In *Figure 13.15*, you can see there are Field names in blue and Field Names in black. The ones in blue are editable directly in the schedule view (as we tried in the sixth step in the *Setting up a schedule* section). Be careful though! Changing the dimensions of an object in a Schedule instead of a 2D or 3D View can create unwanted results.

4. Go back to the Schedule by clicking **OK**.

Wall Schedule							
Element ID	2D Plan Preview	Wall Type	Layer	Height [m]	Thickness [m]	Net Volume [m3]	Position
SW - 001		215 Block Insulated ...	Structural - Bearing	2,930	0,40	6,72	Undefined
SW - 002		215 Block Insulated ...	Structural - Bearing	0,450	0,40	0,66	Undefined
SW - 002		215 Block Insulated ...	Structural - Bearing	2,930	0,40	11,56	Undefined
SW - 002		215 Block Insulated ...	Structural - Bearing	6,450	0,40	15,40	Undefined
SW - 003		215 Block Insulated ...	Structural - Bearing	0,450	0,40	0,93	Undefined
SW - 003		215 Block Insulated ...	Structural - Bearing	2,930	0,40	5,66	Undefined
SW - 003		215 Block Insulated ...	Structural - Bearing	6,450	0,40	13,17	Undefined
SW - 004		215 Block Insulated ...	Structural - Bearing	0,450	0,40	0,64	Undefined
SW - 004		215 Block Insulated ...	Structural - Bearing	2,930	0,40	8,84	Undefined
SW - 004		215 Block Insulated ...	Structural - Bearing	6,450	0,40	15,15	Undefined
SW - 005		215 Block Insulated ...	Structural - Bearing	6,450	0,40	7,57	Undefined
SW - 006		100 Block Double Pl...	Structural - Bearing	6,000	0,13	2,61	Undefined
SW - 006		Stud Partition	Structural - Bearing	6,000	0,10	1,17	Undefined
SW - 007		Stud Partition	Structural - Bearing	6,000	0,10	1,10	Undefined
SW - 008		Stud Partition	Structural - Bearing	6,000	0,10	0,73	Undefined
SW - 009		Stud Partition	Structural - Bearing	6,000	0,10	0,26	Undefined

Figure 13.17 – The resulting Schedule should resemble the one in this screenshot

We can structure and sort the data a little bit more, using the options for Fields to the right. Per Field, there are three buttons, from left to right (*Figure 13.14*):

- The first button lets you decide whether the information should be sorted in ascending or descending order, or not sorted at all. Like in many spreadsheet applications, sorting is done per column (per field) in a hierarchy – first, all data is ordered by the first column (or field), then the second, and so on.

- The sigma is the second button and allows you to determine whether or not the values of a field (column) should be added (the total at the bottom of the column). You can also count the number of values for a field by clicking the button until you get the S1 symbol.

- Adding a flag (the last button) to a Field indicates that (on top of the summations you choose) you want to get a subtotal for all Fields that have a total, grouped for this Field. For example, if you put a flag next to **Home Story**, you will get a total per story for all columns (Fields) for which the summation is activated.

> **Tip**
>
> If you want to get an overview of all available element parameters in **Scheme Settings**, you can access the **Help** function of Archicad through the menu (**Help** > **Archicad Help** – see *Chapter 2*, in the *Getting to know the main UI components* section), and then search for Element Parameters in the search window at the top right, or use the navigation tree on the left-hand side: https://help.graphisoft.com/AC/26/INT/index.htm#t=_AC26_Help%2F060_ElementParameters%2F060_ElementParameters-1.htm.

Formatting a Schedule

When a Schedule is not created based on an existing one, it is not very readable at first. The formatting needs to be updated if we are to see a good result. Broadly speaking, formatting a list is very similar to packages such as *Microsoft Word* and *Microsoft Excel*, and therefore, it should not be too much of a problem for many users to format a list legibly.

Figure 13.18 – Formatting a Schedule – an overview of the most important features

Nevertheless, it is useful to explain the most important features:

- The Schedule style (**1**) determines whether the fields will be on top or on the left (e.g., for a list that lists quantities of materials, you choose *on top*; for a window schedule, you choose *on the left*).

- **Merge Items** (**2**): (Completely) similar items are grouped in one line for a better overview, which is useful for a shorter list with summed quantities of materials. By default, dimensions are ignored when merging, combining the lengths, widths, volumes, and so on of similar elements into a sum total. You can alter this by selecting the column header of a field and clicking the **Merge Items** icon. Choosing the right options here allows you to merge items with various values anyway. This can be useful in cases when, for example, you want to merge items with a different volume and show the total of the volume instead.

Figure 13.19 – Merging various volumes and showing their sum

- **Show Headline**: One or more of the first fields can form a *title row* together, thus providing more structure to the list. Via **Edit…**, the headline settings (**3**) can be set up in detail.

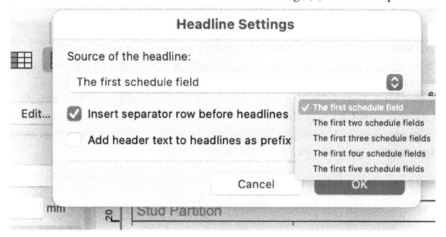

Figure 13.20 – Setting the headlines

- To format a whole table or individual cells/cell types, you have the following options available:

  - Font, font size, and font pen (**4**), or use a favorite setting

  - **Row Heights** (**5**)

  - **Wrap text** (**6**): text wrapping on and off

  - Text formatting and aligning (**7**): Left, center, right, and a final option aligning to the decimal sign

  - A limited number of options for the cell borders and the pen used (**8**)

- Prior to making changes, choose which item you want to change via the **Apply Format Options to drop-down menu**: (**9**).

- A footer (**10**) can be added to tables if desired (when placed on a layout within Archicad).

- Through the gear icon (**11**) there are options available for hiding the different headers, adding a merged header if wanted, and freezing headers (useful when scrolling through long lists). Here, you will also find an **Rebuild Schedule** option, when some values have been altered, for example.

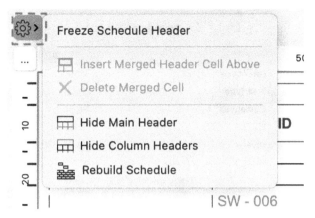

Figure 13.21 – The gear icon in the schedule view

- When a cell is selected, it is (mostly) possible to show the object in question on a plan or in 3D (**12** in *Figure 13.18*). Make sure everything is visible in the (active) 3D window though, as elements in hidden Layers cannot be found in a view when selected from within a Schedule.

- Undoing or redoing changes to the formatting is done via dedicated buttons (**13**).

- The zoom factor of the view can be specified, or set to fit the width of the screen (**14**).

- At the top right, **Scheme Settings** (**15**) can be opened (and, thus, duplicates and/or new lists can be created), as we already know.

- In the top-left corner, the three dots, **...** (**16**), can be found to adjust column widths and row heights to fit the content.

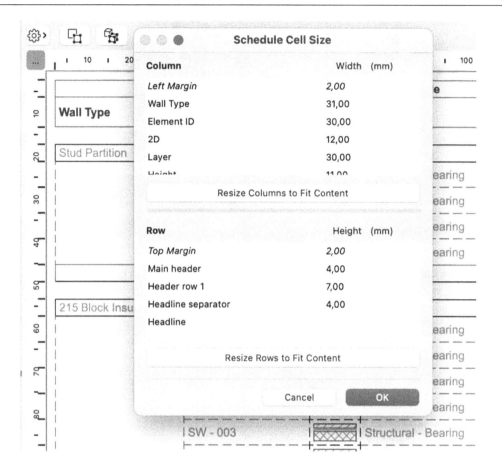

Figure 13.22 – Auto-resizing rows and columns can be helpful at times

Play around with some of the settings, and you will soon get the hang of it. You could try mimicking the formatting of the Schedule, as shown in *Figure 13.17*.

Exporting a Schedule as a spreadsheet

Once you are happy with the result, you can use the view of your Schedule on any layout, as we saw for basic views in *Chapter 7*. However, maybe you would like to share a Schedule in a commonly used spreadsheet format, such as `.xls` or `.xlsx`, so that it can be used in further calculations or other processes. This is easily done in Archicad by simply navigating to **File** > **Save As…** from within the Schedule view and choosing the desired file type from the dropdown. This process can also be optimized, as we will explain in *Chapter 14*.

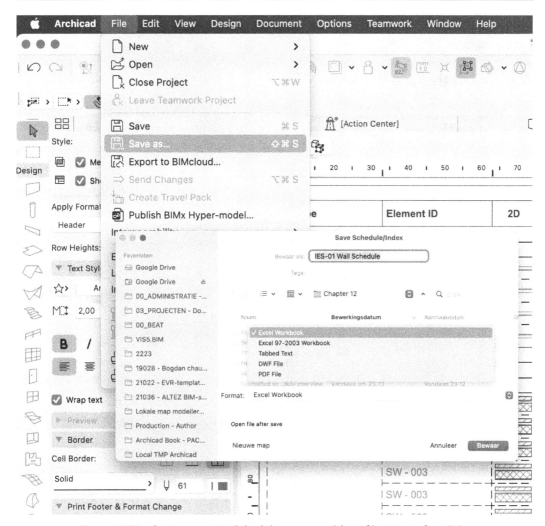

Figure 13.23 – Saving an entire Schedule as a spreadsheet file in just a few clicks

Now that we have learned how to extract valuable data from our model, let's go back to making a graphical representation of the model, this time in 3D!

Graphic Overrides

We already encountered **Graphic Overrides** (**GOs**) when we discussed the **Renovation Filter** (see *Chapter 11*), but they are so much more than that. GOs can give elements another appearance by switching to another pen, fill, color, or surface, based on the properties of the element. They can be used to colorize walls according to their fire rating, to set the color of zones by category or any other property, or indicate which elements are load-bearing. They can also be used to display all elements that lack a required property or classification in red, while setting those elements that are correct to a green transparent hue.

GO combinations

From our template, some GO combinations have already been configured (**Document** > **Graphic Override** > **Graphic Override Combinations**). In this dialog, several override rules are aggregated into a GO combination, allowing us to recall a full configuration in one go.

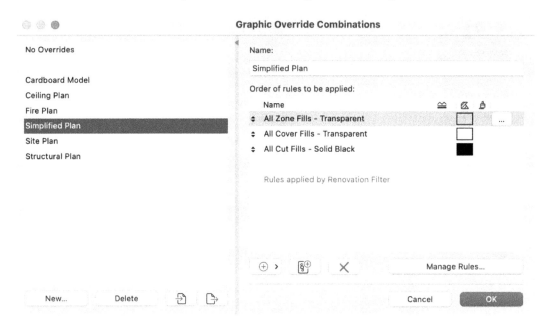

Figure 13.24 – The Graphic Override Combinations dialog

The combinations on the left are just like our Layer Combinations (see *Chapter 14*, in the *Layers and Layer Combinations* section) – a set of configuration rules grouped under a single name to be conveniently activated for our current view. More precisely, when we store a view, we also store the applied GO combination with it.

On the right, you can see the different rules that are applied to this combination. Each combination will have its own set of rules, but the rules can be reused across different combinations. And that is where the magic happens! Let's explain what these rules can do next.

Exploring the rules

GOs are defined by their set of active rules, which control the appearance of elements. Open the **Graphic Override Rules** dialog from the **Graphic Override Combinations** dialog (press the **Manage Rules...** button):

1. The list on the left (*Figure 13.25*, (**1**)) shows all currently available rules, displaying their name, and a small appearance hint using line type, fill, and surface icons. They have a certain logical order in their name, so you may want to follow the same order.

2. On the right, the currently selected rule can be configured. After the **Name** field, we first get a **CRITERIA** panel (*Figure 13.25*, (**2**)). This is very similar to the **CRITERIA** panel in the interactive schedules; you use this to filter the elements in the model to which the rule applies. The current example simply selects all element types. You can add more criteria for more complex filtering – for example, picking multiple element types and removing unneeded element types – but also for looking at properties or any other characteristic of an element. This is where you let the model data steer the visualization.

3. The **OVERRIDE STYLE** (*Figure 13.25*, (**3**)) panel is where you configure how the elements that are picked up by the criteria will look, as a series of optional attribute overrides:

   - **Line Type**: Solid, dashed, and all the other possible types.

   - **Line / Marker / Text Pen**: The pen index (and, thus, the related thickness and color).

   - **Fill Type**: The fill to be used, with a **Hide Skin Separators** option. This is very useful for getting a simplified view of composite elements – for example, during a sketch or preliminary design.

   - **Fill Foreground Pen**: When you want to refine the fill by selecting another pen, regardless of the pen that may have been assigned to the element already.

   - **Fill Background Pen**: This has the same options as the foreground pen, but also gives you the option to set an RGB color instead of relying on an index. This is useful when the color you need is not readily available in one of the available Pen Sets or you want to ignore the actual color from the Pen Set.

   - **Pens / Colors**: Here, you can select to override color, or color and thickness.

   - The last two options relate to the 3D view – **Surface** allows you to select an override surface or select an RGB color and transmittance, and it also gives you the option to override the uncut and cut surfaces.

   - **Hide Model Contours** can help you hide the edges, which is very useful for making certain objects less visible in the 3D view.

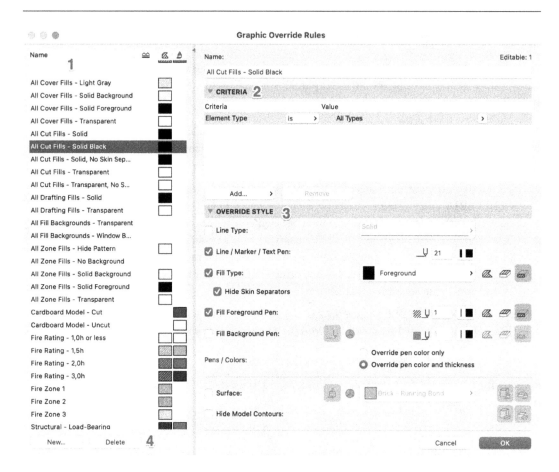

Figure 13.25 – Configuring one rule in the Graphic Override Rules dialog

While understanding how the dialog works and what each part does seems fairly easy, it is good to look at a few examples to get a better understanding of what you can achieve with it.

A Cardboard Model example

The Cardboard Model GO can be used to turn all the surfaces white in a 3D view and turn all lines black (using the 2 pen index, depending on the Pen Set in use), regardless of the applied materials and surface on the models. This GO has two rules – Cardboard Model - Uncut and Cardboard Model - Cut. The uncut rule is displayed in *Figure 13.26*, while the cut rule selects the 22 pen index (a thick black pen in an active Pen Set) and a dark gray surface instead.

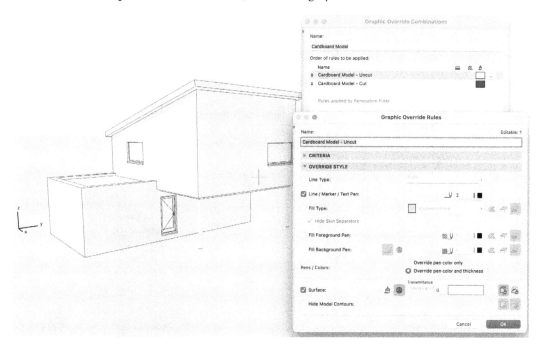

Figure 13.26 – The Cardboard Model GO with the Cardboard Model - Uncut rule configuration

A Simplified Plan example

To display the design as an early sketch, you could use a 1:100 scale, the 03 Plans - Preliminary layer combination, the 03 Architectural 100 Black & White Pen Sets, the 03 Building Plans model view option, and the Simplified Plan GO.

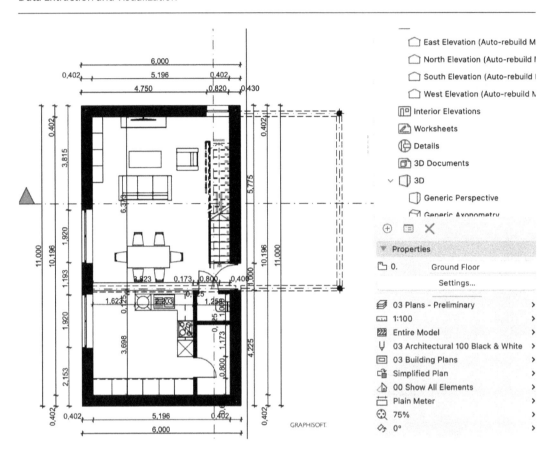

Figure 13.27 – A configured sketch design as a combination of various display options (including the GO)

Remember that GOs are only one of the configurations of a view in Archicad. The full configuration of a view relies on selecting the suitable combination of view options (see *Chapter 14*, in the *Understanding quick options/view filters* section) – layer combination, scale, structure display, Pen Set, model view options, graphic override, and renovation filter. Don't try to solve everything only with a GO.

Some more ideas to try

The possibilities of GOs are nearly endless. You can browse through the various rules that are already available in the template, such as colorizing elements based on their fire rating or setting the color of zones based on their category. This is a much more versatile approach than adding arbitrary materials and surfaces, just for the sake of showing objects in a certain color.

Based on the preceding examples, we invite you to start reflecting on other options, such as the following:

- Showing all self-closing doors in purple, based on their **Element** type, and a custom Boolean property, you could call **Self Closing**, for example, and set it to `true` or `false`

- Making all trees, shrubs, cars, and people transparent in a 3D view, based on their layers

- Making all sanitary and mechanical equipment elements bright red if their manufacturer property is not filled in

Now, let's answer an important question.

Can you hide objects with a GO?

One thing that may not be immediately obvious is that the GOs can only filter elements (using the criteria) and then override their graphic attributes. They cannot really hide objects. Hiding objects can only be done by layers and the **Show Elements in 3D** configurations, explained earlier. However, when you really want the GO to only pay attention to the objects that are relevant to you, you can make their appearance more subdued – for example, by hiding model contours and using a transparent color in 3D, or by using a very thin and light gray (or even white) pen in the 2D graphic settings.

The following example hides the dimensions (using layers) and turns all beams and furniture light gray, using a GO based on `Simplified Plan` but adding two more rules.

The `Furniture turned grey` rule uses **Element Type is Object** and **Layer is Interior - Furniture** as criteria. Then, we select a thin gray pen (e.g., the `101` index) as a **Line / Marker / Text Pen** override. By also setting **Surface** to light gray with a **Transmittance** value of `50%`, the beams and furniture are also displayed as subdued in 3D views, if you apply the same GO.

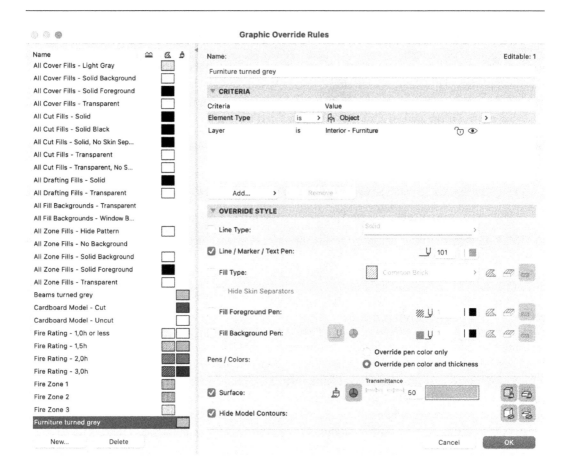

Figure 13.28 – The custom Furniture turned grey rule

For the Beams turned grey rule, we use **Element Type is Beam** as criteria. Since beams and columns reside in the same layer (**Structural - Bearing**), we wouldn't be able to control this from their layer or Pen Set, so we rely on the **Element** type. The rest of the rule is the same as with the Furniture turned grey rule.

Ensure that the rules are put at the top of the override combination so that they are applied first. With the dimensions layer hidden and this GO, we have a more refined sketch design view configured.

Figure 13.29 – Beams and furniture turned gray using two custom rules

So, now you know how you can use GOs and what their limitations are, which brings us to the end of our chapter on how to get the most out of your BIM data and how to use it for graphical representations.

Summary

In this chapter, we discovered that the information in our model (the *I* in *BIM*) can also easily be visualized in tables, using **Schedules**. This helps us to improve the overall quality of the model and prepare models for data extraction. Using these techniques enables us to use our model to prepare a BOQ or other similar lists. We also learned how to communicate in a visual way about these parameters and properties. All the views combined make up our model and its representation. If we want to communicate this information set to other stakeholders, we need to take a look into **Layout Book**, the publishing process, and the attributes involved in our next chapter!

14

Automating the Publication of BIM Extracts

Even with the use of models, we still need to manage the production of derived (2D or 3D) documents, such as floor or ceiling plans, sections, elevations, schedules, and even exports into other formats such as PDF or XLSX. Archicad provides an elaborate publishing workflow that allows you to fully prepare a complete package of deliverables, even when there are last-minute changes in the model. With one click, you can regenerate any number of documents from the model, all in sync and in a wide variety of file formats. This chapter focuses on this overall process.

So, in this chapter, we will further explore how we can create a set of (printed) documents (for an introduction, see *Chapter 7*, in the *Basic export and printing (output)* section), based on our 3D BIM. Finally, it is necessary to know how to get our design into a drawing we can print or send to others as a PDF or DWG file. We will distinguish between printing directly and using a layout, which gives you more control but also requires more preparation.

Archicad has a unique feature called **Publisher**, which allows us to automate this process. But before we can use this Publisher system, we must learn how to set up our **Layout Book**, which contains the **layouts** with **drawings** placed on them. These drawings are in turn based on the views we create in **View Map** using **Quick Options** to determine the look of the views.

Topics treated in this chapter are as follows:

- Understanding **Quick Options** and view filters
- Using the Navigator to switch between **Project Map View Map Layout Book** and **Publisher Set**
- Creating views using the view filters
- Creating layouts in a **Layout Book**
- Publishing documents using the Publisher workflow

Understanding Quick Options/view filters

In various previous chapters, we have created **viewpoints**, such as a section or elevation, and **views** (fixed representations of the viewpoints). Along the way, we already got to know some of the view filters involved in creating a view from a viewpoint, but now it is time to give a detailed and complete overview of all the view filters (or **Quick Options,** as they are also called).

The complete list of view filters (click the right mouse button on a view, then click **View Options…** – shown later in *Figure 14.11*) is as follows:

- **Layer Combination**
- **Scale**
- **Structure Display**
- **Pen Set**
- **Model View Options**
- **Graphic Override**
- **Renovation Filter**
- **Dimensioning**

We will go through them, one by one, refreshing what we have already encountered or explaining the ones that are new.

Layer Combination

Layers are one of the core Archicad attributes, as already discussed a few times throughout this book (see *Chapter 3*, in the *Modeling tools in general* section, for example). They are mainly used to control the visibility of elements, objects, and spaces, but also whether objects are allowed to intersect (see *Chapter 8*, in the *Layer intersection priority* section).

To manage the different layers and set their configuration, we open the **Layers (Model Views)** dialog. It can be accessed through **Options** > **Element Attributes** > **Layers (Model Views)…**, through **Document** > **Layers** > **Layers (Model Views)…**, by using the button in **Quick Options** bar, or by using the shortcut keys *Cmd + L/Ctrl + L*.

The dialog is split into two halves: on the left-hand side, we have the available **Layer Combinations**, and on the right-hand side, we see the **Layers** settings (organized in folders) within the currently selected layer combination from the left side. You can navigate the **Layers** folders at the top-right part. When we select another layer combination, the configuration of the layers can switch, but the list of layers is always the same for each layer combination.

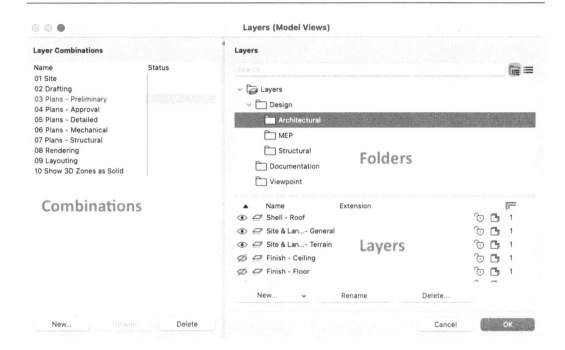

Figure 14.1 – The Layers dialog with its typical triple structure

In the selected combination on the left-hand side in *Figure 14.1* (**03 Plans - Preliminary**), some of the layers on the right-hand side, such as **Finish - Ceiling** and **Finish - Floor**, are turned off. This is indicated by the eye icon just to the left of the layer's name.

We will now go through the different configuration options we have available for each layer in the first column and last three columns (from the left) of the **Layers** dialog:

- **Show/Hide**: Using the eye icon, each layer can be hidden for that layer combination so that all elements in that layer will be hidden in all views that uses this combination. This is a very powerful mechanism to control the visibility of elements in different views. For example, you would hide the layers containing furniture when you created a construction document for the general contractor.

- **Lock/Unlock**: Elements in a locked layer cannot be edited or deleted, nor can new elements be added to a locked layer. This setting is often used to prevent unintentional changes to the design. Layers for grids or a site, or layers containing neighboring buildings, are often locked in most layer combinations.

- **Solid/Wireframe**: This is mostly used in 3D views to alter the display of elements. A typical use case where a wireframe display is applied is for auxiliary layers with objects used solely for the purpose of **solid element operations** (see *Chapter 8*, in the *Using solid element operations (SEO)* section) or to show the building elements as wireframes, while the zones are shown as solids. This is a nice presentation method to explain the spatial layout or the zoning for energy or evacuation in the design.

- **Layer Intersection Priority**: This has been discussed already in *Chapter 8*, and indicates how elements in different layers interact: layers with a different number ignore each other (see the *Layer intersection priority* section in *Chapter 8*). A good use case is when a distinction between new and existing parts of a design is required or when we need to isolate a placeholder model to indicate the neighboring buildings, so they don't intersect or form connections.

You will not be surprised when we say that the configuration of all the layers, including their visibility and other aspects, can become quite complicated. There is no easy way to remember and recall such a configuration each time you need to configure a view, so the layer combination is stored to keep the settings, and the view will recall the chosen layer combination when it is refreshed.

> **Note**
>
> When changing the settings of certain layers and/or layer combinations, we have two options:
>
> - Change the settings of one or multiple Layers within one layer combination:
>   - Select the layer combination you want to alter on the left-hand side
>   - Change the settings of layers on the right-hand side
>   - Click **Update** beneath the combination on the left-hand side
> - Change the settings of one layer in one or multiple layer combinations:
>   - Select the layer for which you want to change the settings on the right-hand side
>   - Make the changes in the layer combinations on the left-hand side (using the same icons as described previously)
>   - No update is needed in this scenario

Scale

It is common practice in the **architecture, engineering, and construction** (**AEC**) industry to display a building on documents at a certain scale. A scale of 1:200 to 1:50 is often used for a general plan. However, to ensure that the details are visible, they are shown to be larger, for example, at a scale of 1:20 or even 1:5. In Archicad, we can set the scale of our views using this specific view filter. The choice does not influence the actual size of the building, it just scales the graphical representation. Several settings can determine the effect of the **Scale** setting on elements in the model. See, for example, *Chapter 6*, in the *Exploring line types* section, when we learned about model size and paper size (as shown in *Figure 6.18*).

Structure Display

The next configuration option for a view is **Structure Display**. Here, you can select how to display elements based on their structural function.

Each 3D element in Archicad can be set to be either **Load-Bearing**, **Non-Load-Bearing**, or **Undefined**. This is very important information for the design and is required by the structural engineer or the contractor.

When we discussed composites in *Chapter 3*, we explained that each composite needs to have at least one Skin that is set to **Core**, indicating the main, typically structural, part of the element.

While setting a Skin to **Core** does not necessarily imply that it is load-bearing, the **Core** Skin and the structural function are used to filter elements and element Skins in the four options of **Partial Structure Display** (**Document** > **Partial Structure Display…**):

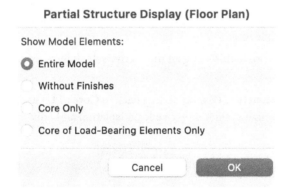

Figure 14.2 – Partial Structure Display options dialog

- **Entire Model**: This is the default. Every element and every Skin is displayed, as expected.

- **Without Finishes**: Every Skin in every composite that is indicated as a **Finish** Skin is hidden. This would typically hide the facade bricks or roof tiles and other finishing Skins of walls, slabs, and roofs.

- **Core Only**: Not only is the **Finish** skin hidden but also the **Other** Skin, leaving only the **Core** Skins visible. This is very useful for indicating a core structural model, without any insulation or coverings.

- **Core of Load-Bearing Elements Only**: This is the final option, and this also hides any element that is not indicated as **Load-Bearing**. This is used to show the net model, disregarding any other element, and can be used to exchange information with a structural engineer or to indicate the structural concept of the design in isolation from any other element.

If you, as an example, switched our current model to the **Core of Load-Bearing Elements Only** option, most of the model would disappear, including, alas, all our walls, beams, and columns. What did we do wrong? Well, until now, we disregarded the **Structural Function** property, and we were not really aware that structural information in our design was still undefined.

By switching to this **Partial Structure Display** option, our model shows what we set (or rather, forgot to set). So, switch back to display the entire model and make the correction to any element that should be load-bearing. Select all walls and switch their **Structural Function** setting to **Load-Bearing**. Do the same for beams and columns and any other load-bearing element.

> **Pro tip**
>
> It is best to do such modifications per element type (e.g., walls, beams, columns, and slabs). You can edit multiple elements at the same time, but when you open the **Element Settings** dialog, it cannot display **Wall Settings** and **Beam Settings** at the same time, since their **Element Settings** dialog has a different layout. For this to work, you would have to use **Edit** > **Element Settings** > **Edit Selection Set…** instead. This specific dialog lets you change a few common attributes and settings across multiple tools.
>
> Alternatively, you could also do this using an interactive schedule, as explained in *Chapter 13*.

Now, we return the **Partial Structure Display** setting back to **Core of Load-Bearing Elements Only** and only our load-bearing elements should be visible, in isolation and only showing their **Core** Skin.

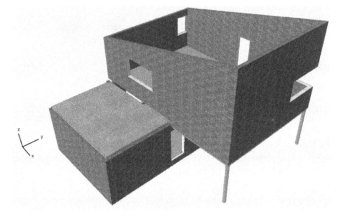

Figure 14.3 – Configured view showing only the core of load-bearing elements

> **Remark**
>
> Note that while the walls are displayed, the windows and doors in those walls are also shown, even if they are not set to be load-bearing! This is a limitation of the visibility control in Archicad; since windows and doors are hosted in walls, they inherit their visibility (and also their layer). To correct this, you can use the **Filter and Cut Elements in 3D** dialog (seen in *Chapter 12*), where you can toggle off the element types you don't need.

Pen Set

The combination of the color and thickness of each pen index was configured and stored in a **Pen Set**, as already discussed in *Chapter 6*.

As part of the view settings, you can select the **pen set** that you want to be used when loading the view. This ensures that the way lines are displayed is consistent across different views and avoids having to configure such a large set of pens for each view separately.

Model View Options

Model View Options (**MVO**) contains a list of combinations on the left side (*Figure 14.4*) and various tabs for each combination on the right side, with specific options for displaying certain elements in Archicad views. The available options affect the objects that have a reference to these options in their GDL script. For example, an object can be shown in several levels of detail (**Full**, **Medium**, or **Low** – see Figure *14.4*), but if a given object does not contain a **Low** (or **Medium** or **Full**) 2D representation, this setting may not have the expected effect for that object. MVO is mostly used to show a predesign in a coarse way versus a detailed, more finished look for a developed design stage (see *Figure 14.5*).

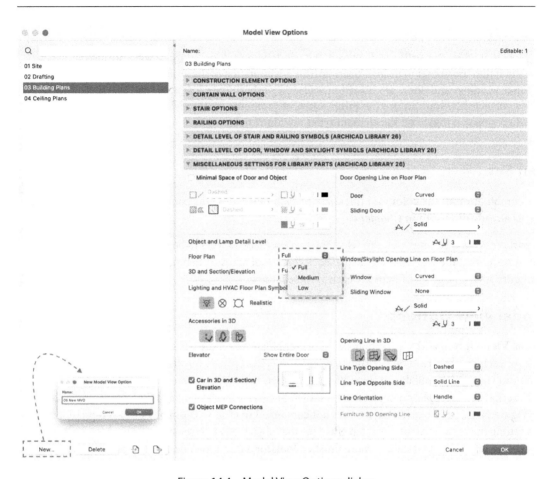

Figure 14.4 – Model View Options dialog

The next figure gives a side-by-side comparison of how the display of objects can be influenced using Model View Options. This gives you global control and is more efficient than having to change each object individually.

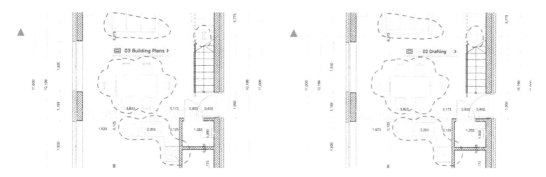

Figure 14.5 – Comparing 03 Building Plans (left) with 02 Drafting (right)

We encourage you to experiment with the available combinations to see the effect and maybe create a combination of your own. Beware though: in the MVO dialog, no *custom* setting is created when editing settings (as is the case for other view filters, such as **Layer Combination** or **Pen Set**). So, when you start experimenting, click **New...** (see *Figure 14.4*) to create a new combination first!

Graphic Override (GO)

As discussed in the previous chapter, the **Graphic Override** combination is a set of configured rules stored under a single name. As part of your view display settings, you can select a GO to be set for the view, so it will be restored every time the view is refreshed.

It is also possible to indicate that no GO has to be set for the view.

Renovation Filter

The **renovation status** of an element was discussed in *Chapter 11*. Remember that **Renovation Filter** is the configuration of how elements appear for that filter, based on their renovation status. Also remember that it was configured as a special GO.

As part of the view settings, you can select the applied **Renovation Filter** for the view.

Dimensioning

As we know from *Chapter 7*, (see the *Understanding dimensions* section), we can set the working units in Archicad for entering data when we model. This does not affect the dimensions displayed by dimension lines and other annotations for showing sizes, angles, and so on. We do have the option, however, to determine which units Archicad has to display, and we can set this choice per view using the **Dimensions** view filter.

Figure 14.6 – The Dimensions dialog

There are many options to set per type (**Linear Dimensions**, angle, radius, spot, and so on) but you quickly get the hang of it, as **Sample** reflects every choice you make immediately. Any edit results in a **Custom** setting you can store as your own setting.

Now that we know about the different view filters, let's see how these can be used to create a view. But before we get to that, we have to take a closer look at the **Navigator** and how it helps us to navigate our model between the different steps in the publishing process. This way, we can use our Navigator to organize our views in a structured way instead of creating views in a slightly random fashion.

The Navigator and the Project Map

Setting up the publication workflow in Archicad requires you to master navigation across a wide variety of viewpoints, views, and other model representations in Archicad. For this, Archicad provides the **Navigator**, which allows you to navigate around the different kinds of representations in your project.

At the top, **Navigator** shows four configuration tabs. We can consider these as the steps in our publishing process (a diagram of this program is shown in *Figure 14.27*). From left to right, we have the following:

- **Project Map**: The index of the entire project, containing every viewpoint of the project (once)
- **View Map**: Contains all defined and configured views, derived from the viewpoints using view filters
- **Layout Book**: Consisting of **subsets** (a bit like chapters) with **layouts** (a bit like pages) on which the views will be placed
- **Publisher**: Contains **Publisher Sets**, which translate various parts of the **Layout Book** and/or the **View Map** to a deliverable (paper print, PDF, XLSX, DWG, interactive BIMx model, etc.)

Next, we will explore each of these tabs in more depth.

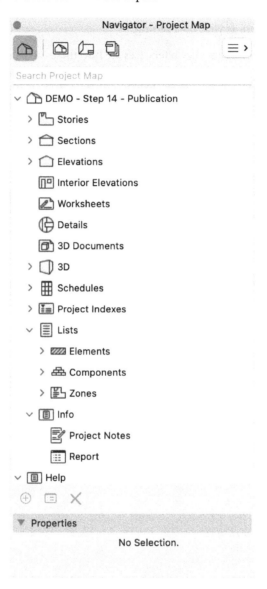

Figure 14.7 – The Navigator palette showing Project Map

Project Map is the first navigation tree or hierarchy that Archicad displays – typically, as a docked palette on the right side of your screen if you follow the default workspace. It displays all available viewpoints in the current project grouped into viewpoint types. Whenever you add a new viewpoint, **Project Map** is updated.

Each time you select an item in the tree, relevant options appear at the bottom of the palette in the **Properties** pane, including a **Settings…** button to open more detailed configuration settings.

A **viewpoint** is a way to display information from the Archicad model. It only tells us what to show and from which position, but not how to display its content. Archicad provides several viewpoints, and we have encountered most of them already:

- **Stories** define the horizontal spatial layout of our building, vertically stacked upon one another. Each story gets its own 2D viewpoint. You can navigate up or down the stories or double-click inside **Navigator** to display them. You can also access the **Story settings** dialog when you click on the **Stories** branch in the tree.

- **Sections** and **Elevations** are also 2D viewpoints but are derived from a vertical cut through the model, as discussed in *Chapter 6*. You also have a branch for **Interior Elevations**, as discussed in *Chapter 12*.

- The **Worksheets** branch contains the independent worksheets, which can be used for non-model-related 2D drawing, also discussed in *Chapter 6*.

- **Details** lists all detail views, which open as a separate viewpoint, as discussed in *Chapter 12*.

- **3D Documents** are 2D drawings projected from the 3D model, allowing them to be annotated and enriched with 2D techniques. We return to them in *Chapter 15*.

- **3D** is the interactive 3D viewpoints, typically split between perspective and orthographic camera projections.

- **Schedules** displays the three types of interactive schedules: **Elements**, **Components**, and **Surfaces**, as discussed in *Chapter 13*.

- **Project Indexes** are dedicated schedules for collecting several kinds of project-related overviews: **Change List**, **Drawing List**, **Sheet Index**, **Transmittal History**, and **View List**. It is especially useful when **Drawing List** and **Sheet Index** are automatically updated and reflect the content of your project output, as it would be annoying to collect this data manually each time you make changes.

- **Lists** gives you access to the ancient listing system, which we don't cover in this course.

- **Info** allows you to access the project notes, which is a dedicated text view for storing any kind of notes or remarks about your project, and the report, which acts as a log of all generated warnings, notifications, and errors. These two viewpoints (project notes and report) are not used that often but can be useful when you have some kind of error or if you need to figure out what Archicad has been doing during long calculations.

- **Help** gives you access to the **About** window and links to the online documentation and the main website of Archicad.

We use **Project Map** all the time, as this gives a good overview of the content of our project and the various representations – especially to switch to schedules, stories, and special dedicated viewpoints.

Organizing the Project Map

At the bottom of the palette, you have a few tool buttons:

Figure 14.8 – Tools for the Project Map

From left to right, we have the following:

- **New Viewpoint** to create a new viewpoint of the currently selected type (e.g., a section or elevation). Depending on the viewpoint type, this button may be disabled (e.g., **Schedules**, **Lists**, and **Help**).
- **Settings** to configure the selected viewpoint or type:
  - When you request the settings of the view type (e.g., sections), this will configure the default settings, which are applied when you next create a new viewpoint. Each type has its own **Settings** dialog.
  - When you request the settings of a story or the **Stories** branch, you simply get the overall **Story Settings** dialog. Similarly, when selecting the settings of the **Schedules** branch, you get the global **Scheme Settings** dialog, where you can pick any of the available schedules, grouped into **Elements**, **Components**, and **Surfaces**.
- **Delete** to delete the currently selected viewpoint. This can have quite considerable consequences in your model, so you get a confirmation dialog, as removing a view effectively removes information from your model. Deleting is undoable, as the dialog that pops up will warn you.

Understanding the logic of the Navigator is crucial in creating a structured model. You don't want to get lost in your own model now, do you? So now that you have learned about the main parts and how your project is indexed in the **Project Map**, let us move on to the **View Map** where we create distinctive representations of our viewpoints: the views.

Creating views and organizing them in the View Map

Switching between viewpoints in **Project Map** only changes the viewpoint. It retains the current **Quick Options**, such as the active layer combination or renovation filter. To store these settings with a viewpoint, we need to use views.

> **Remember**
>
> A View is a Viewpoint with fixed or saved **Quick Options** (view filters):
>
> *View = Viewpoint + Quick Options*

So, whenever you want to return to a particular configuration, you can save it as a view. Double-clicking a view in **View Map** brings you to that viewpoint and it'll apply all the display settings from **Quick Options**.

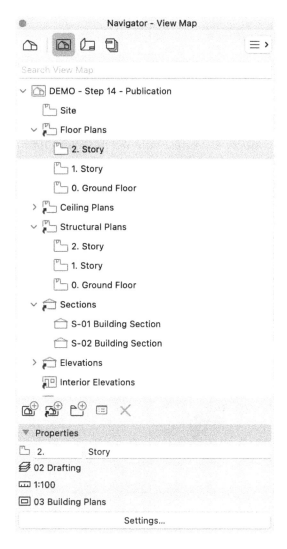

Figure 14.9 – The View Map palette, showing configured views

Each viewpoint can exist only once, but the same viewpoint can be referenced in multiple views; for example, the **1. Story** story exists just once in the model. But you can configure a 1:100 floor plan, displaying the existing renovation filter and have another view at 1:50 scale, showing the new renovation status and possibly with a different layer combination (e.g., to show more annotation and text labels).

Organizing the View Map

There are five dedicated tool buttons available at the bottom of **View Map**:

Figure 14.10 – Tools for the View Map

From left to right, we have the following:

- **Save Current View...** stores the current view and its **Quick Options** as a new view in the currently selected branch of **View Map**.

- **Clone a Folder...** asks you to select one of the folders or branches in **Project Map**. Here, you can create a shared view configuration for all stories, all elevations, or all schedules, for example. This is a very powerful way to control a larger set of views using a single configuration. This will automatically add views when you add new viewpoints later and ensures that all views in the cloned folder use the same settings.

- **New Folder...** just creates an empty, named folder to allow you to group a set of views. Use this to maintain the overview on the **View Map**. It is very common to create a folder for a project milestone or for a dedicated package of related views.

- **View Settings...** opens a settings dialog for the currently selected view or cloned folder.

- **Delete** is obvious, isn't it?

The **View Settings** dialog gives a clear overview of all view filters or **Quick Options** that are applied to the currently selected or created view.

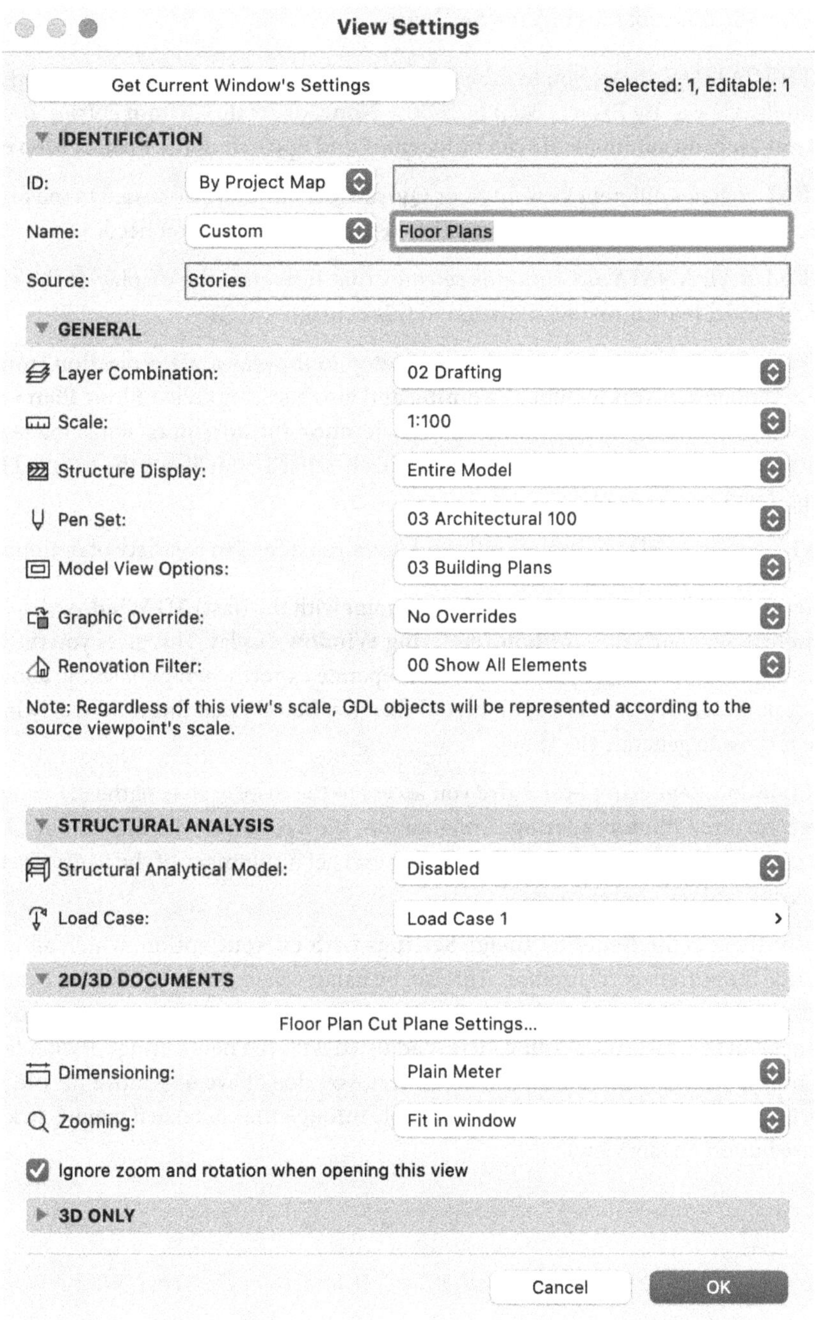

View Settings

Get Current Window's Settings   Selected: 1, Editable: 1

▼ **IDENTIFICATION**

ID: By Project Map ⬍

Name: Custom ⬍   Floor Plans

Source: Stories

▼ **GENERAL**

Layer Combination: 02 Drafting ⬍

Scale: 1:100 ⬍

Structure Display: Entire Model ⬍

Pen Set: 03 Architectural 100 ⬍

Model View Options: 03 Building Plans ⬍

Graphic Override: No Overrides ⬍

Renovation Filter: 00 Show All Elements ⬍

Note: Regardless of this view's scale, GDL objects will be represented according to the source viewpoint's scale.

▼ **STRUCTURAL ANALYSIS**

Structural Analytical Model: Disabled ⬍

Load Case: Load Case 1   ›

▼ **2D/3D DOCUMENTS**

Floor Plan Cut Plane Settings...

Dimensioning: Plain Meter ⬍

Zooming: Fit in window ⬍

☑ Ignore zoom and rotation when opening this view

▶ **3D ONLY**

Cancel   OK

Figure 14.11 – The View Settings dialog showing all the view filters used for the created view

Let's take a look at the different tabs in **View Settings**:

- **IDENTIFICATION** allows you to inherit an ID (short name) and name (descriptive name) based on the view as **By Project Map** or choose **None**, or create **Custom** entries. If you select **By Project Map**, an automatic ID can be assigned and updated as the **Project Map** evolves.

- **GENERAL** contains different view filters or **Quick Options**, as we discussed in the first section of this chapter. You can still switch to another setting or option if you need.

- **STRUCTURAL ANALYSIS** contains settings that influence the display of the structural analytical model (which we didn't discuss in this course).

- **2D/3D DOCUMENTS** configure settings related to the way a 2D projection from the 3D model is configured. This includes **Zooming** and also a setting called **Floor Plan Cut Plane Settings…**, which lets you determine at what elevation the horizontal cut is made to derive 2D representations of the objects (by default, this is set at 1.50m above the finished floor of a building story).

- **3D ONLY** applies to 3D viewpoints only and gives you access to a variety of options:

  - **Generate In**: Allows you to use the 3D viewpoint with the (fast) **3D Window** display or use the potentially much slower **Photorendering Window** display. This gives you the option to place a rendering onto a sheet or even as a separate export when publishing, allowing it to be automatically regenerated upon model changes. Beware that this may hugely impact the time it takes to generate the view.

  - **3D Style** and **Rendering Scene** give you access to the display style of the 3D window or to a preconfigured **Photorendering** configuration. We'll return to visualization in *Chapter 15*. You can also configure the size of the output and get a summary of the various settings for rendering and the engine to use.

  - Finally, there is the **Redefine Image Settings with current** option, which allows you to override the current configuration. This can be handy when you have a decent configuration but made a few changes (e.g., by navigating the camera position or refining some of the settings). In this case, the existing view is adjusted with the new settings. Particularly when you have already added that view onto a sheet, you don't have to remove the old view and insert the new one. This option is also available through the contextual menu (click the right mouse button on any view).

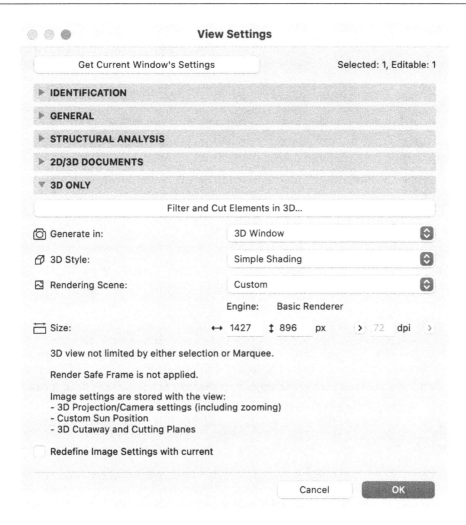

Figure 14.12 – View Settings for 3D viewpoints

Navigating through and modeling in Views

Since each view remembers what is displayed and how it is displayed, we advise you to apply Views not only for publication purposes but also during modeling. You may encounter a situation where, upon modeling, you often have to switch between certain configurations (e.g., with an active marquee, some layers off and others on, a viewpoint and display configuration, or any other (combination of) settings for a certain view). Store this view configuration as a *working view* – that is, a view used for modeling, but not necessarily as part of the output into a published set of documents.

This is especially useful in larger projects, where the time required to regenerate a view may be considerable. In that case, store only a segment of the building to quickly return to; it is quick to regenerate and easier to navigate.

> **Note**
>
> You can model anywhere you want in a View (**View Map**) or in a Viewpoint (**Project Map**). By default, these are all just different representations of your virtual building, as we have discussed before.
>
> Only independent worksheets are not connected to the 3D model. So, if you want to save a certain state of the design, it could be a good idea to create a Worksheet from a Viewpoint or view using the **Worksheet** tool (see *Chapter 6*, in the *Using basic section, elevation, and independent viewpoints* section).

As stated, our Views are preparations for the deliverables in our design process. Next, we will see how we can place Views as Drawings on our virtual paper: the Layouts in our **Layout Book**.

Layouts and the Layout Book

Once we have defined our Views using the view filters and organized them within View Map, we will want to assemble them on a Layout. This layout is the virtual equivalent of a piece of paper on which you want to print your documentation, although it is common practice to use the same paper size (A0, A1, A3, DIN, etc.) for digital printing/export (e.g., PDF) as well.

Drawings, sheets, and masters

The third configuration tab in **Navigator** is **Layout Book**. This is a configuration and navigation tree dedicated to the output into graphical layouts. Again, you get a tree structure with branches, this time containing a variety of layouts (pages) organized within subsets (chapters).

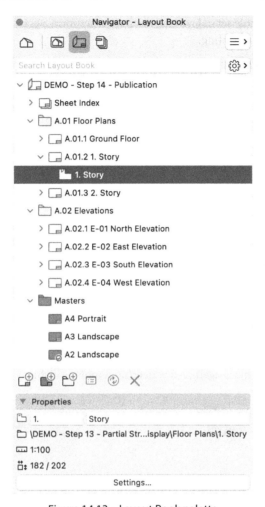

Figure 14.13 – Layout Book palette

Before we can start setting up the **Layout Book** you have to understand a few basic concepts, which we briefly introduced already in *Chapter 7*:

- A **Layout** is a virtual sheet of paper and is always based on a master layout.

- A **Master Layout** is a template sheet that defines the size and printable area and may contain some template graphics, such as a title block, folding marks, or any shared graphic that is required for multiple Layouts:

  - Changes in a Master Layout propagate to all layouts that are based on this master

  - Any graphic element that is placed on a Master Layout is also visible on a Layout, where it becomes part of the non-selectable background graphics (displayed in red to avoid confusion)

- **Layout Book** is the whole set of layouts, which together form a coherent document, ready for publication, sharing, or printing.

- A **Subset** is a group of layouts, giving you some organizational control over the **Layout Book**.

- **Views** are dragged on a layout. This creates a **Drawing**, which references the view and adds some more configuration to the view. The same view may be reused several times, each time creating a new drawing.

- If you try to drag a **Viewpoint** (from the **Project Map**) onto a layout, Archicad asks you to create a view from the viewpoint before it can be turned into a drawing.

> **Newbie warning**
>
> A mistake that we have encountered with many students and new users is adding the drawing directly into a Master Layout. This is not prevented by Archicad since there are scenarios where you may want to, for example, include an overview drawing into the title block, but it is in general the intention that you add your Views onto a Layout, which itself references a master layout.

Now that we know the different parts of the **Layout Book**, let's see how we can create one of our own.

Organizing the Layout Book

As with the previous tree, we have, again, a few dedicated tools available:

Figure 14.14 – Tools for the Layout Book

- **New Layout…** allows you to insert a new Layout in the currently selected position in the tree. When you configure the Layout, you have to indicate the Master Layout it should use. Everything that is on the Master Layout will be inherited on the Layout.

- **New Master Layout…** allows you to insert a new Master Layout. Each Master Layout looks like a Layout, but it directly defines the size of the virtual paper (e.g., A3 landscape, A1 portrait, and the printable area) to indicate the borders onto which no printing is possible.

  The strategy is to define a few basic Master Layouts and to use them across the actual Layouts to get a coherent set of documents (e.g., by having the same title block or using similar page borders or folding marks).

- **New Subset…** is a grouping method to keep a set of Layouts together. This has an impact on the ID and automatic numbering.

- **Settings** allow you to further configure the Layout.

- **Update** – Archicad typically understands when Layouts need to be recalculated after model changes, but you can enforce an update of the currently selected sheet or layout if needed.

- **Delete** – Again, this can be a dangerous operation, so you receive a warning

Working with Layouts and Master Layouts

You use layouts to compose your **Layout Book** and prepare a set of documents for printing or publishing. Anything can be placed onto layouts as long as it is not in 3D. Layouts are a pure 2D environment, much like a traditional CAD drafting system, but with all the integration and coherence of the building model behind it. Here is a list of what you can place on layouts:

- Place views on Layouts to generate Drawings. Drawings have their own dedicated settings, which control what to display (the viewpoint) and how (the view) and add a cropping border and a view title marker.

- You can also place external files onto a layout, such as images or PDF documents (using the **Figure** or **Drawing** tool). This can be a map of the site, the logo of the involved companies, a fragment of a product data sheet, or anything else that you want to include in the layout.

- You can add plain 2D graphics, using lines, arcs, fills, and labels. However, it is not possible to place dimensions or grids onto a layout.

- You can even place GDL objects, as long as they have 2D output. This is mainly used for dedicated 2D symbols. Regular building objects can be placed too, but they are typically not at a suitable scale for a layout and only their 2D representation can be used. Depending on the object library you have available, you would use dedicated 2D objects. Beware that such objects don't become part of the 3D model, so don't use them as part of the building.

So Layouts and Layout Books can contain a variety of elements defined by element settings. But they themselves also have a lot of settings to go through. In the next sections, we will learn about these layout aspects: **Identification**, **ID strategy**, **Drawing Size**, **Drawing Appearance**, **Drawing Frame**, **Drawing Title**, **Title Block**, and **Autotext**. We will explain what all of these mean, how to set them up, and how some of them interact.

Identification

In small projects, it may not be that problematic, but when you have a large project with tens or hundreds of layouts, ensuring consistent and coherent numbering can be a nightmare. In Archicad, you can use automatic drawing identification for both **Drawing ID** (the short name or number) and **Drawing Name** (the descriptive name).

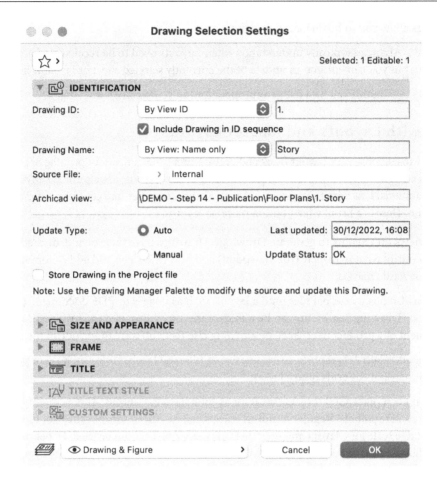

Figure 14.15 – Drawing IDENTIFICATION settings

As you can see, **Drawing ID** can be derived with the help of either **By View ID**, **By Layout**, or **Custom**. You can also see what source this drawing is referencing and can even disable automatic updates, but **Manual** is only recommended for performance reasons, when certain critical drawings would take a very long time to regenerate (e.g., if they contain a complex photo rendering) or when the project (and the accompanying views) are quite large. Otherwise, leave this at **Auto**.

ID strategy across the Layout Book

There are two main strategies for numbering and ID assignment, which have an impact all the way down from the **Layout Book** to subsets, layouts, and drawings (if you ignore manual numbering):

- **Simple sequential layout numbering**: Each layout has its own layout number, assigned by Archicad upon creation of a layout.

- **Automatic layout ID assignment**: This is a configurable ID numbering system, capable of reflecting and automatically updating IDs when the **Layout Book** evolves. Each layout has its own **Layout ID**, also assigned by Archicad. You can configure the numbering rules from the **Book Settings** dialog.

Open the **Book Settings** dialog from the **Layout Book** navigator or from the menu (**Document** > **Layout Book** > **Book Settings…**).

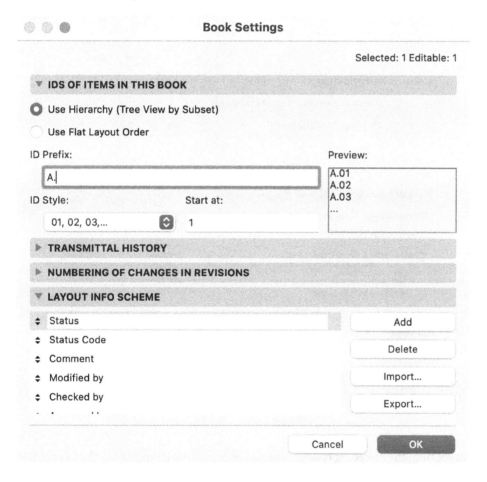

Figure 14.16 – Layout Book Settings dialog

To setup the **Layout Book**, there are a few things we should configure:

- In most cases, you would go for the **Use Hierarchy (Tree View by Subset)** configuration. This derives the layout ID from the order in which the layout is inserted into each subset folder. This way, the ID aligns perfectly with the tree structure in your **Layout Book**.

- **Use Flat Layout Order** simply numbers all layouts in the order of the Navigator, without taking subset IDs into account.

- You can add an ID prefix and select the required **ID Style** and **Start at** details.

Each layout or subset can still be configured as needed using the automatic ID (to ensure the Navigator hierarchy is inherited) or using a custom ID. You can also opt to not include the subset in the ID sequence (e.g., when you have some subsets that fall beyond the regular package of documents to publish).

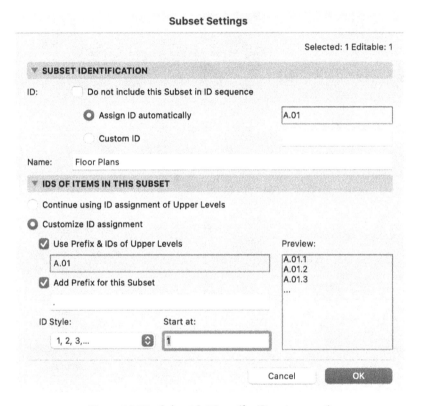

Figure 14.17 – Subset Settings (for ID assignment)

As you can see from the preceding dialog, a subset's only purpose is structuring the **Layout Book**: it is only concerned with ID, naming, and numbering. Subsets don't have any graphical implications on the layouts they contain.

> **Pro tip**
>
> ID management can take some careful thinking and preparation, but once set up (e.g., in your office template), IDs for layouts, subsets, and drawings can all be assigned automatically, ensuring a coherent set of output documents that is ready for publication.

The structure of your **Layout Book** is now set. Let us take a look at how we can add some content to those empty pages, using the views we have prepared in the **View Map**.

Drawing size and appearance

Each drawing is typically a scaled view of the building model. As a result, it will follow the display scale of the referenced drawing (e.g., 1:50). However, you still have the flexibility of magnifying the drawing. This is not recommended for floor plans or sections, but for perspective drawings, it can help resize the drawing to better fit a layout. Most other settings are self-explanatory.

Here, you can also see which **Pen Set** is selected for the drawing, still giving you some flexibility to adjust this (e.g., by enforcing a black-and-white drawing for clean printing).

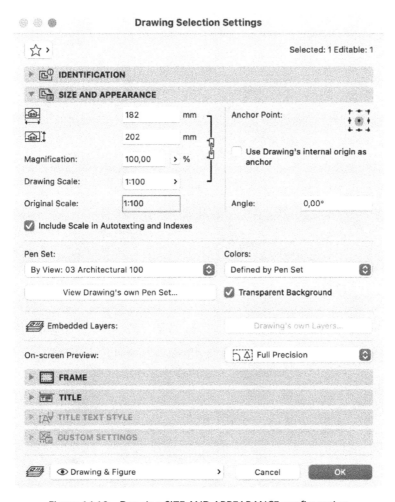

Figure 14.18 – Drawing SIZE AND APPEARANCE configuration

If you want to keep things logical (and understandable for other users perhaps), it is advised to keep the use of these overrides to a minimum. Otherwise, you could end up trying to change the appearance of a view/drawing through **View Settings** using **Quick Options** (which is the default workflow), without these changes being reflected on your layout because of an override (somewhat hidden in **Drawing Selection Settings**).

Drawing Frame

When a Drawing is placed, it not only displays the view that is being referenced but also (optionally) adds a frame to indicate its border. The drawing border is commonly not displayed, but if needed, you can add a printable border using any pen. For reasons of legibility, you can add an offset to ensure that the border does not collapse with the edge of the drawing.

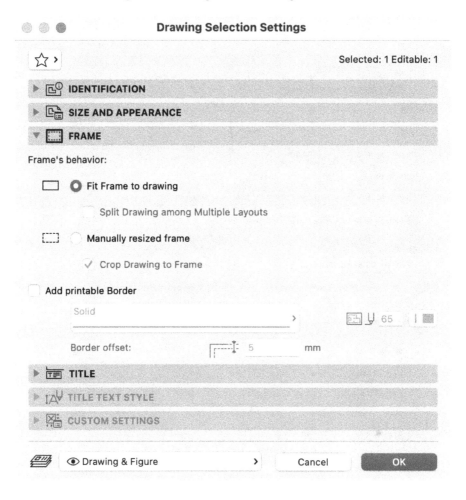

Figure 14.19 – Drawing FRAME configuration

The border of a Drawing is a contour that can be edited, like any other contour in Archicad, using the **Pet Palette**. You already know all of the operations here. The contour of the Drawing crops the underlying view, so if you only need a part of the drawing, you can crop it precisely.

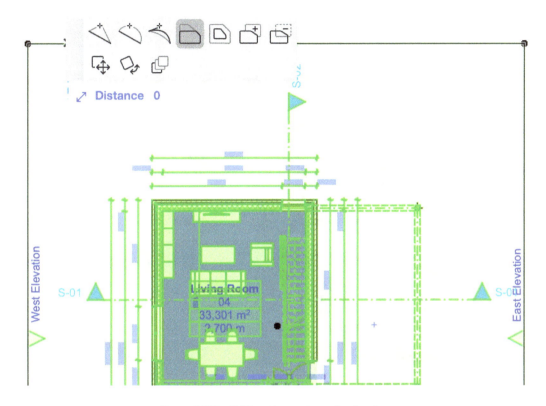

Figure 14.20 – Editing the contour of a drawing

What you may not guess is that the center node of the Drawing has a different Pet Palette: here, you have the option to move the underlying View while keeping the drawing contour in its place. You can use this if you have a cropped Drawing but want to shift the view to show a different fragment (using this with the whole View visible in the Drawing will also crop the view).

Figure 14.21 – Moving the view inside the drawing

> **Pro tip**
>
> When moving Drawings around, you still have the option to use the snapping tools. This can be really helpful when you want to align multiple Drawings based on the building elements. This is commonly used when aligning a Floor Plan with an Elevation on the same Layout so the walls align properly. This improves the graphical quality of your Layout and makes the Layout easier to read.
>
> You can use **Trace & Reference** for tracing one Layout to another and thus aligning Views across multiple sheets.

Drawing title

As already explained, a drawing is a referenced view with added information. You can use **DRAWING TITLE** to indicate on your layout what each Drawing contains by marking its content, such as name, ID, scale, and so on.

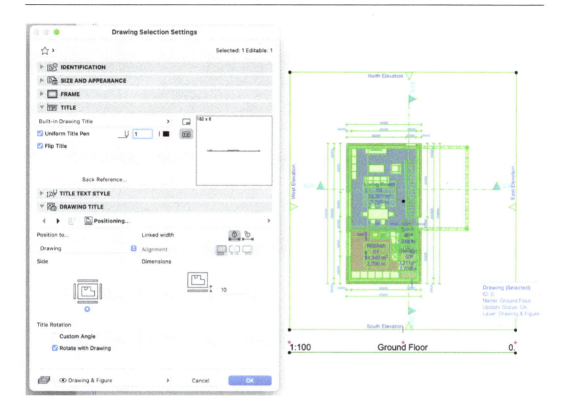

Figure 14.22 – DRAWING TITLE configuration settings

Let's take a look at the different settings for **DRAWING TITLE**:

- Titles are optional. In the **TITLE** panel, you can select between **No Title**, **Built-in Drawing Title,** or a dedicated **Title Marker object** from the loaded library. Here, we stick to **Built-in Drawing Title**, which is plain and simple but works for most cases.

- You can further configure its look using **Uniform Title Pen** and **TITLE TEXT STYLE**.

- The **DRAWING TITLE** panel is again a deep configuration panel with multiple pages:

  - **Positioning…** allows you to anchor and rotate the title to its drawing and set a fixed offset. This helps keep title placement consistent with other drawings.

  - **Title Text…** gives you further options to show the drawing name or use a custom text, show the drawing scale, and show the drawing and/or layout ID. With perspective drawings, we like to hide the scale display as this is a meaningless number for this kind of drawing anyway (a perspective has no scale).

- **Override Text Style…** gives even further options to adjust the text style of the components of **DRAWING TITLE**.

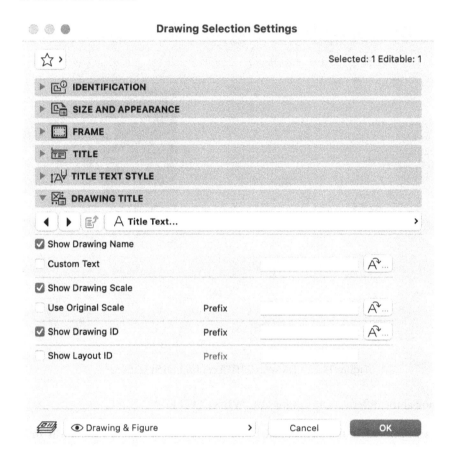

Figure 14.23 – Drawing Title Text configuration

When selecting one of the drawing title objects from the library, you may get other options, allowing you to further refine and control how the drawing title is displayed. But we leave this to you to further explore.

Some differences between Layouts and Master Layouts

In general, a Master Layout and a Layout work in the same way: you can add any 2D graphical element (lines, 2D objects, annotation, etc.) and Drawings, but the main difference lies in the referencing. All graphics that are inherited from the Master Layout are read-only. They cannot be selected or modified in the layout but can be used for snapping.

Figure 14.24 – Hovering over a master item in a layout

Title block and Autotext

A title block is a template block of graphics and text to indicate some information about the project and the elements of the layout, including the architect, the client, and some information about the drawing content and status, such as the revision date, the timestamp of printing or the filename, and so on. As a user, you can choose which information to be displayed by adding graphical elements, 2D objects, and (auto)text to your title block.

A title block is nothing more than a series of lines, fills, and text. It can also contain drawings or figures (e.g., a company logo or a fragment of the project to indicate the location of the current layout).

You can create a title block in multiple ways:

- Draft the title block directly in a Layout. It is strongly advised to do this on a Master Layout and not on the Layout itself so the title block can be reused across layouts. You wouldn't want to repeat this for every single Layout!

- Draft the title block in an independent Worksheet and save it as a view. This has the added advantage that the same sheet can be reused across multiple Master Layouts, ensuring full coherence across the whole **Layout Book** and multiple Layout sizes. That said, an A0 large sheet may have room for a full A4 title block, typically used as the front page after folding. A smaller A3 or A4 layout may only have room for a smaller title block.

Again, by using the referencing system of Master Layouts in a Layout and possibly worksheets into a Master Layout or Layout, you can keep everything synchronized, even when you make an adjustment to the title block.

But wait… some of the information in a title block can differ between Layouts! What about the drawing number? Or the name and number of the sheet and the drawing scale?

This is where **Autotext** is used (also see *Chapter 7*, in the *Associative labeling* section). Each text field can contain one or more **Autotext** fields, which are text fragments that are recalculated based on their context. When you click on the **Autotext** button, you get a huge list of **Autotext** categories with available fields to choose from.

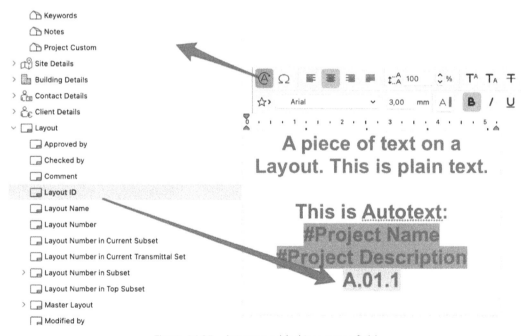

Figure 14.25 – Autotext added into a text field

The **Autotext** fields in the **Layout** category are typically used here, as they adapt to the current active layout. In the preceding example, **#Project Name** and **#Project Description** were picked from the **Project Information** category. Since they have not been filled in already, they are still shown as unresolved fields, indicated by the hashtag (#). The **Layout ID** field, however, picks up the **A.01.1** value for this current layout. If the order and numbering of layouts evolve, this field will adjust automatically, based on the automatic **Layout ID** assignment.

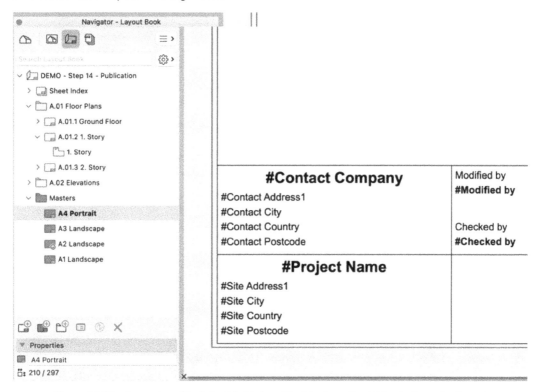

Figure 14.26 – Examples of Autotext fields in a title block on the A4 Portrait master layout

When you use a **Layout** field in a Master Layout (instead of in a Layout), it'll still display the # name since the name can only be resolved in the context of an open Layout and not the Master. Since you are only supposed to print or publish Layouts and not Master Layouts, this is the intended behavior.

Some widely used categories of Autotext that you can use are as follows:

- **Project information** includes details about the site, the building, the client, and so on
- **Layout** as mentioned already
- **Drawing** to reflect on the context of the drawing
- **System** to display the filename or **Last Saved at** timestamp

Not all of these fields make sense in the context of a Layout, though, but the **Autotext** system is used in many other contexts inside Archicad, such as when placing annotations on a Floor Plan or Elevation. Archicad will resolve the **Autotext** field whenever it can. If you keep seeing the # hashtag, the field cannot be resolved in the current context, or will only be resolved when we see the text from another context (e.g., looking at a view referenced in a drawing and placed on a layout).

> **Pro tip**
> When you prepare a template for your office, you can set up office title blocks once and ensure that they can be used easily in each new project. In most cases, the office name and logo can remain static text and images in the title block. Project information is best set as **Autotext** fields, so they inherit the field values from the Archicad project information. This way, even multiple title blocks, possibly configured for different page sizes or layout styles, can reuse the same fields and remain coherent.

When you have completed some Layouts, you could simply print them one by one, or export them to PDF. In the next section, we will greatly improve this process by introducing **Publisher**, allowing us to automate printing and (simultaneously) exporting deliverables in a variety of formats.

Publisher

In this section, we will take a look at managing the whole publishing workflow, from Viewpoint to View to **Layout Book** to **Publisher Set**.

The publishing workflow

We briefly touched upon printing and exporting in *Chapter 7*, but as you may have guessed by now, Archicad goes much further! The **publishing workflow** is a complete pipeline to configure information from the model, displayed as part of Views, referenced in Drawings, and placed in Layouts. The final step of this flow is controlling how everything is published or exported.

Publishing can take on several forms:

- Publishing a **Layout Book** into a multi-page PDF document or booklet
- Sending a series of drawings to a large-format printer
- Generating a package of files for a project milestone

The main idea is straightforward and can be summarized in the following diagram:

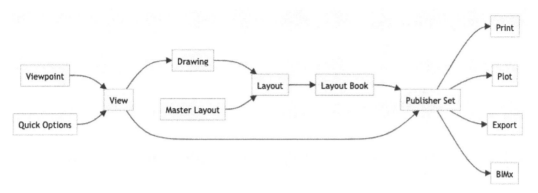

Figure 14.27 – The publishing workflow in Archicad

Figure 14.27 shows how a View is made based on a Viewpoint and **Quick Options**. This View can be placed as a Drawing in a Layout, which is defined by the Master Layout. Layouts are placed inside a **Layout Book**, which is then used within a **Publisher Set** to result in a variety of output formats.

Developing a Publisher Set

The basic goal of setting up your publishing workflow is configuring all required output in one or more dedicated **Publisher Set**. Each set can be configured separately and regenerated upon publication. For each **Publisher Set**, you need to indicate **Publishing method**:

- **Save files**: Update a series of files to be stored either as a single file, a folder structure of files, or a flat file structure. This can be sent to a local file or the **Graphisoft BIM cloud**, if you have access to such an environment.

- **Print** or **Plot**: These are more or less the same and, in most cases, local or network printers are accessible as a printer. If you need to send plotting instructions to a larger-format plotter, this may require additional setup of dedicated plotter drivers. We won't cover this here, but you may have to contact your local plot service provider.

- **Upload BIMx Hyper-model**: This is a dedicated interactive application containing your project, combining the 3D model and a series of drawings, which are linked into the model.

Figure 14.28 – Publisher Set Properties dialog, setup for a single PDF output

When you select **Create single file**, you have two **Format** options:

- **PDF** is a faithful and printable document, which is most common when sharing your output.

- **BIMx Hyper-model** is an aggregated interactive file, containing a real-time 3D version of the Archicad project and related views, such as floor plans and sections, aligned to the model for easy navigation. We return to this option in *Chapter 15*.

If you select **Create a real folder structure**, the structure of the **Publisher Set** will be mimicked to organize everything in files and folders, which is especially useful in large sets with tens or hundreds of files.

You only have to set the output folder location and you are good to go. But now we have to actually define the content of our **Publisher Set**.

From Navigator to Organizer

The easiest method to configure your publication workflow is from **Organizer**. This is an enhanced version of the **Navigator** palette, displaying two navigators side by side (see *Figure 14.26*).

To open this double Navigator, you can use the top-right pop-up menu from the **Navigator** palette or from the menu (**Window** > **Palettes** > **Organizer**). This looks just like the Navigator, so you already know how to use it. The main difference is the fact that you get a second Navigator view on the right side of the palette, allowing you to drag items from one map to the next. We go through the different steps to prepare a **Publisher Set**.

> Remark
>
> You can freely combine a **Navigator** tab on the left and on the right, but you can never select a **Navigator** tab of a lower level on the right side. It is perfectly fine to have two View Maps or Layout Maps side by side if you want. Archicad just disables invalid combinations.

Next, we explain how to use the Organizer for the different steps in the publishing process.

Step 1 – View Editor – from Project Map to View Map

The first configuration of the Organizer is **View Editor**, with **Project Map** on the left and **View Map** on the right. You can easily add new views by either picking a viewpoint in the **Project Map** and dragging it onto the **View Map** or by using the **Save View** >>> button.

We already explained how the **View Map** works, but the Organizer makes it more convenient to prepare a series of views, by configuring each viewpoint with the **View Settings & Storing Options** at the bottom left.

Figure 14.29 – Organizer - View Editor configuration

Step 2 – Layout Editor – from View Map to Layout Book

The second configuration is with the **View Map** on the left and a **Layout Book** on the right, so we get **Organizer - Layout Editor**.

Now you can drag Views (in the tree on the left) into Layouts (in the tree on the right), or more specifically, **Place Drawing** >>>, which references views. Both actions result in one or more Drawings being added to one or more Layouts. This result is quite similar to what we have shown in *Figure 7.44* when dragging a view from within the Navigator onto a Layout, the difference being that, by using the Organizer, you can do this in a more structured way, with a better overview and for multiple items at a time. Since the view is already configured, **View Properties** are not editable in this palette, but you can still access their settings.

Remark

When dragging views onto a layout, Archicad uses a default placement since the layout may not even be opened in the main window. But it is quick and easy to use this way.

Figure 14.30 – Organizer - Layout Editor configuration

Again, there is nothing really new here, so we continue to the Publisher Set next.

Step 3 – Publisher – from Project or View Map to Publisher Set

The final configuration is **Organizer** - **Publisher** and this typically has the **Layout Book** on the left and one of the Publisher Sets opened on the right. In the following example, that set is **1** - **Layouts**, but it could be called anything. Here, the template that we started from was prepared with two sets. You can use the arrow button or the popup to switch between the sets or to go to the top level, where we get an overview of all sets:

- **Layouts**: A set that reflects the **Layout Book** and shows **Sheet Index**, a **Floor Plans** subfolder, and an **Elevations** subfolder.

- **Views**: The second set contains several Views that are collected into a single PDF booklet, but contain only individual views rather than the (full) Layouts, which also have the title blocks and other layout graphics besides the Views (as Drawings).

Figure 14.31 – Organizer - Publisher configuration

When you select the root of the Publisher Set tree, you can see its format has been configured for exporting to PDF and that all branches and leaves underneath will be merged into one PDF file.

You can further refine the **PDF** output from the **Document Options…** button.

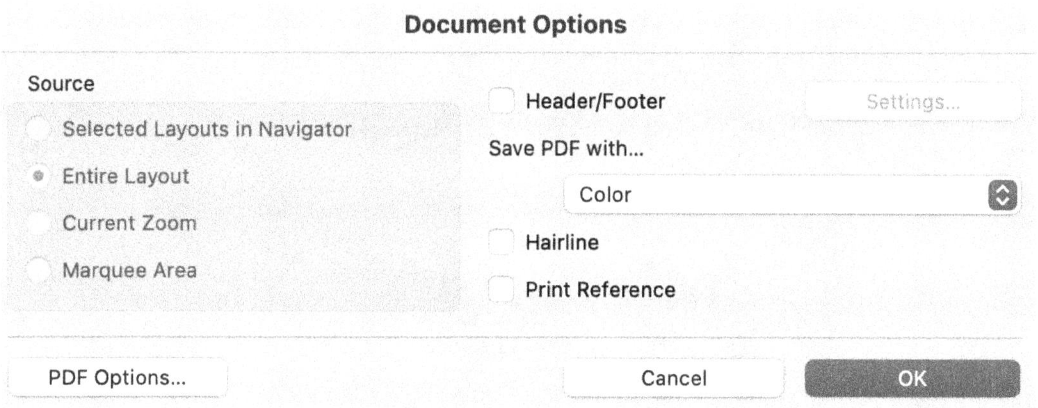

Document Options

Source

○ Selected Layouts in Navigator

◉ Entire Layout

○ Current Zoom

○ Marquee Area

☐ Header/Footer   Settings…

Save PDF with...

Color ⏷

☐ Hairline

☐ Print Reference

PDF Options…   Cancel   **OK**

Figure 14.32 – Document Options configuration for the selected set

Creating a new Publisher Set

We will create a third **Publisher Set**, where we make a small selection of Views and show you output possibilities other than the ones shown in the two included default sets (the first of which is shown in *Figure 14.28*). Since we are collecting both Views and Layouts, the Organizer is the best way to collect them, but it works the same way in the single-pane **Navigator** palette.

Figure 14.33 – Tools for Publisher

Follow these steps:

1. Navigate to the top level in the **Publisher** navigator, where you will see the different available Publisher Sets.

2. From the toolbar, you can create a new **Publisher Set**, duplicate a **Publisher Set** (the currently selected set), set **Publishing Properties…** for the selected set, or delete the selected set (four buttons shown from left to right in *Figure 14.33*). We will now create a new set.

3. Give it a suitable name (e.g., 3 - Client Meeting) and we are good to go. By default, **Publishing method** is set to **Save Files**, but you can switch to another method if needed.

4. In the **Publisher Set Properties** dialog (see *Figure 14.28*), simply point to the desired output folder on your computer or network or in the online BIM cloud and launch the publication. This will display a progress dialog, showing the updates of all Views and Layouts that Archicad needs to process, to ensure that everything is synced from the model.

5. The current set is still empty, so you are not ready for publication yet. Double-click the new **Publisher Set** to open its hierarchy in the Organizer (see *Figure 14.28*). Now, we can start dragging Views and Layouts into the set.

6. From the left panel in the Organizer, go to the **View Map** and select the **0. Ground Floor** view. By default, it is set to PDF format.

7. Do the same for **1. Story** and **South Elevation**. Add any view you like.

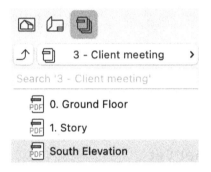

Figure 14.34 – Custom Publisher Set with three views set to PDF

Within an open **Publisher Set**, there are two tools to further organize the set. Let's do this for the set we just created:

Figure 14.35 – Tools for a Publisher Set

To show how you can combine multiple output formats in one **Publisher Set**, we will add a new folder (left button in *Figure 14.33*) using the tools we have explained in this section and name it Images. In this folder, we will drag some more views (using the Organizer), but instead of PDF, we configure their format as JPEG. Archicad displays this with a different icon.

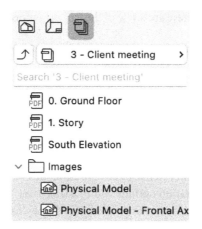

Figure 14.36 – A Publisher Set with multiple file formats mixed

Any item in the set can be further configured:

- **PDF** items have **Document Options** (including zoom, scale, color settings, and arrangement on the page) and **Page Options** to set the output page size
- **JPEG** and other image formats provide a few options for color, space, and compression quality

Depending on the View or Layout you pick, the list of possible formats may vary, but it is extensive.

Figure 14.37 – List of format options for a 2D view

We won't cover them all here, but these are a few of the important ones:

- **DWG**, **DXF**, **DWG**, and **DGN** are vectorial CAD drawings, for software such as **AutoCAD**, **MicroStation**, **BricsCAD**, and **Vectorworks**:

  - Use this when exchanging drawings with other architects, engineers, or contractors who need 2D CAD files but don't use Archicad.

  - Drawings are just the graphics: lines, arcs, text, dimensions, and so on. The model objects and their information are not exchanged this way.

  - These formats come with an extensive configuration, in a so-called **Translator** setup. While explaining all the configuration options within a translator would take a full chapter in this book, the main advantage is that once configured, they can be recalled by the Publisher during export. This ensures every future export is guaranteed to use the exact same settings.

- **JPEG image**, **PNG image**, **GIF image**, **TIFF image**, and **Windows BMP image** are all image formats:

  - Use **JPEG image** for renderings or views from the 3D window when textures or shadows are used

  - Use **PNG image** for clean line drawings, to avoid **JPEG image** compression artifacts (e.g., floor plans, elevations, or 3D documents)

- The **IFC** open standard file format (**IFC** stands for **industry foundation classes**) is used to exchange the actual model (3D geometry and information) with other BIM software. IFC was introduced in *Chapter 13*, in the *Using interactive schedules in Archicad* section, and we will briefly come back to it in *Appendix*:

  - We could write a full book on this format alone, as it is very important in BIM-based information exchange. But it is a deep and complex topic, as Archicad provides extensive configuration options in its **DWG** and **IFC** Translators (see *Chapter 7*, in the *Saving as DWG* section).

  - Use this when exchanging Archicad models with users of **Revit**, **Vectorworks**, **Solibri**, and other software. Walls remain walls, and doors will be doors, but they may not behave exactly the same in other software.

With this in mind, we are able to configure our set of documents, as shown here:

Figure 14.38 – Configured Publisher Set

Notice how the floor plan drawings are published as DWG CAD files in a folder CAD and at the same time, also as PDF files.

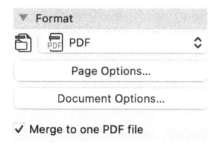

Figure 14.39 – Turning a folder into a multi-page PDF file

Moreover, by grouping the PDF files into a folder, we can set the folder with **Merge to one PDF file** (the checkbox in *Figure 14.39*), so the three views become pages in a single document.

Publishing all your output in one go

At the bottom of the Publisher (when a **Publisher Set** is selected), you can use the **Publish** button to launch the publication of the selected items, the entire set, or even all the sets.

> **Remark**
>
> The first time you do this, you may get a warning dialog to explain that the configuration of this set is not complete. In this case, you still need to set the address or path for the output to be stored.

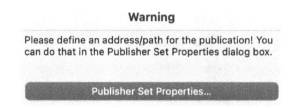

Warning

Please define an address/path for the publication! You can do that in the Publisher Set Properties dialog box.

Publisher Set Properties...

Figure 14.40 – Warning when the Publisher Set is not yet fully configured

The **Publisher Set Properties** dialog asks you to set the **Publishing** methods, as discussed earlier, and then set the type of output as single or multiple files and the related folder structure.

Now, Archicad can start generating the different files in the set. In this example, we have eight steps, but the last three steps will lead to only a single file, our PDF booklet.

Publishing... (8/8 done)

	Name	Publisher file	Status	Size
✓	Physical Model	/Users/stef...l Model.jpg	Saved	215 KB
✓	Physical Model -...	/Users/stef...ometry.jpg	Saved	197 KB
✓	0. Ground Floor	/Users/stef...Floor.dwg	Saved	50 KB
✓	1. Story	/Users/stef...Story.dwg	Saved	47 KB
✓	South Elevation	/Users/stef...vation.dwg	Saved	24 KB
✓	0. Ground Floor	/Users/stef...ut/PDF.pdf	Saved	67 KB
✓	1. Story		Saved	61 KB
✓	South Elevation		Saved	101 KB

Note: Double-click on the item in the list for details.   Close

Figure 14.41 – The publishing steps

In a small model, this happens very quickly, but rest assured that large projects with large Publisher Sets can take a considerable calculation time. However, even then, this is several times faster than trying to remember all the individual files you have to configure and export.

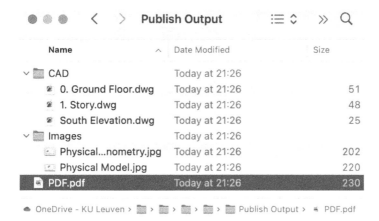

Figure 14.42 – Example of the output folder with different published files

Also notice that the folder names simply reflect the names in the set. You have full control over how they are named and which file formats they contain.

This brings us to the end of the section on the **Publisher Set**, in which we explained the intelligent and efficient system Archicad provides for creating an automated flow in publishing your BIM to multiple output formats in an organized way. With this, we also finish our chapter on the publication of BIM extracts. Let us briefly summarize what we have learned here, before moving on.

Summary

In this chapter, we completed our exploration of the Archicad output.

First, we dove deep into the different options we store in views: Layer Combinations, Pen Settings, Graphic Overrides, the Renovation Filter, and the Partial Structure Display.

We learned about the powerful Navigator and how it helps us to find all our viewpoints in the **Project Map**, the **View Map** with configured Views, the **Layout Book** with sheets, Layouts, and Master Layouts, and finally, the **Publisher Set**.

The publishing workflow is a good example of the power of BIM inside Archicad; all output is collected with in a single publishing operation so you don't have to remember all the different filenames, formats, and the sometimes very extensive export settings.

Within a single set, you can set up any combination of documents, such as a PDF booklet created from different layouts, individual images from a 3D window, or CAD drawings and models generated from selected Views for sharing with others who may not be using Archicad.

In the next chapter, we will move forward on our Archicad journey by exploring the various visualization techniques on offer, from 3D documents over photo renderings to interactive BIMx hyper-models.

15

The Various Visualization Techniques in Archicad

By now, you have a very clear and broad view of the modeling capabilities of Archicad, as well as an overview of how it manages the information in your model and how the software provides workflows for sharing the derived drawings and data schedules with other parties using the publication flow explained in the previous chapter.

It is now but a small step to use this fully developed model, which of course includes the material definitions added via the Building Materials, to create some more advanced images. Within this chapter, we can further explore the possibilities Archicad provides for the typical 2D views, add some options for graphical representations in 3D, and even render an image for a photorealistic view of our project. Beyond that, there are also ways to use your model as a starting point in other imaging software, of which we will give a short overview and some general advice.

In this chapter, the following topics are covered:

- Creating more graphical or less technical 2D Views
- Creating graphical 3D representations, using 3D Documents
- Creating both rendered still images and rendered animations
- Using external rendering software
- Creating (interactive) real-time visualizations with Graphisoft **BIMx Hyper-models** or external gaming engines

This chapter finishes our journey into **Building Information Modeling** (**BIM**) and how we can use Archicad for this. In more than one way, visualization is the cherry on the cake – it offers you the possibility to make your design stand out in presentations and shows you how you can communicate about a design in a variety of ways. Let's start by showing you how to enhance your 2D and 3D output!

Creating compelling images with Archicad

In the world of architecture, construction, and building design, almost every project must be presented at some point. The images used for presentations often have a different look and approach compared to the standard technical construction drawings such as floor plans, sections, and elevations. They are generally more expressive and less technical, making them more appealing to a client and not specifically aimed at construction professionals. In this section, we will explore how we can further develop these kinds of presentation images using Archicad Views, and we will even add some new inventive workflows for creating these kinds of images, using techniques we have already learned.

Getting creative with 2D and 3D Views – preparation

In *Chapter 12*, we have explained how views such as Sections and Elevations can be set up. The focus in that chapter was creating a clear, technical drawing aimed at communicating the design with a construction company. In the following section, we will combine such Views with other techniques such as **Image Fills** (see *Chapter 6*, in the *Using Fills for 2D drafting and annotating* section) and **Graphic Overrides** (see *Chapter 13*, in the *Graphic Overrides* section).

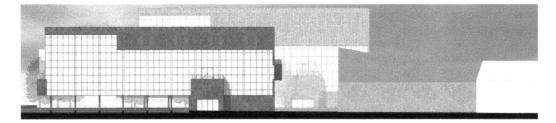

Figure 15.1 – Elevation from Chapter 6 (Figure 6.28) – it is time to learn how to create it

Before we start, we advise you to add some exterior elements to your model similar to what is shown in *Figure 15.2*. Otherwise, the design is floating in space and lacks context.

Figure 15.2 – Creating some context for the design will improve the final visualization

In our example, we have done the following (see *Figure 15.2*):

- We added a Mesh for the terrain, including a slight slope, using the technique described in *Chapter 10*, (in the *Site model using the Mesh tool* section).

- We also added a driveway using slabs and a Roof for the sloping part and then combined the Slab with the sloping terrain using some solid Element Operations, as described in *Chapter 8*, in the *Using Solid Element Operations (SEO)* section.

- Next, we created a pond or small swimming pool and a deck or terrace – again, using Slabs and Solid Element Operations.

- Finally, we added a few objects (trees, a car, etc.) to complete the scene. There are many useful objects waiting for you in the Archicad library.

Figure 15.3 – Gravitation in Archicad

> **Tip**
>
> To place trees or any other object at the right height on top of a sloping terrain, you should use **Gravitate to Mesh** (see *Figure 15.3*). When placing your object in 2D, this toggle will make the object drop to the correct height.

Now that the design has been staged, we can start creating some visualizations. We will start by taking our 2D presentation skills to another level, reproducing an elevation similar to the one shown in *Figure 15.1*.

Adding texture to Floor Plans and Elevations using Graphic Overrides

Although **Elevation Settings** allow you to choose **Surface - Texture Fill, shaded** to display the texture applied in the settings of the **Surface** attribute, this does not always give you the result you want. In *Figure 15.1*, we see a mix of graphical elements (the trees, the windows, etc.), photorealistic images (the clouded sky), and a few things in between. Achieving such an image is done by combining several techniques that were already introduced throughout this book:

- Creating an Elevation View (*Chapter 12*)
- Creating a Layer Combination (*Chapter 14*)
- Using a Graphic Override (*Chapter 13*)
- Adding 2D objects (*Chapter 6*)
- Placing Drawings on a Layout (*Chapter 14*)

Let's break this process down into simple steps, which we will apply to the project we have modeled and documented in *Chapters 3* to *14*:

1. Open **South Elevation** and save this as a view in **View Map**:

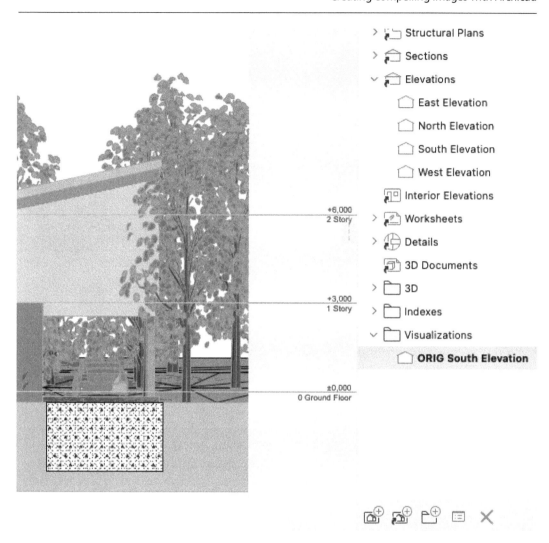

Figure 15.4 – We called this ORIG South Elevation because we will
need this version later on (ORIG stands for Original)

2. Limit the elevation vertically by going to the **Elevation Selection Settings** dialog and in the
    GENERAL section choosing **Limited** for **Vertical Range**. Set the upper value to 12000 and
    the lower value to -500 so the trees are fully showing and we see just a little bit of the bottom
    of our project.

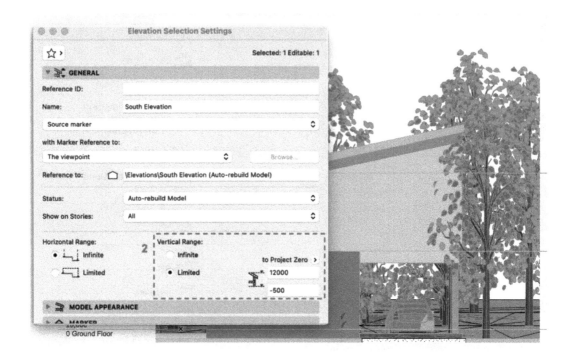

Figure 15.5 – Setting Vertical Range for an Elevation and the result in the background

3. Create a new layer combination and call it `11 Elevations for presentation`, in which the layers containing the car and the trees (**Site & Landscape - General** and **Site & Landscape - Terrain**) are *hidden*. We are going to replace the 3D projected trees and the car with 2D images. This will create a more balanced, graphical image. The graphical representation of our 3D Mesh is somewhat technical in appearance, showing all the triangular faces. Unfortunately, there is no nice way of hiding all these ridges, so let's simply hide the shape in our elevation by hiding its Layer in this new Layer Combination.

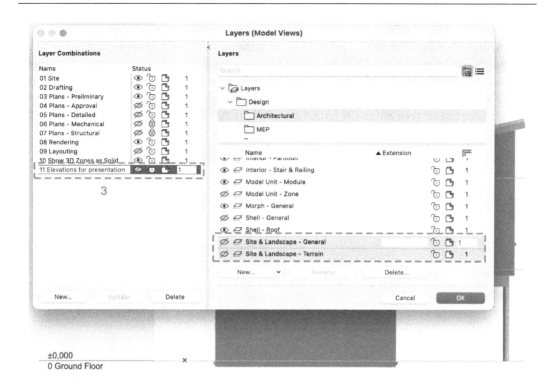

Figure 15.6 – The mesh is typically placed on the Terrain layer, and trees
and other props are on the General layer for Site & Landscape

4. Turn off transparency by going to **Elevation Settings** > **MODEL APPEARANCE** > **UNCUT ELEMENTS** > **Transparency** as, for an elevation, it can be confusing to see through the exterior windows. Elevations rarely show transparency in conventional architectural drawings.

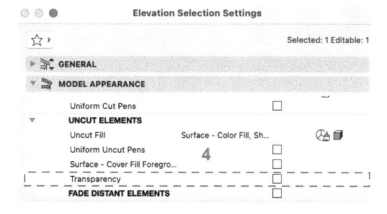

Figure 15.7 – For elevations, it is common to turn transparency off

5.   Add a few appropriate Image Fills for brick and wood in the project through **Options** > **Element Attributes** > **Fills**. Click **New…** (**a**) to add a fill to the project, give it an appropriate name (**b**), and choose **Image fill** (**c**). Click **Browse…** (**d**) to search for a fitting image. Such an image could be chosen from **Archicad Library 26** (**e**) or you can add (**f**) one you downloaded from the internet. Make sure to set **Pattern Unit size: [mm]** (**g**) in accordance with the source of the image and set your Fill to **Use with: Cover Fill** (**h**). As this is a real-life texture, you should select **Scale with Plan (Model Size)** (**i**). Finish by clicking **OK**.

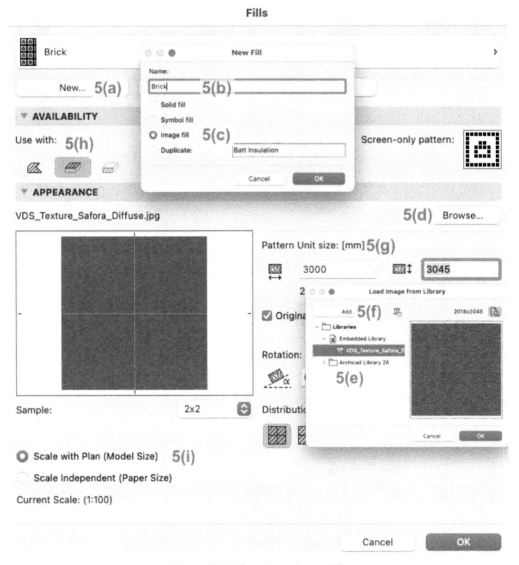

Figure 15.8 – Creating an Image Fill

> **Tip**
>
> Using good texture images makes all the difference in your final image. They can be found on many websites (paid and free options). The brick texture we used in *Figure 15.8* is available from the manufacturer's website of *Vandersanden* (`https://www.vandersanden.com/en/texture-generator`). They have good-quality customizable textures and even provide an Archicad plugin to support you in creating the right look for your images. In general, you should look around for sites offering *seamless* textures, which can be tiled without ugly borders. Beware, however, that large texture files will start slowing down Archicad eventually. Unless you need an extreme close-up, textures seldom need a width or height of more than 2,000 pixels.

6. Create a Graphic override that shows textures for the opaque parts of the exterior walls. In our example, we have a brick finish on the ground floor and wooden cladding on the upper floor. You could do this through the **MODEL APPEARANCE** tab in **Elevations Settings** as well, but this is a binary kind of choice; every element or object is shown in the same way – there is no room for more nuance.

Figure 15.9 – Creating a Graphic Override Combination, with two similar
rules based on the composite name used in the external walls

7. Save the view with a different name in your **View Map** (thus applying the edits made in *steps 2* to *6*).

8. Although the resulting elevation view may look quite barren, it is time to place it on a layout. In our example, we are using a simple, *empty* A4 landscape Layout, without any general project information or text on the Master Layout, as we are only focusing on producing this image.

9. To finish the image, we will stack several elements (drawings, but also an object and a fill) on top of each other, as shown in *Figure 15.10*.

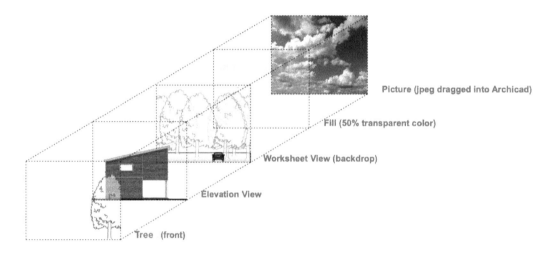

Figure 15.10 – Combining several drawings and other elements on a
layout allows us to create and fully control our own style

When these parts are placed in the right order (from the bottom to the top of the stack – in that order), this doesn't need any further configuration, but you can always adjust **Display Order** through the context menu to rearrange the stack. For example, if you placed the clouds as the last element, this image would cover all the other parts in the stack. To rectify this, select the clouds image, use the right mouse button for the context menu, and choose **Display Order** > **Send to Back**.

The following elements are added (in the *right* order):

A. A **figure**: This tool can be used to add bitmaps to your project (see *Chapter 6*, in the *Adding a figure* section). If you drag the cloud image directly into your Layout, this tool is automatically used.

B. A semi-transparent **fill**: We chose a light yellowish color to temper the appearance of the clouds. You can experiment and, for instance, apply **Gradient Fill**. Obviously, the background pen is set to transparent (**ø**), or the fill blocks out the clouds entirely.

C. A **Drawing** referring to a **Worksheet**: This Worksheet contains symbolic 2D representations of several elements; the hill is shown as a gradient fill and 2D objects are used for the trees and the car. Correct placement is achieved by using **Trace & Reference** with **ORIG South Elevation** as a reference. 2D objects can be made transparent by setting **Fill Type** to a semi-transparent fill (**50%** in our example) and setting **Fill Background Pen** to ø. A car symbol is found in **Archicad Library 26** under **1.7 2D Elements 26** (see *Chapter 6*, in the *2D objects* section). Save the Worksheet as a view before placing it onto the Layout.

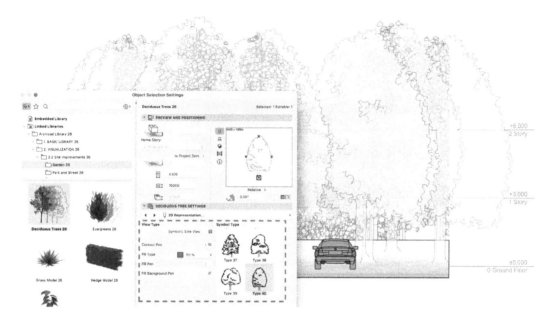

Figure 15.11 – Tracing the complex 2D representations of 3D objects with simplified transparent 2D symbols

> **Note**
> Using this Worksheet creates a cut in the automatic updating process of the final image: the elements traced on the Worksheet are not updated when editing their counterparts in the 3D model. So, preferably, use this technique solely for static elements in the scene or at the last stage of a design phase!

D. A drawing referring to the Elevation view created in *step 7*.

E. An **object** (tree) placed directly on the layout. This tree is (partly) in front of our building. You can simply copy and paste one of the trees from the worksheet. The instance on the layout will become a fill and a group of lines and no longer an object, so trimming it along the edges of the frame of your final image should be easy: use **Pet Palette** for the fill (**Subtract from Polygon**) and the **Trim** function (*Cmd + LMB* or *Ctrl + LMB*) for the lines. Don't forget to suspend groups (*Alt + G*). You can choose to set it up differently – for example, more or less transparent or with a different pen for the contour line, and so on.

10.   Your final image is ready. Compare it to your starting point (*Figure 15.4*).

Figure 15.12 – The finished image may look like it was photoshopped but is fully produced within Archicad

As you may have noticed, techniques like those described in the preceding steps require a lot of the skills we have introduced throughout this book, and thus you will need some experience in these basic skills before being able to create images like these off the cuff. We are confident, however, that any Archicad user can produce similar results with practice, patience, and the steps provided in this book, without requiring separate editing in software such as Photoshop, Illustrator, or Affinity.

Now that we have mastered some extra possibilities using 2D views, there is one last type of 2D document to look at: 3D documents. Confusing, huh? Let's see in our next section how a tool called **3D Document** results in a two-dimensional image (actually, we prefer the term 2½D – somewhere in between 2D and 3D…).

Taking it a step further with 3D documents

Using these techniques directly, or by combining several techniques, there is a huge range of visualization styles you can apply. We want to explicitly introduce **3D Document** here.

3D Documents are 2D projections of any 3D View. They can be enriched with 2D annotations such as text and dimensions (which is not possible on the 3D source View), but remain linked to the model and can be refreshed. This is different from the 3D source View, a 3D render, or a screenshot from the 3D View, which do not allow us to annotate in 2D. So, in a way, you can think of it as 2½D with benefits of both view types.

Creating a 3D Document

Any 3D View or Viewpoint can be turned into a 3D Document. From **Project Map**, there is a branch called **3D Documents**. Use the + button to create a new 3D Document in this branch. It will be generated from the currently open 3D window. You can also use *RMB* on any existing 3D view and create a new 3D Document from within the context menu.

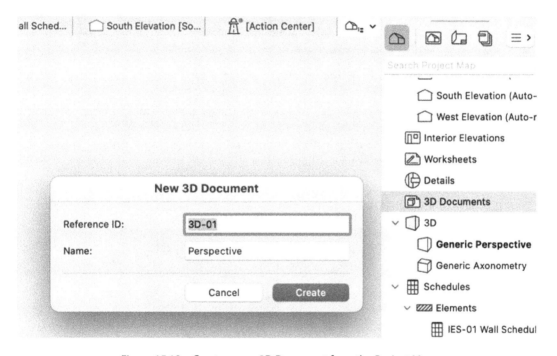

Figure 15.13 – Create a new 3D Document from the Project Map

The 3D document uses the same style as sections or elevations, which is based on a **Vectorial Engine**. It supports colors and pens but does not display textures.

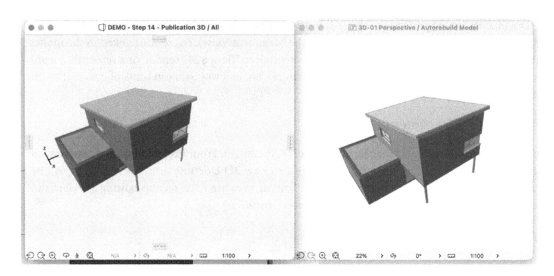

Figure 15.14 – 3D document created from a 3D window

This 3D Document behaves like a regular 2D window, supporting annotation and grids and snapping when drafting, but it also has similar settings as the Sections and Elevations we covered in *Chapter 12*, in the *Configuring Sections and Elevations in detail* section. However, when you add dimensions based on nodes of elements, the real 3D distance will be picked up.

Figure 15.15 – Annotating a 3D document

You can return to the source 3D window from the 3D Document by picking **Open Source View** from the context menu (*RMB*) of the Viewpoint in the Project Map or on the View's background.

You can also update the 3D Document (**Rebuild** and **Rebuild from Model** from the context menu) and even select **Redefine 3D Document based on current 3D Window**, which can be used if you want to adjust the perspective or camera position, for example. Archicad will warn you that this is not undoable (it cannot be reversed).

Within **3D Document Selection Settings** (*Figure 15.16*), you have very similar options as in **Section Settings** and **Elevation Settings** on how to project the 3D window into this 2D display: the display of **CUT** and **UNCUT ELEMENTS** and the display of **SUN** and **SHADOWS** can all be set to your needs and preferences (see *Chapter 12*, in the *Configuring Sections and Elevations in detail* section).

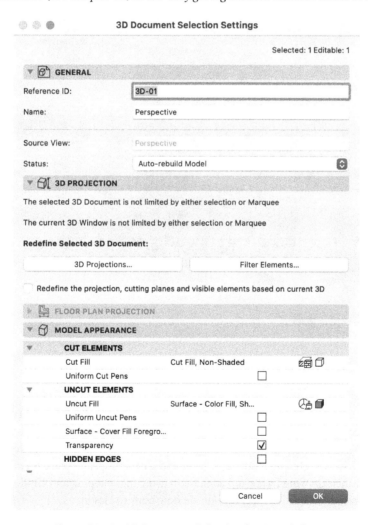

Figure 15.16 – 3D Document Selection Settings dialog

Now that we have learned the technical aspects of 3D documents, let's get creative with them for visualizing our project.

Creative use of the 3D document for visualization

Here are two examples of how 3D documents can be applied:

- Prepare a view with **3D Cutaway** or the **Marquee** tool, as we explained in *Chapter 12*. When you create a 3D document from this view, it will display the Composite structure of Wall, Slab, and Roof structures in a 2D style, with the correct fill patterns as in a regular 2D section or elevation. We like this method to create open 3D sections, which display the composition of elements cleanly. Such a drawing would take an enormous effort in a regular 2D or even 3D CAD system.

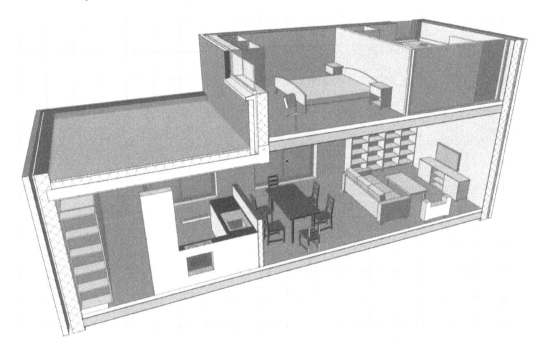

Figure 15.17 – 3D Document from a view with two cutaways

- To create an exploded 3D axonometric view, we need to create 3D documents from different stories and compose them on a layout. To display only a single story, use the **Filter and Cut Elements in 3D** option (**View** > **Elements in 3D View** > **Filter and Cut Elements in 3D**). Prepare two Views, each with a different **Filter and Cut** setting, and place them onto a Layout. You can hide their drawing frame and align them using the snapping and Guides. And to finish things off, draw a few dashed lines to indicate how both stories fit.

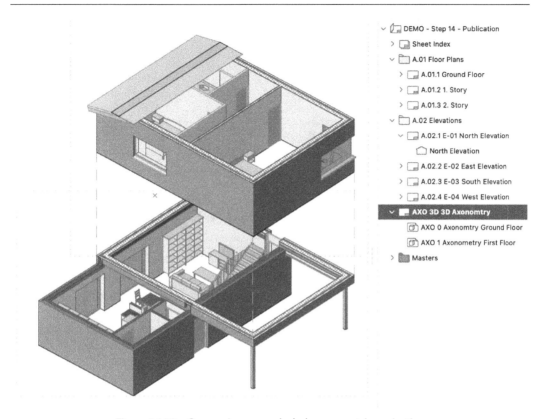

Figure 15.18 – Composing an exploded axonometric projection
using two 3D documents aligned on a layout

This second example takes a little more effort than having it in a single view, but you retain full control, and the views will be updated with model changes. Again, not something you would typically consider doing in regular CAD software.

Of course, you can even go further with the previous techniques and combine them with what was shown for creating the elevation view at the beginning of this chapter. For example, you can use the **Marquee** tool to create complex cutouts for your exploded axonometric projection, add some graphic overrides, combine the result with 2D views, and maybe mix in a suggestive background, as shown in the following examples.

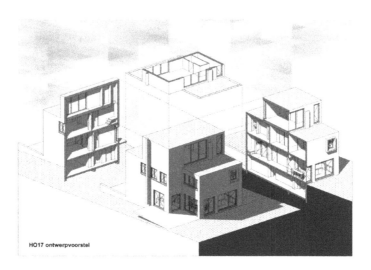

Figure 15.19 – Example 1 (©Flotus/studiov2 - arch. Pieter Vandewalle)

The preceding figure shows a combination of 3D documents, with a background image and graphic overrides for the roofs and site. The following figure combines a background and a 2D tree to set the atmosphere, while the 3D images provide insight into the overall spatial design.

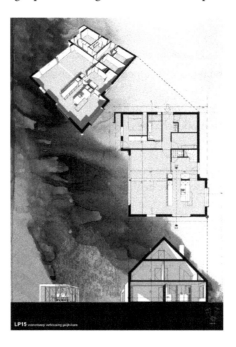

Figure 15.20 – Example 2 (©Flotus/studiov2 - arch. Pieter Vandewalle)

As shown in these examples, without any external software and without rendering, Archicad already provides a powerful workflow for creating stunning and compelling images for your design presentation. Nevertheless, it is very interesting to also know the basics of the software's rather impressive built-in rendering capabilities. So, let's move on to the next section to do so!

Creating rendered images

The techniques displayed so far mainly rely on regular projection and attributes. However, there is also a dedicated system for the generation of **Photorealistic Images** or **Renderings**. In fact, Archicad has always provided means to generate a 3D view with lighting and shadows. This really got a boost with the integration of the **Cineware** engine, which is borrowed from the *Cinema 4D* software by *Maxon*. This is a professional and dedicated visualization software, and Archicad has gained the whole rendering functionality to allow realistic lights, materials, shadows, textures, reflections, and the distribution of light using *Global Illumination* techniques.

This is where your computer will start using its full power to generate highly realistic images of your Archicad model.

Since Cineware is a very extensive system, we cannot cover all its details. Luckily for us, *Graphisoft* has decided to implement this using a preset system. You typically pick one of the presets and then, if needed, still have access to all its underlying expert settings should the need arise.

Rendering requires three main parts – geometry, surfaces, and light – and we have complete control over all of them.

The whole book until now focused on building a model, so we assume that the preparation of geometry has been well covered by now. Only remember that the more geometry the system needs to deal with, the more calculation time you should expect. Rendering a skyscraper is more challenging than a single-family house.

Surfaces for rendering

The next step is the preparation of the surfaces. **Surfaces** are attributes in Archicad that are used in **Building Material** definitions and **Model Overrides**. They can be set up through **Options** > **Element** > **Attributes** > **Surfaces**.

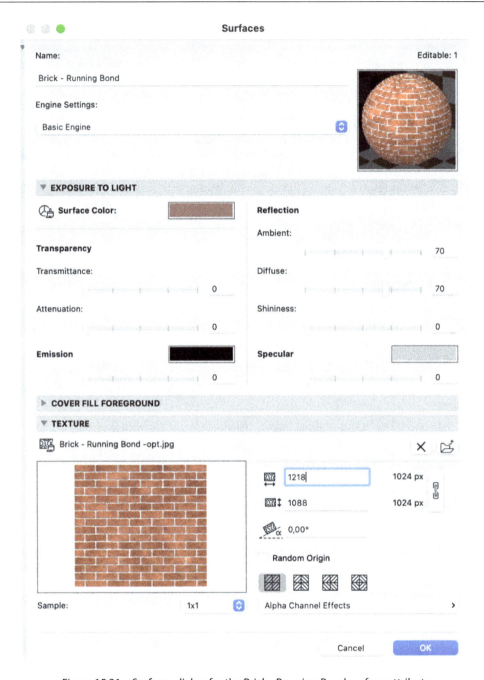

Figure 15.21 – Surfaces dialog for the Brick - Running Bond surface attribute

Within the **Surfaces** dialog box, there are a few important settings to configure:

- **Surface Color** sets the core color of the surface, which is what you see in the Viewport or in shaded views if no texture is applied.

- **Transmittance** defines how transparent the material surface will be. Most Building Materials are completely opaque, but for glass materials, this should be increased.

- **Shininess** and **Specular** are used to indicate the reflectivity intensity and color of the surface. Keep it at zero for dull and rough materials. For metal, plastic, glass, and other more reflective materials, increase **Shininess** and adjust the **Specular** color. Plastic and glass typically have a bright white or light gray **Specular** color, while metals take on their own color (e.g., yellowish for gold).

- **Texture** is where you can optionally add an image texture onto the surface. In the default Archicad library, there is a whole range of textures available that can be used, and most surfaces in the template already have a texture assigned. The addition of a photographic texture already contributes a lot toward having realistic surfaces as is. Just be careful to also set the **Horizontal** and **Vertical** sizes of the texture: this tells Archicad how far the texture should be stretched when being applied to an element. In the preceding example, based on the size of individual bricks and the width of the mortar, the horizontal size was set to 1218 [mm] and 1088 [mm] in the vertical direction.

Pro tip for textures

Depending on the image (bricks, planks, tiles, etc.) you need to adapt the texture size. Check with the manufacturer documentation when you need to closely mimic a real material. Getting this wrong destroys the results! You don't want such bricks to be only 3 mm high or 3 m long. There is more to this but ensure that these are at least set correctly. When you are in doubt, just look at some of the example surfaces in the template and start from their values. Again, you could use good textures provided by manufacturers, as mentioned in *step 5* in the *Adding texture to Floor Plans and Elevations using Graphic Overrides* section.

As discussed before, ensure that all Building Materials that are assigned to your elements have their surface attribute properly configured, including their color, transparency, and texture.

What may not be directly obvious is that, due to the availability of multiple 3D engines, surfaces also have multiple definitions. The main surface definition we just discussed influenced how surfaces appear in the 3D window and in sections or elevations. In the context of rendering, an alternative definition of a surface can be configured. Luckily for us, this has been configured already for all out-of-the-box surfaces, and if you have access to an additional **Surface Catalog** (e.g., via your maintenance license), these are all set up to work well with no need for further tweaking.

When you create a custom surface (e.g., by selecting another color and texture), you need to ensure that the rendering configuration of the surface follows along. You can do this from the **Surface Attribute Settings** dialog (**Options** > **Element Attributes** > **Surfaces**).

Within this dialog, you can switch the display of the surface using the **Engine Settings** dropdown. By default, **Basic Engine** is shown, and this is what you usually see in the viewport. Switch to the **Cineware and Redshift by Maxon** engine instead.

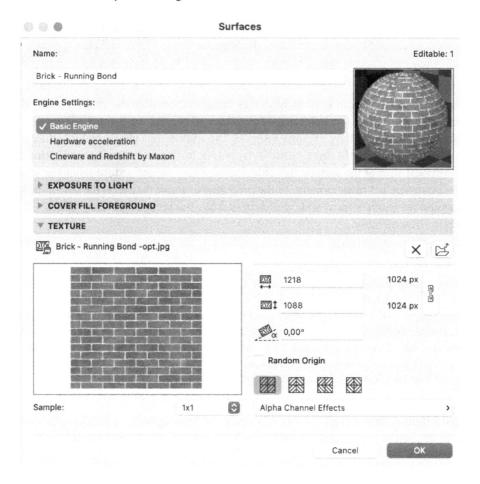

Figure 15.22 – Switching to the Cineware and Redshift by Maxon engine

After you switch, the **Surfaces** settings dialog displays **CINEWARE SETTINGS**, which really expand the amount of control you have over the surface, replacing the **EXPOSURE TO LIGHT** and the **TEXTURE** sections of the dialog with more extensive options. Instead of having a few color values and a single texture, as we had before, you now have multiple channels, each contributing to the surface style.

When you create your own surface, we suggest starting from **Basic Engine** and using the **Match Settings…** (*Figure 15.23*) popup to select **Update Cineware Settings (from Basic)**, so you get a good starting point. However, since there are many additional settings with Cineware, don't call **Match Settings…** for the already configured surfaces!

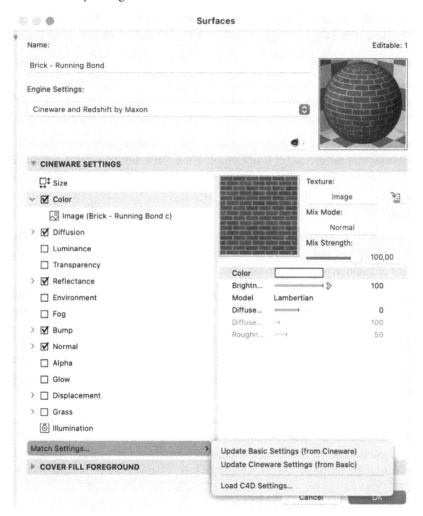

Figure 15.23 – Cineware settings for a surface: matching settings

We mention the most important ones only:

- **Size** is the first setting, and impacts the scaling of texture maps. Get this absolutely right or your whole surface will look out of place. Here you indicate, just like with the Basic Engine, the **Horizontal** and **Vertical** size of the texture: this tells Archicad how the texture should be stretched when being applied onto an element.

- **Color** is the basic display of the material, as a combination of an RGB color value and a texture map (a seamless image of a material surface). This is the most important part of your surface, so ensure the right image is loaded here.

- **Diffusion** is an extra layer that interacts with the surface. In most of the Cineware materials, this is a grayscale version of the texture used in **Color Channel**, but set to **Affect Specularity** – for example, making the mortar layer in the brick wall duller and reflecting less light.

- Use the **Transparency** channel for glass and other transparent or translucent surfaces.

- **Reflectance** gives more in-depth control over how the surface will react to light. This will dictate the difference between a dull and rough material (such as brick or concrete) or a shiny, polished, and reflective material (such as varnished parquet, polished stone, or inox and aluminum).

- **Bump** and **Normal** are two methods to mimic surface undulations or perturbations. This gives the impression that the geometry is not simply a flat plane, but that it is much more detailed, with mortar lines, cracks, small dents, and other surface details reacting to light. This requires a special texture map, which is typically provided in the advanced surface catalog or in third-party texture libraries.

- **Displacement** takes **Bump** and **Normal** literally up to an even higher level by having the rendering engine generate actual geometry instead of simply simulating the bumps. This is a very heavy calculation and only makes sense in very specific advanced situations.

- The same applies to the **Grass** channel, which is, again, a rendering effect to add 3D grass into the geometry. This works nicely on a site mesh object but can explode rendering times considerably. Use this only when it really adds value to the rendering.

In most of these channels, textures maps can be used. But it goes even deeper. Instead of a single texture map, this can become a combination of layers of textures and other effects (e.g., by mixing multiple textures and blending them with procedural effects).

We don't want to push you this far, but due to the integration of the Cineware engine, this depth of control is fully available inside Archicad. Luckily, it is well covered in the excellent Graphisoft Help Center (at the moment of publishing, version 26 of the software): `https://help.graphisoft.com/ac/26/int/index.htm#t=_AC26_Help%2F140-1_CineRenderSurfaces%2F140-1_ CineRenderSurfaces-1.htm&rhsearch=cinerender%20surfaces&rhsyns=%20.`

Lighting

Without light, we would all be in the dark. So for proper renderings, we need to configure which light sources should be taken into account. In Archicad, there are a few sources of light:

- The **Sun** is the main source of light and is well presented in Archicad. You can set the position of the project and the date and time of day, which allows you to have an accurate rendition of the direction of the **SUN** and **SHADOWS**. In the context of rendering, the sun is a light source with a high light intensity and strong, parallel light rays casting shadows. The sunlight has a bright color, which is typically slightly yellow, but which may vary based on the season and the time of day: orange and murky in the morning and evenings, bright yellow or almost white at noon, and orange or slightly purple before dawn.

You can control the sun orientation from the **Location Settings** dialog (**Options** > **Project Preferences** > **Location Settings**). Here, you indicate the geographic location of the project and **North Angle**. This in turn ensures that Archicad will generate an accurate and realistic sun trajectory for renderings or views, including a correct placement of shadows.

Figure 15.24 – Location Settings dialog, resulting in an accurate sun position

- Alongside the sun, there is also **Sky**, which provides a background (color, gradient, clouds, etc.) and acts as an environmental light source. When you are on the shaded side of a building, you are not in complete darkness, far from it. In fact, due to the way our atmosphere scatters the light from the sun, reflected rays from the sun also reach shaded areas, brightening and tinting these areas. The way the sky illuminates the project is defined as part of the render settings, which are discussed in the next section.

- The next source of light comes from Archicad **Lamp** objects. In the library, there is a collection of parametric objects, which also contain a light source. This can be used to simulate light luminaires and light fixtures, combining the geometry of the light armature and the light-casting properties of the lamp itself.

You position lamps like any regular object, but they also have a dedicated set of properties that control their light source, including brightness, light color, falloff, and shape of the light cone. We prefer to position them first in a floor plan view and then switch to a section view, to check their vertical position.

The following is the **Pendant Lamp** selection from the library (*Figure 15.25*). It is intended as an interior light fixture, with both the geometry of the fixture and the light source itself contained in a **Lamp** object:

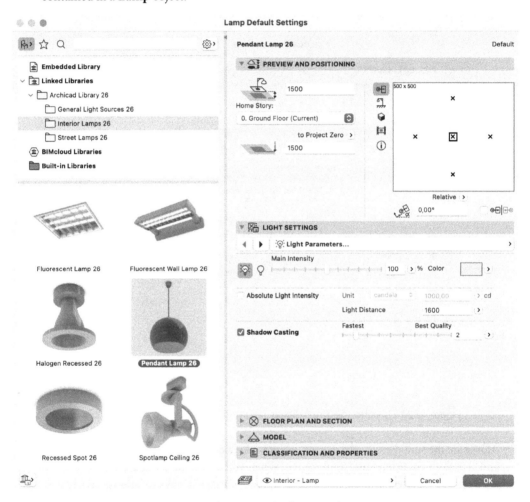

Figure 15.25 – Pendant Lamp displaying Light Parameters

Some points of attention in this dialog are as follows:

- Notice the **Main Intensity** and the **Shadow Casting** options. The first is an easy way to quickly configure the strength of a lamp, and the second is a checkbox: when switched off, no shadows will be calculated, which would normally be created by the lamp light hitting other objects.

- Other pages in the object editor are comparable to regular library objects, with a variety of options to set the geometry, dimensions, colors, and other aspects parametrically.

> **Pro tip**
>
> In addition, there are a few dedicated Archicad lamps for special effects. It would lead us too far to explain them all, but there is one trick we want to share with you: for high-end interior renderings, it is often difficult for the exterior sunlight to reach sufficiently deep into the building, even when light bounces off walls, floors, and ceilings. Most of the light is blocked by the walls anyway. In that case, you can apply the **Window Light** lamp object (*Figure 15.26*). This adds an array of smaller light source objects, which you can position inside a window and which will replicate the sunlight orientation and light color. This way, more light enters the interior directly, relying less on light bounces. The result? A cleaner image, more even and softer light spreading, and much less noise and artifacts in the rendering.

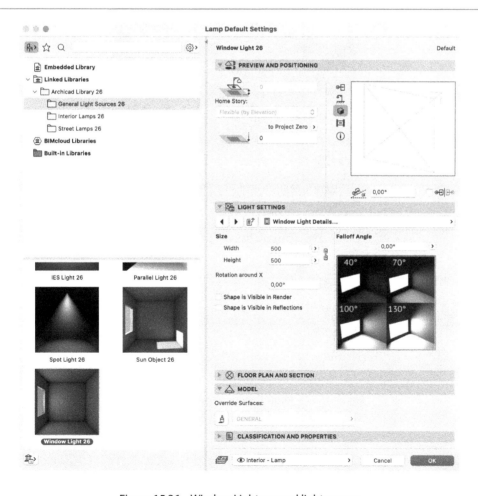

Figure 15.26 – Window Light general light source

Materials and light should be set by now, so let's move on to the **PhotoRendering** part itself and the configurations needed to create a good render.

PhotoRendering Settings

Here comes the fun part: generating renderings! You do this from the **PhotoRendering Settings** dialog (**Document** > **Creative Imaging** > **PhotoRendering Settings**). This opens a separate modeless palette, which you can leave open while tweaking the rendering settings.

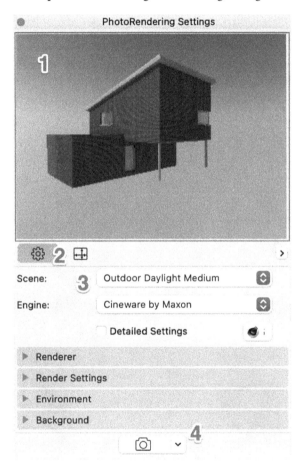

Figure 15.27 – PhotoRendering Settings dialog

As a user, you don't immediately have to know how to configure the engine since you are presented with a list of so-called **Scenes** (i.e., preset configurations for a variety of purposes). Let's have a look at the most important parts of this dialog, shown in *Figure 15.27*:

1. At the top of the dialog, you get a rendering preview window. Click on it and a small-size preview rendering is generated, which helps you to assess the current settings without the burden of long rendering times. It is simplified and not as clean as the final output, but this way, the system can cut a lot of corners and give fast feedback. Use this to your advantage: test often before you run the full rendering. The most important part is that the view composition is correct and that the light balance is right, and for this, the preview is more than enough.

2. The toolbar displays settings and a sizing tab. The **Settings** wheel icon gives you access to the rendering configuration in the panels underneath, while the **Size** tab simply lets you set the output image size and resolution.

3. At its simplest, you pick a **Scene** preset from the dropdown and ignore everything else. You can open the different panes to refine it or to see what has been configured. If you have some experience with high-end rendering, you may also toggle **Detailed Settings**, which show even more specialized and sometimes rather obscure options.

4. Once you are pleased with the preview, you use the large **PhotoRender Projection** button at the bottom of the palette (the one with the camera icon) to launch a full rendering, which will take much longer than the preview, alas. Start with fairly small rendering sizes first (e.g., 800x600), before you commit to large sizes, which take much longer: double the size is four times the amount of pixels! You can follow the rendering process at the bottom of the screen and even cancel it if you don't want to wait for the result.

That's, in a nutshell, all there is to it. You'll get an image that gets rendered, sometimes fast, sometimes painstakingly slow, depending on your project size and chosen preset.

Let's dive a bit deeper into the use of scenes to learn how these can help you get good images.

Using Scenes

From the **PhotoRendering Settings** palette, you can open the **Select and Manage Scenes…** dialog.

This dialog (*Figure 15.28*) gives an extensive list of rendering configurations that you can access, but also add to, rename, or delete. There are two main folders, which are further subdivided: **Photorealistic Scenes** and **Sketch Scenes**. Let us start with exploring **Photorealistic Scenes**. **Sketch Scenes** are covered later in this chapter.

Photorealistic Scenes are focused on the *Cineware* and *Redshift* engines and are intended to generate accurate and realistic renderings, taking into account materials and lighting.

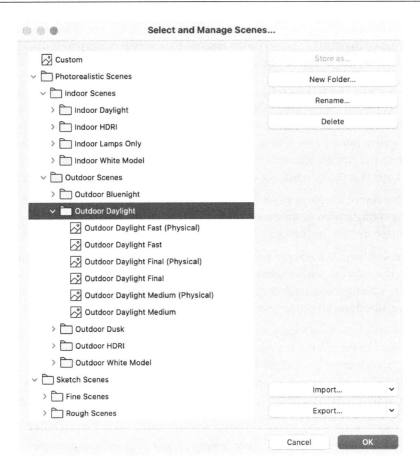

Figure 15.28 – Select and Manage Scenes dialog

We explain a few of the common presets and why you would want to use them. Notice that *Graphisoft* used a naming convention to help you decide on the best usage:

Preset Name	Intended usage
Outdoor versus **Indoor**	This balances the brightness of the scene. Our human eye easily compensates for brightness differences by opening our pupil when going inside, but we have to give a hint to the rendering system. Indoor scenes bump up the brightness considerably.
Fast - Medium - Final	This controls the balance between rendering speed and rendering quality. Alas, it is either "fast and rough" or "slow and nice." Start testing using **Fast** presets and only switch to the **Final** preset when you are pleased with the composition, materials, and light balance.

Preset Name	Intended usage
HDRI versus **Daylight**, **Bluenight**, **Dusk**, or **Physical Sky**	This presents alternative methods of environmental lighting: using **HDRI**, an image with a *High Dynamic Range* is used as an environment map, with the brightness and color of pixels illuminating the scene. These images are typically captured on location, and well-prepared HDRI images can simulate a real location (e.g., the beach on a summer day or a concert hall or museum space).
	Daylight, **Dusk**, and **Bluenight** allow you to switch the time of day and have a fitting sky background as a bonus.
	Physical Sky, on the other hand, uses a numerical model to simulate any place in the world, including a presentation of a cloudy, overcast, or clear sky.
Lamps Only	These are presets where only **Lamp** objects contribute. This can be used for evening or night renderings in an interior.
	It is recommended to tweak lamp intensities using a **Lamps Only** preset; otherwise, the brightness of the sun will make their contribution very hard to distinguish.
White Model and **Cardboard Sepia**	These rendering presets override all materials with a clean white or sepia color and ignore all textures (apart from glass). This is a beloved option for architects, as it mimics a physical model or maquette. It works well to present an early design concept before materials and finishes are fully specified.

Table 15.1 – Overview of Scene preset names and intended uses

Let's look at a few examples. For such a small scene and using a small image size, you don't have to wait long for the results (5-10 minutes max). Outdoor scenes are typically faster than indoor scenes, which need higher settings to get a clean and noise-free result.

Figure 15.29 – Outdoor Daylight Medium (Physical) showing shadows from trees

Figure 15.29 shows the general lighting capabilities and how shadow casting performs in Archicad, while *Figure 15.30* shows the water part and some windows and their reflections in more detail.

Figure 15.30 – Outdoor Daylight Medium (Physical) showing reflections (water and windows)

While the previous two images are more realistic, *Figure 15.31* shows how Archicad is also capable of rendering more schematically – showing a physical model-like representation of the project.

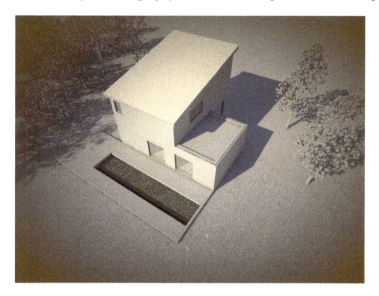

Figure 15.31 – Outdoor Cardboard Sepia Fast in an overview results in a cardboard model look

Interior images have a different approach and atmosphere, but using the predefined scenes, Archicad also delivers a good-quality rendering for such images, as shown in *Figure 15.32*.

Figure 15.32 – Indoor Daylight Medium (Physical)

The **Indoor** example took considerably longer (several minutes instead of seconds) since we needed to select a better setting to get good results. To improve rendering times, you can use a slightly faster setting. Adding **Window Lights** as mentioned earlier in the *Lighting* section also helps to obtain a cleaner light distribution in the model, especially if there is limited direct sunlight entering the interior.

You can always tweak an existing preset and save it under a new name. Using the scenes preset manager, you can export or import presets for use in other projects. Apart from switching between presets, you can also switch from standard to **Redshift by Maxon**. This is a more recent addition to the Archicad render engine's arsenal and is a powerful alternative to the Cineware engine. Access is, however, limited to subscription-license holders and also depends on your hardware for compatibility reasons. Rest assured that the Cineware standard engine is a very capable and professional system, so don't feel left out if Redshift is not available for you.

> **Pro tip**
>
> Apart from the built-in rendering engines, there are a few third-party plugins available for Archicad, which will interface with an external rendering engine directly. It does bring the advantage of not having to export and import the Archicad project into another file format, but you are still in a fairly constrained environment of software, which is mainly oriented to model authoring and not to rendering and visualization.

This finishes up the section on photorealistic renderings, but we still have to learn how the other type – Sketch renderings – are created, so let's move on to the next section!

Conceptual images with sketch rendering

Visualization by architects and designers evolves with trends, tastes, hypes, and software tools. There are times when photorealistic rendering fits – for example, when you want to impress clients or give a life-like representation of the lighting, materials, and mood of a design.

At other times, however, realism may be misleading. Showing articulate textures, bumps, and light reflections may give the client the impression that the design is fully resolved, where, in reality, you may still be developing the outline of spaces, the position and angle of a roof, or the distribution of windows in a facade. Regardless of the reason why you may want to skip realism, Archicad offers an alternative rendering engine catering to conceptual rendering styles, mimicking more of a hand-drawn look, or offering a variety of graphical styles that don't pretend to be realistic. Architects have always sketched and still do in a digital context. The **Sketch Rendering** engine in Archicad is a nice toolset to generate such images straight from the model.

While you can, again, go into great depth with settings and configurations, we will use the same presets system in the **PhotoRendering Settings** dialog as before. Here are a few examples to get you started, but feel free to try some others as well.

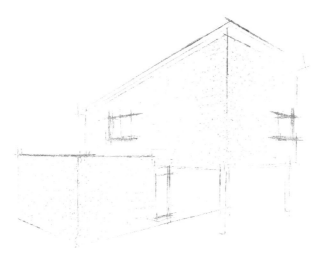

Figure 15.33 – Bamboo Sketch Rendering of the project (without environment)

In sketch rendering, you can mimic certain drawing materials and techniques, such as pencils and extended lines at the corners in *Figure 15.33*, and a more crayon-like feel in *Figure 15.34*.

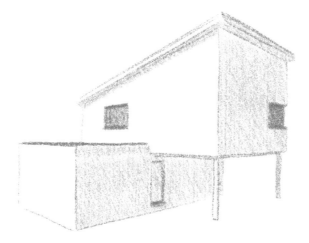

Figure 15.34 – Darjeeling Sketch Rendering of the project (without environment)

These Sketch renderings typically don't take too long and can be used to represent early design variants or can spice up a dull layout sheet. You may have to play around with the best resolution of the image, depending on the thickness of the sketch lines and edges.

In some cases, it also helps to simplify the 3D window (e.g., by hiding trees or furniture layers before rendering) to not overload the view with useless detail.

Animation in Archicad

Let us start by saying that Archicad is not animation software. If this is your objective, you are better off with tools such as *Blender*, *3ds Max*, *Artlantis*, or *Cinema 4D*. However, two types of animation are prevalent in architectural visualization: the simulation of light changes over time and an animated walk-through or fly-through of the design. And for both, there is a solution inside Archicad.

Animated sun and shadow study

In a **Shadow** or **Sun Study**, we mimic the path of the sun over the course of a typical day. This allows the user to see the impact of the building on its surroundings (the shadow it is casting) or see the direct light that falls through windows and curtain walls in the interior.

The idea is that you animate the time of day, which in turn, automatically adjusts the position of the sun. Depending on the chosen day of the simulation, you may see the sun path for a winter day (low sun, with light piercing deep through the windows) or a summer day (high sun position, short but harsh shadows). This helps to see the impact of overhangs, facades extending above windows, or exterior shading devices and louvers.

This type of animation is straightforward to set up in Archicad. Open the **Create Sun Study** dialog (**Document** > **Creative Imaging** > **Create Sun Study**) to configure the animation.

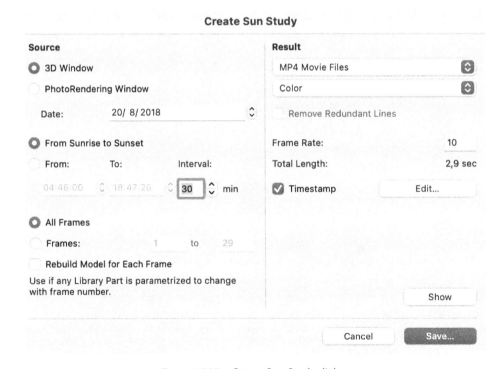

Figure 15.35 – Create Sun Study dialog

In this dialog, several items need our attention:

- The **Source** window can be either **3D Window** or **PhotoRendering Window**. Using **3D Window** is very fast but be sure that you have selected a 3D style where shadows are displayed! Otherwise, you get a useless sun study.

- Pick a date and set the time or let Archicad define the time as **From Sunrise to Sunset**. Easy!

- You can directly render to a movie file in **MP4** or **Quicktime** container format, but there is also the option to generate individual images (see *Figure 15.36* for the results) or even some 3D model formats that support animation. Rendering to images can be useful if you need to further edit the movie or need to test the best movie compression format, without requiring a re-render.

- Based on the **Interval** duration, the number of frames is defined. Combined with **Frame Rate**, you can find the **Total Length** duration of the animation. 30-minute intervals at 10 frames per second gives a 2.9-second movie.

- Finally, press **Save…** to export the movie, but not before you test the animation at least once using the **Show** button. This will show the movement in the 3D window without launching the actual rendering output.

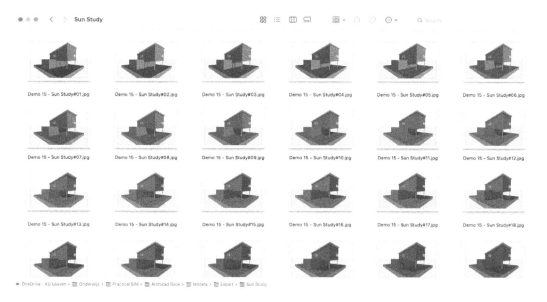

Figure 15.36 – Sun study exported as individual images

A sun study is an ideal way to optimize your design, even in a preliminary stage, or convince a client of certain design decisions (position and size of windows, adding sun screens or canopies to dampen the sunlight, etc.).

Our second form of animation – the fly-through – is more aimed at presenting the design as a final product. Let's go ahead and see how we can create such an animation in Archicad.

Fly-through

In a **fly-through** or walk-through, the position and orientation of a camera are animated over time, giving the impression of the user moving or even flying through the design. The result can give a nice view of how you would navigate around the building, although the movement is fixed in advance, using a static navigation path.

To create a fly-through, you need to define a path for the camera to follow, with Archicad calculating all the steps in between the main positions. Do this as follows:

1. Go to the plan view and activate the **Camera** tool in the **Viewpoint** section of the toolbox.

2. Place new cameras, which will become part of a **camera path**.

3. Each camera can be further edited using the graphical handles, similar to how you'd edit a spline: an arrow line points at the movement direction and the length of the handles define how strongly they pull the path into the indicated direction.

4. Use the palette to refine the properties of each camera (e.g., **Camera Z** height and/or camera **Target Z**, **View Cone**).

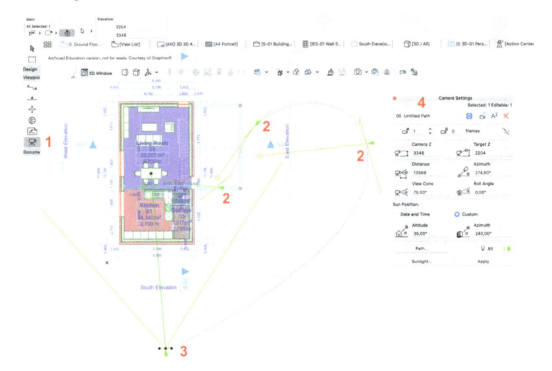

Figure 15.37 – Graphically editing the path for the camera to follow

From the **Path…** button, a **Path Options** dialog opens allowing you to further indicate name, path, and display options.

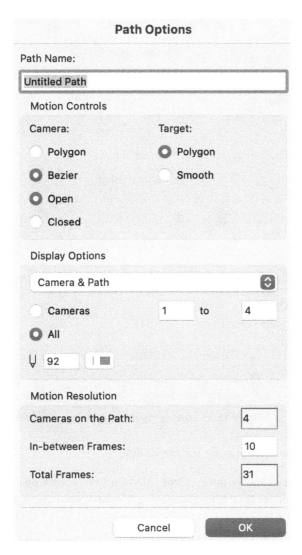

Figure 15.38 – Path Options dialog

Depending on the shape and smoothness of the path, this can appear nice and gentle but also rough and even erratic. Having a smooth and fairly basic movement is highly recommended to not disturb the viewer or cause nausea. You don't want to give the impression of a theme park ride on a roller coaster.

To preview or render the animation, open the **Create Fly-Through** dialog (**Document** > **Creative Imaging** > **Create Fly-Through**).

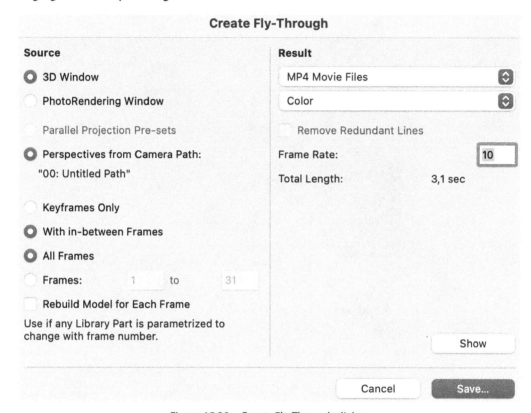

Figure 15.39 – Create Fly-Through dialog

The dialog has some similar options as for the sun study:

- You can opt to generate the fly-through from **3D Window**, mimicking exactly what you see in the Archicad viewport, or from **PhotoRendering Window**, which needs to calculate a rendered image for each frame.

- You have similar **Result**, **Save**, and **Show** options as with the sun study. Click **Show** to get a quick preview, before you launch the actual rendering.

Rendering into a movie file and how to optimize the calculations

Both types of animation require you to render a series of images, which are then compiled into a movie file to be played to the user.

> **Warning!**
> We have to warn you of potentially very long rendering times, especially when we combine animation with the photorealistic rendering engine.

Each image is a separate rendering, potentially taking minutes or even hours to render. That may take so much time and require so much CPU power that you can't use your computer for several hours. Here are some tips:

- Set the image size as small as you can for the intended purpose. You may not really need a 4K rendering if a rendering size of 1,280x720 pixels (HD) or 1,920x1,080 (Full HD) pixels is sufficiently usable. 4K is not 2 but 4 times the number of pixels compared to Full HD!

- Keep the frame rate reasonable. Don't aim for 60 frames per second for a movie. 30, 25, 24, or even 20 frames per second could be enough. Again, 30 frames are half the amount of rendering time as opposed to 60 frames per second.

- Keep the rendering quality to the basics. You can limit the amount of oversampling or the amount of Global Illumination bounces. Finding the right balance between realism and rendering times is an art form and probably one of the most challenging tasks when rendering animations.

- Think about whether the illumination changes throughout the animation. This is, by definition, the case in a sun or shadow study, but the lighting would remain static in most walk-throughs. Static lighting can benefit from all types of caching mechanisms from the rendering system, such as shadow position and light bounces. Reflections, however, need to be recalculated each time the camera position changes. So, both methods have some optimizations and some aspects that increase rendering times.

- Disable all layers you don't need. Having a movie displaying all furniture and plants and people and cars is nice, but maybe not really needed for a more basic movie.

Having a faster CPU or more CPU cores will greatly improve rendering. A dedicated workstation with a multi-core CPU and high amounts of memory and high clock speed is really beneficial for rendering, much more than it is for modeling or drafting.

And maybe, if you really want to take it to the next level, Archicad is not the best environment for animated movies, so we will discuss some alternatives in the next section.

Using external imaging software

Since Archicad is not really a dedicated animation software, it makes sense to be able to export the 3D model's geometry, alongside the used surfaces, into other software. This is what professional animation experts would do due to the increased control and a much smoother and more efficient workflow.

Model conversion

Luckily for us, Archicad has a good set of supported export formats that can be used by various third-party animation and visualization software applications.

Here is a list of some of the supported 3D export formats and for which purpose they should be considered:

Format	Description	Considerations
COLLADA DAE	Open format for 3D animation software	Fairly well-structured scene, including lights, cameras, and materials, but all geometry will be meshes.
FilmBox FBX	Autodesk format for 3D animation software	Comparable to COLLADA. This requires the *Twinmotion* add-in to be installed.
Artlantis ATL	Native Artlantis format	Only useful if you use *Artlantis*, but in that case, it works really well and allows you to update the Archicad model while retaining the Artlantis setup.
SketchUp SKP	Popular 3D format	Mesh-based format and basic material and single texture, supported by many 3D applications.
3D Studio 3DS	Old 3D Studio format	Meshes and textures are supported, but there are inherent limitations in mesh size and texture naming. This is the native format of the old *3D Studio for MS DOS*, not the format for the current *3ds Max*.
Wavefront OBJ	Older 3D format	Widely supported, but recovering materials (from an accompanying MTL file) varies per software. The model structure is limited by the export (flat list of objects or merged by material). Be aware that the *y* axis is up in this format, instead of *the z axis*.
DWG/DXF	AutoCAD format, supporting solids and mesh geometry	Objects become blocks and layers are used to give structure, but don't expect full surface transfer. More useful for CAD software.

Table 15.2 – A list of 3D export formats and their general purpose

The list of supported 3D formats is actually much longer, including a few file formats for rendering software that have been deprecated for a long time, such as *Piranesi*, *Electric Image*, and *Lightscape*. We advise you to stick to the previous options and preferably go with DAE or FBX if a native conversion is not available.

Exporting to CAD formats such as DWG/DXF or Rhino 3DM is used to transfer geometry for further modeling rather than visualization.

Exporting to **Industry Foundation Classes (IFC)** contains the full BIM model and all its information but is typically not compatible with visualization software.

Without going too much into detail, the main experience we have suggests that static geometry, typically mesh-based, is perfectly usable in 3D rendering software as long as the front and back sides of faces are properly oriented. For walls, slabs, and other volumetric objects, Archicad generates clean geometry. For glass panes, you may have to assign a dual-sided material to avoid these planes becoming invisible in the rendering when looking from the back.

In addition, ensure that you set up both systems to use the same units. While having a model out of scale may seem not to be that big of a problem for rendering, it actually has a huge impact on light behavior (light fall-off) and the scale of props and accessories that you may want to introduce later.

Preparing an Archicad View

To ensure you only send what you need, you are strongly advised to prepare a view for the 3D export. In that view, ensure that you only enable those Layers that contribute to the visualization and check whether the other display options are well suited. The default Archicad 26 template provides such a Layer Combination.

Any object that you want to replace in the rendering software (e.g., trees, shrubs, cars, etc.) are best left off inside Archicad. Those would otherwise contribute to a large export file and are not at all set up for high-end rendering anyway. In fact, visualization specialists often have their own dedicated furniture libraries, including cushions, bed sheets, plants, and all kinds of accessories to spice up the scene, for which the more technical Archicad objects are no match. Focus on the architecture and leave such details for the visualization library.

Material conversion or replacement

While geometry is typically usable, materials are less so. Either the file formats don't contain much information about such materials, or they are not translated in full by the receiving software.

Here we encourage you to focus on setting basic materials in Archicad and focus on getting them named clearly. You have to assign a texture with the right size so the texture coordinates are reusable in the receiving software. Even then, the best effect is obtained by replacing the material with an equivalent software-compatible material, taking full advantage of the material system of the rendering software. For example, ensure that an Archicad facade brick is named clearly and its texture size (horizontal and vertical) is set to the right scale, but don't worry too much about the rendering settings if you move to a different rendering engine.

The worst you can do is not assign materials with surfaces, as this will lead to lots of additional work in the rendering software: re-assigning materials, possibly by splitting meshes face by face and re-assigning texture coordinates. That said, for most objects, a box or cubic projection gives a reasonable result.

Third-party software examples for rendering

In a book on Archicad, we don't have much room to start explaining how other software works, as each such explanation can fill a complete book series by itself. But we can give you a few pointers to a quick and efficient workflow when using external rendering software systems.

> **Beware!**
>
> We have to warn you about using such dedicated specialist software, though. In stark contrast with software such as *Artlantis*, the number of options and settings for rendering, lighting, materials, and camera in most of these systems is staggering. You have extreme control, but it also means you need to really dedicate time to learn about all the options to get an optimal result. It's a profession in itself.

Maxon Cinema 4D

The Cineware engine inside Archicad is actually derived from *Maxon Cinema 4D*, a high-end rendering and animation software, which is also part of the *Nemetschek* group to which *Graphisoft* belongs.

This brings not only an advanced rendering system into Archicad but also allows us to export the Archicad project directly into a Cinema 4D file, ready to be further refined and expanded to a full animation.

You can export the current view to Cinema 4D from the **PhotoRendering Settings** dialog, as the second option alongside the **PhotoRendering Projection** button.

Since the Archicad surfaces contain both the internal engine and the Cineware structure, this ensures that materials retain their full setup upon export. So, rather than having to replace or recreate the materials, visualization artists can focus on enriching the project with more detailed and optimized assets, such as trees, cars, furniture, and even people. There is also a full animation system that gives full control over lights (which can be turned off or on), and doors that can swing open when the camera passes through.

Cinema 4D has a complete modeling system as well, which is easier for modeling organic objects, but we recommend focusing on having the core architectural geometry managed inside Archicad and only adding things related to visualization in Cinema 4D. For example, walls or floors modeled in Cinema 4D are not synchronized into the Archicad model and hence do not appear in any extracts or documents derived from your BIM. You also don't have the in-depth parametric control as in the Archicad objects for doors, windows, and beams, for example.

Moreover, upon rendering an animation, there are more techniques available, including caching of the Global Illumination, rendering to individual frames, and advanced control over keyframes to perfect camera movements, effects, and practically everything else.

Artlantis

This is also a popular option for Archicad users. It is a dedicated architectural visualization software from the French company *StudioBase2* (they used to be called *Abvent*), which has a long tradition of providing easy-to-use software for architects. The interface is oriented toward quick scene organization: dragging materials from its library on top of existing materials, positioning lights, camera, and accessories into the scene, and having a rendering configuration dialog with a streamlined set of options. It supports still images and movies, but also 360° images and even network rendering.

Archicad has a direct export to the Artlantis *ATL* format, and it supports a clever workflow where the Archicad export can be updated while still retaining the setup of the visualization inside Artlantis.

Another nice advantage of Artlantis is the very quick rendering preview, which can speed up the guesswork of balancing light and materials before you launch a dedicated rendering.

Autodesk 3ds Max

Autodesk has two advanced visualization systems with *3ds Max* and *Maya*. For architectural visualization, 3ds Max is more widely applied, not least because it has better support for working with CAD and BIM models than previous external solutions mentioned. While integration with *Revit* is easier (as both are part of the Autodesk software collection), it is still possible to bring an Archicad model into 3ds Max.

While you can use the old *3D Studio for MS Dos* format (*3DS*), this will give you a lot of additional work to recreate and reconfigure all materials. You get better results using the *COLLADA DAE* format or the *FBX* format, for which you need to have the *Twinmotion* add-in installed for Archicad.

Specialists often apply the *V-Ray* rendering system, which is a third-party external rendering plugin for 3ds Max and is very popular for architectural visualization, with an efficient workflow for global illumination animations and an elaborate material system. However, understand that due to the nature of model and format differences, you need to do more effort to convert the Archicad material definition into a 3ds Max or V-Ray definition. Again, as we explained before, ensuring that the Archicad surfaces are mapped correctly takes a lot of additional effort away in the conversion.

Beware, though, that 3ds Max is a Windows-only software, so macOS users are better served with Maya or Cinema 4D instead.

Blender

Finally, we don't want to leave you without mentioning *Blender*. This is an open source and cross-platform animation and visualization software. It has wide support for importing 3D models, including COLLADA and FBX, which are recommended.

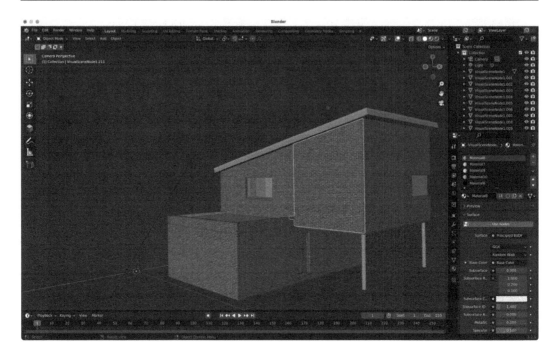

Figure 15.40 – Archicad model converted into Blender using the COLLADA DAE format

As you may notice in the preceding figure, at least the geometry and the materials are supported, giving you the basic color and texture as with Archicad. From here on, you can use all the advanced possibilities Blender has to offer.

The software has quite an unusual interface, but it is very complete and has a very active community of users. The same warning applies, though: it is specialist software with lots of settings and options to control, so don't expect to dive in and have fantastic results by simply hitting the **Render** button. But it is nice to know that there is an option that is completely free and can lead to professional results if you learn how to use it properly.

If you also install the *BlenderBIM* add-in, you get support for the **IFC** open standard. This can be used to also transfer the model information. It is not oriented to rendering, so apart from their basic color, don't expect much in terms of the conversion of Archicad surfaces.

Any recommendations?

While we can't recommend one third-party software over another due to their varied capabilities, here are some pointers for you to keep in mind:

- Having an internal visualization engine inside Archicad makes it easy to integrate visualization tasks with the rest of your workflow. This is where Cineware or third-party engines have their place.

- Dedicated architectural visualization software, such as Artlantis, with a workflow to update models after initial export, make the external tool accessible and still well integrated.

- Expert tools such as Cinema 4D or 3ds Max or Blender take the most effort and have a less smooth workflow, but they are the best option if you need high-end animation and full control over everything – at the cost of a more expensive license and more effort, obviously.

We warned you at the beginning of the *Animations* section about long rendering times when it comes to animated visualizations, and this also goes for external applications. More recently, external applications using game engine technology have been developed toward visualizing architectural designs. Archicad has a similar technology with *Graphisoft's BIMx Hyper-models*. Let's take a look at these solutions as well.

Real-time visualization

There are times when a full animation just takes too long and still does not bring enough added value to the project. Maybe the client doesn't have the budget, or the design is still evolving. In this case, we can apply an alternative technique: converting the Archicad project into a **real-time interactive visualization**. Much like a 3D game, you can navigate freely around the model instead of watching a pre-rendered movie. You will sacrifice a little on the graphical side, but with current hardware (desktop and mobile), it is still a convincing display of the design, albeit not entirely photorealistic. But, more importantly, it can be experienced by the user without the need for an Archicad license.

There are two main approaches available to you: using the *BIMx* real-time engine, available with Archicad, or using an external system, such as a game engine or dedicated visualization software.

BIMx, an interactive Graphisoft app

BIMx is a dedicated application that is included with Archicad and allows you to export the Archicad 3D model and also include a set of layouts, which are anchored to the model. The user can then browse the project in real time, including switching between 3D and 2D views and seeing information that is attached to objects in the scene.

The BIMx export can be displayed in a separate desktop viewer, but there are also viewers for iOS, Android, and the web. A typical use case is sharing a design proposal with a client who normally doesn't have Archicad available. They can visit the project as if it were a 3D game and even collect comments to be shared with the architect.

Creating a BIMx project

Remember the publishing methods that were available in *Chapter 14*: the Archicad Publisher can be used to export the model in one go.

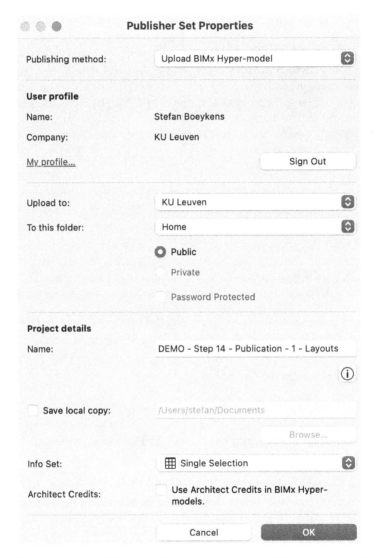

Figure 15.41 – Publishing the Archicad project into a BIMx Hyper-model

Publishing your project as a BIMx format and opening it in the desktop or mobile environment of your choice will provide you with an interactive model to walk through or fly around. You can also create live sections in the 3D window or consult the 2D layouts included when added to the Publisher Set in *Figure 15.41*. You can even show the 2D documents as an overlay in the 3D environment (see

Figure 15.42). A **BIMx Hyper-model** remembers the links (the blue icons on the 3D window in *Figure 15.42*) between these different views and documents. It is like having a read-only, interactive copy of your Archicad model – ideal for sharing with clients and using in presentations!

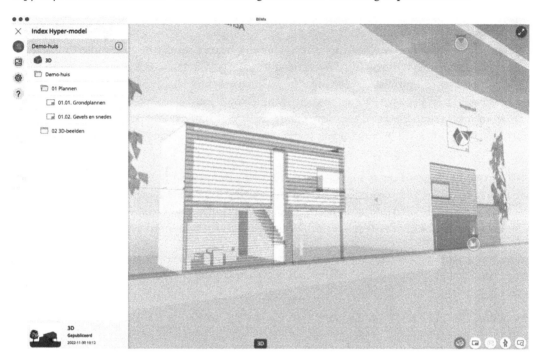

Figure 15.42 – BIMx lets you show your 2D sections as an overlay in 3D

As good an integration as it is, BIMx is somewhat limited in creating photorealistic images and has no capabilities for showing a real animation, such as doors that swing open. For this, again, we could turn to external applications, which are covered in the next part.

External game engine applications

While Archicad has a very capable real-time system included with BIMx, sometimes, external engines may be applied. These days, it is possible to reach levels of visual quality and fidelity of lighting and materials using modern game engines. We still remember the time when we were pleased with pre-rendered images that took several minutes or even hours, while the current generation of game engines can render high-resolution, detailed images in real time: instead of hours, each image takes less than 1/30th or even 1/60th of a second, allowing you to walk or fly around your model interactively, while still having the full range of lighting, material texturing, shadows and a whole set of other tricks to fool us into something that appears close to reality.

Two of these engines have become hugely popular due to their accessibility in terms of features and price, and obviously, due to the graphic quality they can achieve.

In an introductory course on Archicad, we don't have room to elaborate on this in detail, but we want to mention the main options, nonetheless, giving you something to work towards:

- The **Unity Game Engine** is very popular, not only for gaming but also for many other interactive applications. This is a cross-platform engine that allows you to create a project, comprised of multiple scenes, in which you can organize interactivity through custom scripting. There is an extensive asset store where content, scripts, and whole game systems can be added, ranging from small and free to large and, well, more expensive. Unity requires payment of a license fee for companies based on your global yearly revenue, which may limit its use in a professional architectural or engineering company.

- **Unreal Engine** by **Epic** is a high-end game engine that can be used freely for visualization purposes. Once you start publishing a game, there is a royalty payment required. There are countless courses and tutorials available, but, as with Unity, expect to spend considerable time learning about such a system, especially when you want to add more interactivity.

In both cases, Archicad models can be imported into the game engine from an exported 3D model. You get good results using FBX or DAE formats, which are recommended, as with the external 3D rendering software.

After adding the Archicad 3D model, there are a few steps to take:

- Ensure the geometry is loaded and is using the right scale.

- You may have to revise the materials to ensure they make optimal use of the game engine shaders.

- Add cameras and lights to help the visualization.

- For the best visual quality, the lighting system can be pre-calculated or baked onto the geometry, making it look realistic and still perform well in real time.

- Finally, you need to add interactivity. Here are some examples:

  - Letting doors swing open or lights toggle on when you get nearby

  - Having an on-screen user interface to toggle elements on/off or to select different material options

  - Animating a camera for a turntable-style visualization

  - Adding an interactive first- or third-person player that can be used to freely run around the scene

The end result can be previewed in the game engine editor but eventually has to be published to a self-running executable or game before it can be shared. These systems typically support multiple platforms and devices, such as desktop, mobile, and even web browsers. For publishing to gaming consoles, you need to enter into a separate license agreement with publishers such as Sony, Microsoft, or Nintendo. But if you go this far, you are probably not working as an architect anymore.

> **Pro tip**
>
> There is no upper limit as to how far you can go. You can add all kinds of animations, including doors that swing open when the player approaches, pop-up info panels that display information for elements that are picked by the user, elevators you can call when pressing a button, sounds that play in a concert hall, people and animals walking around in the hallways, undulating water in the swimming pool, or TV screens that play a movie.

Dedicated visualization software

Developing a game – or rather, an interactive project – in Unreal or Unity can be very rewarding and reach an almost unlimited level of freedom when you add custom scripting, but the effort and experience required may not be practical for your office or your clients.

In that case, it may be more pragmatic to look for dedicated real-time visualization software. And luckily for us, there are a few options available that work well with Archicad. Next, we'll look at one such option.

Twinmotion

Twinmotion (`https://www.twinmotion.com`) is directly dedicated to architectural real-time visualization. It is based on Unreal Engine, but rather than diving deep into scripting, blueprints, and custom assets, it is presented as a streamlined albeit very large package of pre-optimized scripts, assets, and a direct connection to the Archicad software (and a few others as well).

With Twinmotion, you establish a connection by exporting your Archicad project into a Twinmotion scene. That scene is loaded into the Twinmotion software and, from then on, you enrich the scene by adding assets and interactivity. Here are a few of the many possibilities that are available:

- Adding trees, shrubs, bushes, and grass, but also water, planes, flying birds, and so on, all fully animated and highly detailed – something that simply isn't feasible inside Archicad

- Adding furniture, such as benches, couches, tables, and closets, but also lighting armatures, cars, and even animated people, and so on, with the lights actively illuminating the scene in a realistic way, but without the long calculation times of a rendering tool such as Cineware

- Tweaking a day/night light cycle to mimic the light effect and the path of the sun throughout the day, like the sun study that Archicad provides

- Playing with the sky and atmosphere and even simulating the weather by adding rain or snow or, on the contrary, choosing a bright sunny day

Twinmotion presents this as a streamlined workbench interface, with the whole scene interactively displayed in the main 3D view and panels of assets to be dragged into the project.

Even more important, and very much in line with a BIM-based workflow, the Archicad model is not simply copied into Twinmotion but is referenced. This implies that you can continue to update the model and have the changes automatically synchronized. You keep the model as the main source but use Twinmotion to add interactivity and enrich it with animated assets.

This one-click synchronization is very effective and works with a few other BIM software tools as well.

To get precise information on pricing, please go to the Twinmotion website. There are free offerings to try out the software or if you are an educational user.

There are alternatives to Twinmotion, such as *Enscape* and *Lumion*, which work in a similar way. As Graphisoft has included Twinmotion with Archicad within the SSA package up until December 31, 2021, we saw it fit to explain rendering with game engines using this example, but you should explore options for yourself, using a trial or educational license of the different software, or comparing their specs.

And with this overview, we have come to the end of this chapter on visualization. Let's summarize what we have learned!

Summary

In this final chapter of the book, we used our model to start making some impressive visuals. Not only is Archicad unmatched in creating 2D and 3D Drawings and combining them into presentation images but it also has a very decent built-in rendering engine, which allows us to make good and quite realistic photographic renderings of our project with little to no effort or in-depth knowledge.

Should this not suffice, we have also explained how you can export the model and use the geometry in dedicated rendering software or real-time rendering applications that use the latest game engine technology. This topic alone is worth a whole book and can be explored much further, and we actually hope that we leave you wanting to get to know even more about the possibilities of BIM and how Archicad provides you with tools and workflows to get the most out of this process!

Appendix: Some Final Tips and Tricks

We hope this book has helped you reach a good level of understanding of Archicad and how to use it to improve your BIM workflow. Throughout the previous chapters, we have tried to share our knowledge on a variety of topics to help architects and students of architecture start to use the software as a basis for their design workflow. We've done this by modeling a project along the way – of which we provided downloadable versions per chapter. These downloads can be used as a check for assessing your own result(s) or to help you get back on track, should you have gotten lost somewhere along the road. In this chapter, we will finish our journey, look back at what we have learned so far, and look ahead to what can still be learned after finishing this book. The topics covered are as follows:

- Links to some of the more important intermediate results for the project we have modeled throughout the book

- Some strategies on what to do when you do get stuck when using Archicad

- Tips on how and where to get some additional help on Archicad

- An overview of topics that we did not cover in this book

With this chapter, our introductory story into the world of BIM using Archicad comes to an end, but we hope this journey opens a lot of new directions for you to follow. Let's start by looking back at what we have modeled in the previous chapters and where to find the intermediate results for our project.

What to do when you get stuck

This book is aimed at reaching an intermediate level in using Archicad, in a step-by-step way. However, from our teaching experience, we know that anyone learning to use a new software or application – no matter how well they follow the steps or do the exercises – will reach a point where they get the feeling, *I thought I knew this, and I'm pretty sure it should work this way, but it doesn't*. This sense of frustration is completely normal and mostly occurs when you try to apply newly learned techniques within your own projects. To help you overcome this feeling, and any difficulty you might have in using Archicad in your own projects, we would like to provide some tips and tricks.

Remember the on-screen feedback

At the very beginning of the book (*Chapter 2*, in the *Understanding Archicad on-screen feedback* section), we showed you how Archicad helps the user to do what the software expects or needs as input. When reviewing the steps you take in a certain action, or when if something went wrong or you got unexpected results when using a tool, the questions asked would then be as follows (in this order):

1. *What input is needed (a node, selecting an edge, etc.)?*

    Carefully read the message in the Status Bar (e.g., **Enter First Node of Line**):

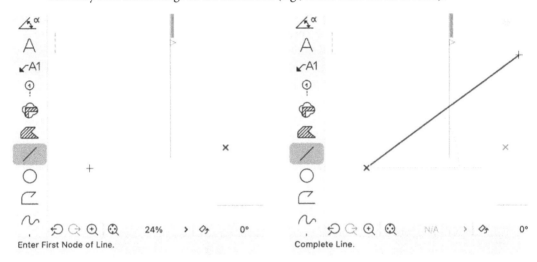

Figure 16.1 – The Status Bar in action (see Figure 2.3 as well)

2. *Are you providing the correct input (are you clicking a node, edge, intersection, etc.)?*

    Look at the **Intelligent Cursor** cycle for feedback:

Figure 16.2 – Some shapes of the Intelligent Cursor (see also Figure 2.6)

3. *Are you entering precise input or rather estimating?*

    Look at **Tracker** and provide exact input if needed. Combine this with **Snap Guides** and **Pet Palette** for maximal efficiency:

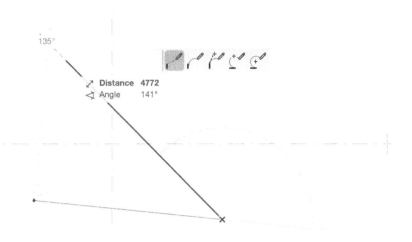

Figure 16.3 – Tracker, Pet Palette, and Snap Guides all help for modeling accurately (see Figure 2.7 as well)

4. *Are you selecting the correct element(s)?*

    Carefully watch for preselection in blue and/or cycle through selection options using the *Tab* key while hovering! This is called **selection cycling** and comes in handy once you start having overlapping elements, such as the wall and floor slab in the example in *Figure 16.4*:

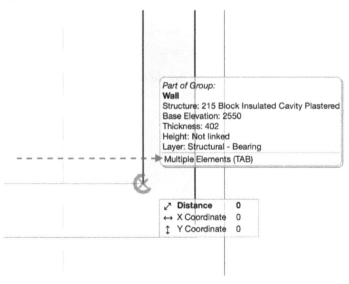

Figure 16.4 – Use selection cycling to select the desired element of multiple (overlapping) elements

5. *Are you entering the correct value(s) in dialogs?*

    Most Archicad dialogs have an interactive preview of the element, showing you the influence of certain values (e.g., a bottom offset of 0 for a wall results in a preview where the wall picture starts at the line representing its Home story).

Going through all the steps again

When students report a problem and state, *The command is not working, although I am doing all the steps you showed us,* this often has the same solution. Although few people believe in the so-called *healing by the laying on of hands* attributed to saints and gods, miraculously, the problem is often simply solved while we are standing next to the student, who claims they are doing the exact same things as before. The explanation for this is simple: having a teacher watching your every move forces you to concentrate on every step of the process, resulting in success!

So before creating a shrine and hanging our picture next to your computer screen, you should first try to meticulously go through *all* the provided steps once more.

Getting help

We know that, sometimes, all the previous points might still not help you solve the issue at hand. Maybe you are overlooking something, or maybe a setting (view filter, attribute, etc.) is influencing the result. If you can't fix the problem by yourself in a reasonable time, it is time to get some help.

Don't get us wrong – we are big fans of Niels Bohr, the Danish physicist, who once said, *An expert is a person who has made all the mistakes that can be made in a very narrow field.*

Learning from your own mistakes is one thing, but there is nothing wrong with asking for help. Luckily for Archicad users, help is available in abundance! Next, we provide some interesting links, both official Graphisoft sites and some unofficial but equally interesting sites run by Archicad fans, geeks, and gurus.

Official Graphisoft sites

Here are some of the official Graphisoft websites that can help you:

*   *Graphisoft Learn*: This is an online platform containing video courses and e-books, with free and paid options. (`https://learn.graphisoft.com/home`)

*   *Graphisoft Community*: This is the central hub for sharing knowledge on all things Archicad. Both users and Graphisoft staff/developers are active on this website. Search through over 330,000 posts or start a new conversation with a (yet) unanswered question. Beginners are welcomed with open arms! (`https://community.graphisoft.com`)

*   *Graphisoft User Guides*: Here, you'll find all the manuals for all the software developed within the Graphisoft group, including Archicad, BIMx, and others. (`https://graphisoft.com/resources-and-support/user-guides`)

- *Graphisoft GDL Center*: This is unfortunately not covered in this book (see the next section), but the **Geometric Description Language** (**GDL**), which lets users and developers create fully parametric objects, is well documented here. (`https://gdl.graphisoft.com`)

- *Graphisoft BIMcomponents*: In the odd case you couldn't find the right object in the Archicad 26 library, you can try this website, which has a vast collection of objects, ready to be inserted into our favorite BIM authoring tool. (`https://bimcomponents.com`)

External resources

Some interesting Archicad gurus we would recommend looking up are the following:

- **Shoegnome**

  American Seattle-based architect *Jared Banks*, *AIA*, has been an Archicad-enthusiast since the early 2000s and started the **Minnesota Archicad User Group** in 2009. He has been blogging on BIM and his favorite software since 2010. He shares a free template and gives some very good insights into BIM in general, and Archicad specifically (his blog posts on the **Pen Set** gave us the insights we used in this book on the subject):

  `http://www.shoegnome.com`

  `https://www.youtube.com/@Shoegnome`

- **Eric Bobrow**

  As a true Archicad guru (coming from an environment at the crossroads of architecture and information technology), Eric not only created a very unique way of using Archicad through his *MasterTemplate* but has also trained numerous users in the software for over 20 years through his *Best Practices* course, and has recorded numerous interesting talks with professional users all over the world:

  `https://bobrow.com`

  `https://www.youtube.com/@EricBobrow`

- **Jeroen de Bruin**

  Jeroen works as a senior BIM product specialist at **KUBUS**, the Archicad distributor for the Netherlands and the Dutch-speaking part of Belgium. Besides doing an excellent job in helping to develop workflow and template solutions for KUBUS (often sharing this knowledge with tips and tricks through social media), Jeroen also has expertise in GDL. With his own company, **Masterscript**, he has been developing great objects throughout the years that are worth checking out:

  `https://www.youtube.com/user/kubusinfo` (Dutch spoken) `http://www.masterscript.nl`

- **Nathan Hildebrandt**

  Skewed is a company led by Australian architect Nathan Hildebrandt (FRAIA, BARCH (Hons), and BBE (Arch St.)), which provides expertise to assist people and companies in solving problems with Archicad and BIM in general, without just following industry trends. They organize events and training, and also provide consultancy. As an Archicad specialist, Nathan has talked on numerous occasions about topics such as OpenBIM, visualization with Enscape, the digital transition, and so on. His BIM-focused podcast *The Digital Transition* is also worth mentioning!

  `https://www.skewed.com.au`

  `https://www.skewed.com.au/podcasts`

What we didn't cover

An introductory book about Archicad cannot possibly cover everything this software has to offer. The Archicad manual PDF is well over 1,000 pages long and is replaced with online help, which is a complete inventory and documentation for every tool, menu, dialog, and integrated add-in that Archicad provides.

When you think about Archicad as a design tool for architects, we have covered most of what you need to be ready for effective modeling, presentations, layouts, and even visualization. However, we couldn't cover some of the more advanced functionalities, which all relate to information management, parametric design, automation, and multi-disciplinary design and collaboration.

That said, here is a brief list of things we encourage you to research further, using the resources mentioned. There is enough content here to cover a few more books about Archicad:

- Archicad supports collaborative projects, with well-established support for OpenBIM standards. There are **Industry Foundation Classes** (**IFC**) (`https://www.buildingsmart.org/standards/bsi-standards/industry-foundation-classes/`) as part of the open data scheme and data format to exchange models with other project partners, independent of the software they may use. There is also support for the **BIM Collaboration Format** (**BCF**) (`https://www.buildingsmart.org/standards/bsi-standards/bim-collaboration-format-bcf/`), which is used to manage comments and issues, linked to models and shareable across software platforms. For good coordination and geospatial reference, a dedicated survey point was also added from version 25.

- Advanced **classification** and **property management** are available in Archicad. To manage the information part of BIM, Archicad has a best-in-class system for managing custom element properties, which relates them to the classification of elements, allowing you to fine-tune how each type of element carries the right information. Classifications help to bring more structure to models and are usable in **Schedules**, **Graphic Overrides**, and with export to **IFC**. This is required in collaborative projects to ensure you transfer the right amount of information to other project partners.

- **Change management** and **collision detection** also relate to collaborative work. So, you can mark up all model changes and manage them with your co-workers. With **collision detection**, Archicad can figure out overlaps between elements to prevent design or modeling errors and (again) to help with collaborative design.

- Creating designs with repetitive elements or putting together multiple buildings in masterplan projects is perfectly supported by the **Hotlinked Modules** system within Archicad. You could check that out.

- We could also dive deeper into the vast extent of the Object library, third-party libraries, third-party surfaces, and other templates, but we advise you to also contact your local reseller to see how they can support you to align with local standards and working methods.

- The Archicad library is extensive, but sometimes you need more. You can create your own, fully parametric objects, using either the GDL, the visual dataflow system called **PARAM-O**, or the brand-new **Library Part Maker**. The subject of GDL is so vast and extensive that we could write a whole book about this alone. In fact, some books have already been published about GDL, not even counting the *GDL Reference Guide* from **Graphisoft**.

- There is a whole **energy evaluation** system included with Archicad, helping you to assess the energetic performance of your design and to calculate the CO_2 emissions and related energy costs.

- Structural engineers now have an elaborate system to manage **Structural Analytical Model** directly inside Archicad. You can define axes, nodes, constraints, and load combinations, ready to be shared with structural calculation software, via the open **SAF** file format.

- Mechanical engineers shouldn't feel left out either, with the Archicad **MEP Modeler** extension providing a whole library of parametric ducts, pipes, and other MEP elements. The Archicad product name is not only about architecture these days.

So far, this book assumed that you were working alone. But rest assured that Archicad can also be used in direct collaboration within a design team and co-workers:

- The **Teamwork** system allows you to share a model via a local server or via **Graphisoft BIMcloud**, and enables collaborative modeling – multiple people working on the same model in parallel, with a clever system of reserving and releasing elements, so everything is synced, while at the same time, protecting the objects you are working on from being changed by somebody else.

Even then, if the feature set of Archicad is not enough, you can look out for **add-ons**: external plugins that expand the Archicad feature set. There are add-ons for additional modeling tasks, integration with other systems, or integrating additional file formats or rendering engines. Let's list some of the plugins we like and use often and where you can find the most common add-ons:

- Some specific add-ons focusing on IFC include connections to **Solibri** for model-checking via IFC, the **Design Checker** powered by **Solibri.Inside**, and an improved interoperability add-on for **IFC Model Exchange** between Archicad and **Revit**.

- A special mention goes to the connection of Archicad with the **Grasshopper** visual programming environment via **Rhinoceros 3D**, bringing advanced parametric design into Archicad. This last add-on allows you to create a parametric script that communicates directly with Archicad and can be used to create objects programmatically instead of manually.

- You may not be aware of this but a large set of Archicad features is provided already via such add-ons, including the **Cineware** rendering and the **Ecodesigner** energy evaluation.

- Some functionality may be dependent on your Archicad license. Subscription (SSA/Forward) customers receive a few additional tools and add-ons to extend Archicad even more and to reward these license holders.

- Check out the Graphisoft add-on downloads at `https://graphisoft.com/resources-and-support/downloads?section=add-on`.

And finally, for the real Archicad advanced users, you can extend Archicad yourself using one of the **Application Programming Interfaces** (**APIs**) (`https://archicadapi.graphisoft.com`):

- The **Archicad Software Development Kit** provides software libraries for creating your own add-on, which can talk to most of the Archicad features and data structures. Add-ons are written in C++ and compiled into a dynamic APX library, and allow you to add your own commands.

- A more recent addition is the **JavaScript Object Notation** (**JSON**) API (`https://archicadapi.graphisoft.com/JSONInterfaceDocumentation`). This covers a subset of Archicad features and can be used by external software to communicate with a running Archicad instance. You can use it to manage classifications, properties, currently selected elements, and parts of **Layout Book**.

- Ultimately, this JSON API has also been wrapped into the **Archicad Python library** (`https://pypi.org/project/archicad/`). So if you feel more familiar with Python, you can start writing scripts that communicate with a running instance of Archicad.

Summary

So that's it – the end of this book on Archicad and how to use it in your design workflow when implementing BIM. We wrapped it up by briefly looking back into what we learned and looking forward to some interesting roads you can travel along next! Who knows, maybe our roads will cross again in one way or another…

Index

G

H

I

J

K

Kitchen Layout object
configuring 152, 153

L

labels
associative labeling 244-246
exploring 243
object-based label types 246
Layer Combination 462
Layer Intersection Number 273
layers 107
Layers dialog
Layer Intersection Priority 464
Lock/Unlock 463
Show/Hide 463
Solid/Wireframe 463
Layout Book 471, 480, 482
appearance configuration 488
Autotext 494-496
drawing frame 488, 489
drawings 480, 482
drawing size 487
drawing title 490, 491
identification 483
ID strategy 484
layout 481
layouts, versus master layouts 493
layouts, working with 483
master layout 481
master layouts, working with 483
masters 480
organizing 482
palette 481
setting up 485, 487

sheets 480
subset 482
title block 493, 494
views 482
left mouse button (LMB) 40
Library Container file 150
Library Part Maker 569
license types 7
developer license 8
educational license 8
full commercial license 7
migrating between 10
pay-per-use license 8
subscription license 8
temporary trial license 8
licensing 7
Line tool 184
Line Types
categories 189, 190
exploring 189
using, as introduction to Archicad
attributes 187, 188, 189
scaling 190, 191
Linework Consolidation 426
lists, Archicad 430
limitations 430
localized versions 10

M

Maxon Cinema 4D 554
**mechanical, electrical and plumbing
(MEP) design 20**
MEP Modeler extension 569
Merge Elements connection
overlaps, removing with 290

Z

www.packtpub.com

Subscribe to our online digital library for full access to over 7,000 books and videos, as well as industry leading tools to help you plan your personal development and advance your career. For more information, please visit our website.

Why subscribe?

- Spend less time learning and more time coding with practical eBooks and Videos from over 4,000 industry professionals

- Improve your learning with Skill Plans built especially for you

- Get a free eBook or video every month

- Fully searchable for easy access to vital information

- Copy and paste, print, and bookmark content

Did you know that Packt offers eBook versions of every book published, with PDF and ePub files available? You can upgrade to the eBook version at packtpub.com and as a print book customer, you are entitled to a discount on the eBook copy. Get in touch with us at customercare@packtpub.com for more details.

At www.packtpub.com, you can also read a collection of free technical articles, sign up for a range of free newsletters, and receive exclusive discounts and offers on Packt books and eBooks.

Other Books You May Enjoy

If you enjoyed this book, you may be interested in these other books by Packt:

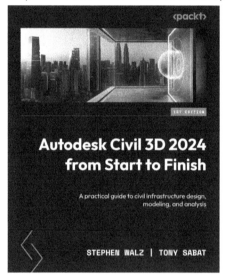

Autodesk Civil 3D 2024 from Start to Finish

Stephen Walz, Tony Sabat

ISBN: 978-1-80323-906-4

- Understand civil project basics and how Autodesk Civil 3D helps achieve them
- Connect detailed components of your design for faster and more efficient designs
- Eliminate redundant workflows by creating intelligent objects to handle design changes smoothly
- Collaborate with distributed teams efficiently and produce designs swiftly and effectively
- Optimize 3D usage and decision-making, using a model-based approach on the impact of your designs and accelerate your career

3D Printing with SketchUp

Aaron Dietzen

ISBN: 978-1-80323-735-0

- Understand SketchUp's role in the 3D printing workflow
- Generate print-ready geometry using SketchUp
- Import existing files for editing in SketchUp
- Verify whether a model is ready to be printed or not
- Model from a reference object and use native editing tools
- Explore the options available for adding onto SketchUp for the purpose of 3D printing (extensions)
- Understand the steps to export a file from SketchUp

Packt is searching for authors like you

If you're interested in becoming an author for Packt, please visit authors.packtpub.com and apply today. We have worked with thousands of developers and tech professionals, just like you, to help them share their insight with the global tech community. You can make a general application, apply for a specific hot topic that we are recruiting an author for, or submit your own idea.

Hi!

We are Ruben Van de Walle and Stefan Boeykens, authors of *A BIM Professional's Guide to Learning Archicad*. We really hope you enjoyed reading this book and found it useful for increasing your productivity and efficiency in Archicad.

It would really help me (and other potential readers!) if you could leave a review on Amazon sharing your thoughts on this book.

Go to the link below or scan the QR code to leave your review:

`https://packt.link/r/180324657X`

Your review will help us to understand what's worked well in this book, and what could be improved upon for future editions, so it really is appreciated.

Best wishes,

Stefan Boeykens

Ruben Van de Walle

Download a free PDF copy of this book

Thanks for purchasing this book!

Do you like to read on the go but are unable to carry your print books everywhere?

Is your eBook purchase not compatible with the device of your choice?

Don't worry, now with every Packt book you get a DRM-free PDF version of that book at no cost.

Read anywhere, any place, on any device. Search, copy, and paste code from your favorite technical books directly into your application.

The perks don't stop there, you can get exclusive access to discounts, newsletters, and great free content in your inbox daily

Follow these simple steps to get the benefits:

1. Scan the QR code or visit the link below

https://packt.link/free-ebook/9781803246574

2. Submit your proof of purchase

3. That's it! We'll send your free PDF and other benefits to your email directly

www.ingramcontent.com/pod-product-compliance
Lightning Source LLC
Chambersburg PA
CBHW060920060326
40690CB00041B/2731